KNOTT'S
HANDBOOK FOR
VEGETABLE GROWERS

KNOTT'S

HANDBOOK FOR VEGETABLE GROWERS

SIXTH EDITION

GEORGE J. HOCHMUTH

Soil, Water, and Ecosystem Sciences
University of Florida
Gainesville
USA

REBECCA G. SIDEMAN

Department of Agriculture, Nutrition & Food Systems
University of New Hampshire
Durham
USA

This edition first published 2023

© 2023 John Wiley & Sons Ltd

Edition History

John Wiley & Sons, Inc. (5e, 2007); Wiley (4e, 1997); Wiley (3e, 1988); Wiley (2e, 1980); Wiley (1e, 1957)

The right of George J. Hochmuth and Rebecca G. Sideman to be identified as the authors of this work has been asserted in accordance with law.

Registered Offices

John Wiley & Sons, Inc., 111 River Street, Hoboken, NJ 07030, USA

John Wiley & Sons Ltd, The Atrium, Southern Gate, Chichester, West Sussex, PO19 8SQ, UK

Editorial Office

The Atrium, Southern Gate, Chichester, West Sussex, PO19 8SQ, UK

For details of our global editorial offices, customer services, and more information about Wiley products visit us at www.wiley.com.

Wiley also publishes its books in a variety of electronic formats and by print-on-demand. Some content that appears in standard print versions of this book may not be available in other formats.

Library of Congress Cataloging-in-Publication Data has been applied for

ISBN 9781119811077 (Paperback); ISBN 9781119811176 (Adobe PDF); ISBN 9781119735977 (e-Pub)

Cover Design: Wiley

Cover Image: Picadilly Farm, Winchester NH, USA, courtesy of Rebecca Sideman

Set in 10.5/13pt NewCenturySchoolbookLTStd by Straive, Pondicherry, India

SKY10035937_090622

Knott's Handbook for Vegetable Growers,
sixth edition dedicated to
Dr. Donald N. Maynard

The Knott's Handbook for Vegetable Growers *has been in production since 1956 when Dr. James E. Knott envisioned the need for a handy source of useful vegetable production information. This current edition is the sixth edition over the 65 years since Dr. Knott's first book. Dr. Don Maynard has been instrumental in the creation of three of those editions over the years, the most recent being the fifth edition. Unfortunately, Dr. Maynard passed away on 27 November 2016.*

Dr. Maynard was born on 22 June 1932, in Hartford, CT, and grew up on a small family farm on the Connecticut River in Connecticut. He obtained his bachelor's degree in Plant Science in 1954 and a master's degree in Horticulture from North Carolina State University. He began his career as an instructor of Olericulture at the University of Massachusetts, where he completed his doctorate in Botany in 1963. He rose through the professorial ranks at The University of Massachusetts and, after 23 years, moved to the University of Florida, in Gainesville, FL, as Chair of the Vegetable Crops Department. He was a tireless supporter of research, teaching, and extension work in support of the vegetable industry in Florida. He completed his distinguished career as a vegetable extension specialist at the Gulf Coast Research and Education Center in Bradenton, Florida, retiring 2003.

Dr. Maynard received many awards and honors over his career, from every organization he worked with. He was a Fellow of the American Society for Horticultural Science (ASHS). He received the Distinguished Alumni Award from the University of Connecticut and the Outstanding Alumnus Award from North Carolina State University. Honors for his writing, research, and extension successes include the ASHS Marion Meadows Award, the ASHS Environmental Quality Research Award, and the ASHS Outstanding Extension Publication Award. He also was a two-time recipient of the ASHS Outstanding Extension Education Award. Other awards in recognition of his contributions to research and extension include the Southern Region Extension Publication Award, six times. He received the Wilson Popenoe Award for Vegetable Crops from the Interamerican Society. At a Cucurbitaceae Conference, he received the Lifetime-Achievement Award for Crop Production. In 2021, Dr. Maynard was elected to the American Society for Horticultural Science Hall of Fame, which recognizes horticulturists who have distinguished themselves professionally to an outstanding level.

Dr. Maynard was an inspiration to countless horticulturists around the world, and this sixth edition of the Knott's Handbook for Vegetable Growers *is being dedicated to Dr. Maynard for his tireless work on behalf of vegetable growers and specifically for his support of the* Knott's Handbook for Vegetable Growers.

George J. Hochmuth and Rebecca G. Sideman

CONTENTS

PREFACE TO THE SIXTH EDITION

The availability of information and the speed with which we can access it have changed radically in recent decades, and technological advances in all aspects of our lives are accelerating at a pace that is hard to comprehend. Nevertheless, it can be difficult to find relevant, accurate information in a digestible format, when it is needed.

Our goal in this sixth revision of the *Knott's Handbook for Vegetable Growers* is to provide up-to-date information on vegetable crop production in handy format for growers, as a textbook for students, and as a reference book for extension personnel, crop consultants, and all those concerned with commercial production, handling, and marketing of vegetables.

New technical information has been included on the status of the vegetable industry, water management with drip irrigation and soil moisture sensors, fertigation scheduling, petiole sap testing, plant tissue testing, fertilizer recommendations from several regions in the United States, worker protection and food safety standards, post-harvest and handling topics, new computer-based technologies to improve farm efficiencies, and protected agriculture production systems. We have added and updated many references that provide additional information on vegetable production.

The authors are grateful for the splendid assistance of the editors at Wiley: Rosie Hayden, Kerry Powell, and Rebecca Ralf. The authors are also grateful to Drs. Oscar A. Lorenz and Donald N. Maynard, whose leadership was critical to the second through fifth editions.

We hope our sixth edition of The *Knott's Handbook for Vegetable Growers* will continue to be a timely and useful reference as Dr. J. E. Knott envisioned when the Handbook was first published in 1956.

GEORGE J. HOCHMUTH,
University of Florida, USA

REBECCA G. SIDEMAN,
University of New Hampshire, USA

ABOUT THE COMPANION WEBSITE

This book is accompanied by a companion website.

https://www.wiley.com/go/hochmuth/vegetablegrowers6e

Resources on the website include:
- Figures from the book, in Powerpoint slides
- Tables from the book, in PDF format

PART **1**

VEGETABLES AND THE VEGETABLE INDUSTRY

Knott's Handbook for Vegetable Growers, Sixth Edition. George J. Hochmuth and Rebecca G. Sideman.
© 2023 John Wiley & Sons Ltd. Published 2023 by John Wiley & Sons Ltd.
Companion website: https://www.wiley.com/go/hochmuth/vegetablegrowers6e

2

BOTANICAL NAMES OF VEGETABLES AND COMMON NAMES OF VEGETABLES IN NINE LANGUAGES

TABLE 1.1. BOTANICAL NAMES, COMMON NAMES, AND EDIBLE PARTS OF PLANTS USED AS VEGETABLES

Botanical Name	Common Name	Edible Plant Part
Class Polypodiopsida		
Equisetaceae	HORSETAIL FAMILY	
Equisetum arvense L.	Horsetail	Young strobili
Dennstaedtiaceae	BRACKEN FERN FAMILY	
Pteridium aquilinum (L.) Kuhn.	Bracken fern	Immature frond
Osmundaceae	ROYAL FERN FAMILY	
Osmunda cinnamomea L.	Cinnamon fern	Immature frond
Osmunda japonica Th.	Japanese flowering fern	Immature frond
Parkeriaceae	WATER FERN FAMILY	
Ceratopteris thalictroides (L.) Brongn.	Water fern	Young leaf
Polypodiaceae	COMMON FERN FAMILY	
Diplazium esculentum (Retz.) Swartz.	Vegetable fern	Young leaf
Class Liliopsida		
Alismataceae	WATER PLANTAIN FAMILY	
Limnocharis flava (L.) Buchenau	Yellow velvetleaf, Sawah-lettuce	Young leaf, petiole, floral shoot
Sagittaria sagittifolia L.	Hawaii arrowhead	Corm
Sagittaria trifolia L. (Sieb.) Ohwi	Chinese arrowhead	Corm

3

TABLE 1.1. BOTANICAL NAMES, COMMON NAMES, AND EDIBLE PARTS OF PLANTS USED AS VEGETABLES (Continued)

Botanical Name	Common Name	Edible Plant Part
Amaryllidaceae	ONION FAMILY	
Allium ampeloprasum L.	Leek, kurrat, elephant garlic	Bulb and leaf
Allium cepa L. Aggregatum group	Shallot	Pseudostem and leaf
Allium cepa L. Cepa group	Onion	Bulb
Allium chinense G. Don.	Chinese onion, Rakkyo	Bulb
Allium fistulosum L.	Welsh onion, Japanese bunching onion	Pseudostem and leaf
Allium macrostemon Bunge	Chinese garlic, Japanese garlic	Leaf
Allium proliferum Schrad. Ex Willd.	Tree onion, Egyptian onion	
Allium sativum L.	Garlic	Bulb and leaf
Allium schoenoprasum L.	Chive	Leaf
Allium scorodoprasum L.	Spanish garlic	Leaf and bulb
Allium tuberosum Rottl. ex Spreng.	Chinese chive	Leaf, immature flower
Allium victorialis L.	Alpine leek	Bulb, leaf
Araceae	ARUM FAMILY	
Alocasia macrorrhizos (L.) G. Don.	Giant taro, alocasia, ape	Corm, immature leaf, petiole
Amorphophallus paeoniifolius (Dennst.) Nicolson	Elephant yam	Corm
Colocasia esculenta (L.) Schott	Taro, dasheen, cocoyam	Corm, immature leaf
Cyrtosperma merkusii Schott.	Gallan, giant swamp taro	Corm
Xanthosoma brasiliense Engl.	Tahitian spinach	Immature leaf
Xanthosoma sagittifolium (L.) Schott	Tania, arrowleaf elephant's ear	Corm and young leaf

4

Scientific name	Common name	Part
Asparagaceae	LILY FAMILY	
Asparagus acutifolius L.	Wild asparagus	Shoot
Asparagus officinalis L.	Asparagus	Shoot
Asphodelaceae		
Hemerocallis spp.	Daylily	Flower
Cannaceae	CANNA FAMILY	
Canna indica L.	Indian canna, arrowroot, edible canna	Rhizome
Cyperaceae	SEDGE FAMILY	
Cyperus esculentus L.	Rushnut, chufa, nutsedge	Tuber
Eleocharis dulcis Hensch.	Water chestnut, Chinese water chestnut	Corm
Dioscoreaceae	YAM FAMILY	
Dioscorea alata L.	White yam, water yam	Tuber
Dioscorea bulbifera L.	Potato yam, aerial yam	Tuber
Dioscorea cayenensis Lam.	Yellow yam	Tuber
Dioscorea dumetorum Pax.	African bitter yam	Tuber
Dioscorea esculenta (Lour.) Burk.	Lesser yam	Tuber
Dioscorea polystachya Turcz.	Chinese yam	
Dioscorea rotundata Poir.	White Guinea yam	Tuber
Dioscorea trifida L.f.	Indian yam	Tuber
Iridaceae	IRIS FAMILY	
Tigridia pavonia (L.f.) Redouté	Common tiger flower	Bulb
Liliaceae		
Lilium spp.	Lily	Bulb

TABLE 1.1.   BOTANICAL NAMES, COMMON NAMES, AND EDIBLE PARTS OF PLANTS USED AS VEGETABLES (Continued)

Botanical Name	Common Name	Edible Plant Part
Marantaceae	ARROWROOT FAMILY	
Goeppertia allouia (Aubl.) Borchs. & S. Suárez	Sweet corn root, Guinea arrowroot	Tuber
Maranta arundinacea L.	West Indian arrowroot	Rhizome
Musaceae	BANANA FAMILY	
Musa paradisiaca L.	Plaintain	Fruit, flower bud
Poaceae	GRASS FAMILY	
Bambusa spp.	Bamboo shoots	Young shoot
Dendrocalamus latiflorus Munro	Bamboo shoots	Young shoot
Pennisetum purpureum Schum.	Elephant grass, napier grass	Young spear
Phyllostachys spp.	Bamboo shoots	Young shoot
Saccharum edule Hassk.	Sugarcane inflorescence	Immature inflorescence
Setaria palmifolia Stapf	Palm grass	Young plant
Zea mays subsp. mays	Sweet corn	Immature kernels and immature cob with kernel
Zizania latifolia (Griseb.) Turcz. ex Stapf.	Water bamboo, cobo	Swollen shoot/stem
Pontederiaceae	PICKERELWEED FAMILY	
Monochoria hastata (L.) Solms.	Hastate-leaved pondweed	Young leaf
Monochoria vaginalis (Burm.f.) C. Presl	Pickerel	Young leaf
Taccaceae	TACCA FAMILY	
Tacca leontopetaloides (L.) Kuntze	East Indian arrowroot	Rhizome

6

Zingiberaceae | GINGER FAMILY

Alpinia galanga (L.) Willd. — Greater galangal — Floral sprout and flower; tender shoot, rhizome

Curcuma longa L. — Turmeric — Rhizome

Curcuma zedoaria (Christm.) Roscoe — Long zedoary — Rhizome

Zingiber mioga (Thunb.) Roscoe — Japanese wild ginger — Rhizome, tender shoot, leaf, flower

Zingiber officinale Roscoe — Ginger — Rhizome and tender shoot

Class Magnoliopsida

Acanthaceae — ACANTHUS FAMILY

Justicia insularis T. And. — Tettu — Young shoot, leaf, root

Rungia klossii S. Moore — Rungia — Leaf

Aizoaceae — CARPETWEED FAMILY

Mesembryanthemum crystallinum L. — Ice plant — Leaf

Tetragonia tetragoniodes (Pall.) O. Kuntze — New Zealand spinach — Tender shoot and leaf

Amaranthaceae — AMARANTH FAMILY

Alternanthera philoxeroides (Martius) Griseb. — Alligator weed, Joseph's coat — Young top

Alternanthera sessilis (L.) DC. — Sessile alternanthera, Rabbit-meat — Young top

Amaranthus spp. — Amaranthus, pigweed — Tender shoot, leaf, sprouted seed

Atriplex hortensis L. — Orach — Leaf

Beta vulgaris L. ssp. *cicla* (L.) W.D.J. Koch — Chard, Swiss chard — Leaf

Beta vulgaris L. ssp. *macrocarpa* (Guss.) Thell. — Garden beet — Root and leaf

7

TABLE 1.1. BOTANICAL NAMES, COMMON NAMES, AND EDIBLE PARTS OF PLANTS USED AS VEGETABLES (*Continued*)

Botanical Name	Common Name	Edible Plant Part
Blitum bonus-henricus (L.) Richb.	Good King Henry	Leaf
Celosia spp.	Cockscomb	Leaf and tender shoot
Chenopodium quinoa Willd.	Quinoa	Leaf, seed
Bassia scoparia subsp. *scoparia*	Mock cypress, burning bush	Tender shoot
Kali komarovii (Iljin.) Akhani & Roalson	Komarov Russian thistle	Leaf and young shoot
Salsola soda L.	Salsola, soda saltwort	Leaf and young shoot
Spinacia oleracea L.	Spinach	Leaf
Suaeda glauca Bunge	Common seepweed	Young stem, leaf, plant
Apiaceae	CARROT FAMILY	
Angelica archangelica L.	Garden angelica	Tender shoot and leaf
Angelica keiskei (Miq.) Koidz.	Japanese angelica	Tender shoot and leaf
Anthriscus cerefolium (L.) Hoffm.	Chervil	Leaf
Apium graveolens L.	Celery, celeriac	Petiole, leaf, root, seed
Arracacia xanthorrhiza Bancroft	Arracacha, Peruvian carrot	Root
Centella asiatica (L.) Urban	Asiatic pennywort	Leaf and stolon
Chaerophyllum bulbosum L.	Tuberous chervil	Root
Coriandrum sativum L.	Coriander, Chinese parsley	Leaf and seed
Cryptotaenia japonica Hassk.	Japanese honewort	Leaf
Daucus carota L. subsp. *sativus* (Hoffm.) Schübl. & Martens	Carrot	Root and leaf
Foeniculum vulgare var. *azoricum* (Miller) Thell.	Fennel	Leaf

8

Foeniculum vulgare var. *dulce* Fiori	Florence fennel	Leaf base
Glehnia littoralis F. Schm.	Coastal glehnia	Leaf, stem, root
Hydrocotyle sibthorpioides Lam.	Hydrocotyle, lawn marsh pennywort	Young shoot and leaf
Myrrhis odorata (L.) Scop.	Garden myrrh, sweet cicely	Leaf, root, and seed
Oenanthe javanica (*Blume*) DC. subsp. *javanica*	Chinese celery, water dropwort	Leaf and tender shoot
Pastinaca sativa L.	Parsnip	Root and leaf
Petroselinum crispum subsp. *crispum*	Parsley	Leaf
Petroselinum crispum var. *tuberosum*	Turnip-rooted parsley, Hamburg parsley	Root and leaf
Petroselinum crispum (Mill.) Fuss	Italian parsley	Leaf
Sium sisarum L.	Skirret	Root
Araliaceae	ARALIA FAMILY	
Aralia cordata Thunb.	Spikenard, Japanese asparagus	Tender shoot
Aralia elata (Miq.) Seeman	Japanese aralia	Young leaf
Asteraceae	SUNFLOWER FAMILY	
Arctium lappa L.	Edible burdock	Root, petiole
Artemisia dracunculus L.	French tarragon	Leaf
Artemisia vulgaris L.	Mugwort	Leaf
Aster scaber Thunb.	Aster	Leaf
Bidens pilosa L.	Bur marigold, Black-jack	Young shoot and leaf

9

TABLE 1.1. BOTANICAL NAMES, COMMON NAMES, AND EDIBLE PARTS OF PLANTS USED AS VEGETABLES *(Continued)*

Botanical Name	Common Name	Edible Plant Part
Chrysanthemum spp.	Edible chrysanthemum	Leaf and tender shoot
Cichorium endivia L.	Endive, escarole	Leaf
Cichorium intybus L.	Chicory, witloof chicory, Belgium endive	Leaf
Cirsium dipsacolepis (Maxim.) Matsum.	Gobouazami	Root
Cosmos caudatus Kunth	Cosmos	Leaf and young shoot
Solanecio biafrae (Oliv. et Hiern) C. Jeffrey	Sierra Leone bologni	Young shoot and leaf
Crassocephalum crepidiodes (Benth.) S. Moore	Hawksbeard velvetplant	Young shoot and leaf
Cynara cardunculus subsp. *cardunculus*	Cardoon	Petiole
Cynara scolymus L.	Globe artichoke	Immature flower bud
Emilia sonchifolia (L.) DC.	Emilia, false sow thistle	Young shoot and leaf
Enydra fluctuans Lour.	Buffalo spinach	Young shoot and leaf
Farfugium japonicum (L.) Kitamura	Japanese farfugium, Leopardplant	Petiole
Fedia cornucopiae (L.) Gaertn.	Horn of plenty, African valerian	Leaf
Galinsoga parviflora Cav.	Galinsoga, gallant soldier	Young shoot
Gynura bicolor (Roxb. Ex Willd.) DC.	Gynura, red-vegetable	Young leaf
Helianthus tuberosus L.	Jerusalem artichoke	Tuber
Lactuca indica L.	Indian lettuce	Leaf
Lactuca sativa L. var. *asparagina* Bailey	Asparagus lettuce, celtuce	Stem
Lactuca sativa L. var. *capitata* L.	Head lettuce, butterhead lettuce	Leaf
Lactuca sativa L. var. *longifolia* Lam.	Romaine lettuce, leaf lettuce	Leaf
Launaea taraxacifolia (Willd.) Amin ex C. Jeffrey	Wild lettuce, African lettuce	Leaf

Scientific name	Common name	Part used
Petasites japonicus (Sieb. & Zucc.) Maxim.	Butterbur	Petiole
Smallanthus sonchifolius (Poepp. & Endl.) H. Rob	Yacon strawberry	Root
Scolymus hispanicus L.	Golden thistle	Root and leaf
Scolymus maculatus L.	Spotted golden thistle	Leaf
Pseudopodospermum hispanicum subsp. *hispanicum*	Black salsify	Root and leaf
Sonchus oleraceus L.	Milk thistle, sow thistle	Leaf
Blainvillea acmella (L.) Philipson	Brazil cress	Young leaf
Acmella ciliata Kunth	Guasca	Young leaf
Acmella uliginosa (Sw.) Cass.	Getang	Young leaf and flower shoot
Acmella paniculata (Wall ex DC.) R.K. Jansen	Getang	Young leaf and flower shoot
Struchium sparganophora (L.) Kuntze	Bitter leaf	Young shoot
Taraxacum officinale Weber ex Wiggins	Dandelion	Leaf, root
Tragopogon porrifolius L.	Salsify, vegetable oyster, Jerusalem star	Root and young leaf
Tragopogon pratensis L.	Goatsbeard, meadow salsify	Young root and shoot
Gymnanthemum amygdalinum (Delile) Sch.Bip. ex Walp.)	Bitter leaf	Young shoot
Basellaceae	BASELLA FAMILY	
Basella alba L.	Indian spinach, Malabar spinach	Leaf and young shoot
Ullucus tuberosus Caldas	Ulluco	Tuber
Boraginaceae	BORAGE FAMILY	
Borago officinalis L.	Borage	Petiole
Symphytum officinale L.	Common comfrey, boneset	Leaf and tender shoot

TABLE 1.1. BOTANICAL NAMES, COMMON NAMES, AND EDIBLE PARTS OF PLANTS USED AS VEGETABLES (Continued)

Botanical Name	Common Name	Edible Plant Part
Symphytum × uplandicum Nyman	Russian comfrey	Young leaf and shoot
Brassicaceae	MUSTARD FAMILY	
Armoracia rusticana Gaertn., Mey,, Scherb.	Horseradish	Root, leaf, sprouted seed
Barbarea verna (Mill.) Aschers	Upland cress	Leaf
Brassica carinata A. Braun	Abyssinian mustard	Leaf
Brassica juncea (L.) Czernj. & Coss. var. *capitata* Hort.	Capitata mustard	Leaf
Brassica juncea (L.) Czernj. & Coss. var. *crassicaulis* Chen and Yang	Bamboo shoot mustard	Stem
Brassica juncea (L.) Czernj. & Coss. var. *crispifolia* Bailey	Curled mustard	Leaf
Brassica juncea (L.) Czernj. & Coss. var. *foliosa* Bailey	Small-leaf mustard	Leaf
Brassica juncea (L.) Czernj. & Coss. var. *gemmifera* Lee & Lin	Gemmiferous mustard	Stem and axillary bud
Brassica juncea (L.) Czernj. & Coss. var. *involuta* Yang & Chen	Involute mustard	Leaf
Brassica juncea (L.) Czernj. & Coss. var. *latipa* Li	Wide-petiole mustard	Leaf
Brassica juncea (L.) Czernj. & Coss. var. *leucanthus* Chen & Yang	White-flowered mustard	Leaf

Scientific name	Common name	Part used
Brassica juncea (L.) Czernj. & Coss. var. *linearifolia*	Line mustard	Leaf
Brassica juncea (L.) Czernj. & Coss. var. *longgepetiolata* Yang & Chen	Long-petiole mustard	Leaf
Brassica juncea (L.) Czernj. & Coss. var. *megarrhiza* Tsen & Lee	Tuberous-rooted mustard	Root
Brassica juncea (L.) Czernj. & Coss. var. *multiceps* Tsen & Lee	Tillered mustard	Leaf
Brassica juncea (L.) Czernj. & Coss. var. *multisecta* Bailey	Flower-like leaf mustard	Leaf
Brassica juncea (L.) Czernj. & Coss. var. *rugosa* Bailey	Brown mustard, mustard greens	Leaf
Brassica juncea (L.) Czernj. & Coss. var. *strumata* Tsen & Lee	Strumous mustard	Stem
Brassica juncea (L.) Czernj. & Coss. var. *tumida* Tsen & Lee	Swollen-stem mustard	Stem and leaf
Brassica juncea (L.) Czernj. & Coss. var. *utilis* Li	Penduncled mustard	Young flower stalk
Brassica napus L. subsp. *rapifera* Metzg. Ex Sinskaya	Rutabaga	Root and leaf
Brassica napus L. subsp. *napus*	Vegetable rape, canola, oilseed rape	Leaf and young flower stalk
Brassica napus L. var. *pabularia* (DC.) Alef.	Siberian kale, Hanover salad	Leaf
Brassica nigra L. Koch.	Black mustard	Leaf
Brassica oleracea L.	Cabbage	Leaf
Brassica oleracea L. var. *acephala* DC.	Kale, collards	Leaf

TABLE 1.1. BOTANICAL NAMES, COMMON NAMES, AND EDIBLE PARTS OF PLANTS USED AS VEGETABLES (Continued)

Botanical Name	Common Name	Edible Plant Part
Brassica oleracea L. var. *alboglabra* (L.H. Bailey) Sun	Chinese kale	Young flower stalk and leaf
Brassica oleracea L. var. *botrytis* L.	Cauliflower	Immature floral stalk
Brassica oleracea L. var. *costata* DC.	Portuguese cabbage, tronchuda cabbage	Leaf and inflorescence
Brassica oleracea L. var. *gemmifera* DC.	Brussels sprouts	Axillary bud
Brassica oleracea L. var. *gongylodes* L.	Kohlrabi	Enlarged stem
Brassica oleracea L. var. *italica* Plenck.	Broccoli	Immature flower stalk
Brassica oleracea L. var. *medullosa* Thell. Marrow	Marrow stem kale	Leaf
Brassica oleracea L. var. *ramosa* Alef.	Thousand-headed kale, perennial kale	Leaf
Brassica oleracea L. var. *sabauda* L.	Savoy cabbage	Leaf
Brassica oleracea L. var. *viridis* L.	Borecole, Portugese kale	Leaf
Brassica rapa subsp. *oleifera* (DC.) Metzg.	Spinach mustard, tendergreen mustard	Leaf
Brassica rapa L. subsp. *chinensis* (L.) Hanelt	Pak choi; Bok choy, Chinese mustard	Leaf
Brassica rapa L. var. *narinosa* (Bailey) Olsson	Broad-beaked mustard, Chinese savoy	Leaf
Brassica rapa L. var. *pekinensis* (Lour.) Olsson	Chinese cabbage, Napa cabbage	Leaf

Brassica rapa L.	Turnip, Turnip green	Enlarged root, leaf
Bunias orientalis L.	Hill mustard	Leaf
Capsella bursa-pastoris (L.) Medikus	Shepherd's purse	Young leaf
Cardamine pratensis L.	Cuckoo flower	Leaf
Crambe maritima L.	Sea kale	Petiole and young leaf
Crambe tatarica Sebeók	Tartar bread plant	Petiole and young leaf, root
Diplotaxis muralis (L.) DC.	Wallrocket	Leaf
Eruca vesicaria subsp. *sativa* (Mill.) Thell.	Rocket salad, arugula, Italian cress	Leaf
Lepidium meyennii Walp.	Maca, Peruvian ginseng	Root
Lepidium sativum L.	Garden cress	Leaf
Nasturtium officinale W.T. Aiton	Watercress	Leaf
Raphanus sativus L.	Rat-tail radish	Immature seed pod
Raphanus sativus L.	Radish, Daikon	Root
Sinapis alba L.	White mustard	Leaf and young flower stalk
Eutrema japonicum (Miq.) Koidz.	Wasabi, Japanese horesradish	Rhizome, young shoot
Cabombaceae	WATER LILY FAMILY	
Brasenia schreberi Gmelin	Watershield	Young leaf
Cactaceae	CACTUS FAMILY	
Opuntia ficus-indica (L.) Mill.	Prickly pear, Barbary fig	Pad, fruit
Campanulaceae	BELLFLOWER FAMILY	
Campanula rapunculus L.	Rampion	Root and first leaf
Capparaceae	CAPER FAMILY	
Capparis spinosa L.	Capper, Finders rose	Flower bud
Gynandropsis gynandra L.	Cat's whiskers, spiderwisp	Leaf, young shoot, fruit
Platycodon grandiflorus (Jacq.) A. DC.	Chinese bellflower	Leaf

TABLE 1.1. BOTANICAL NAMES, COMMON NAMES, AND EDIBLE PARTS OF PLANTS USED AS VEGETABLES (*Continued*)

Botanical Name	Common Name	Edible Plant Part
Convolvulaceae	BINDWEED FAMILY	
Calystegia pubescens Lindl.	Rose glorybind	Root
Ipomoea aquatica Forssk.	Water spinach, kangkong	Tender shoot and leaf
Ipomoea batatas (L.) Lam.	Sweet potato	Root and leaf
Crassulaceae	ORPINE FAMILY	
Sedum sarmentosum Bunge	Sedum	Leaf
Cucurbitaceae	GOURD FAMILY	
Benincasa hispida (Thunb.) Cogn.	Wax gourd	Immature/mature fruit
Citrullus lanatus (Thunb.) Matsum & Nakai	Watermelon	Ripe fruit and seed
Citrullus amarus Schrad.	Citron, preserving melon	Fruit
Coccinia grandis (L.) Voigt	Ivy gourd, tindora	Fruit, tender shoot, leaf
Melothria sphaerocarpa (Cogn.) Schaef. & Renner	White-seeded melon	Fruit and seed
Cucumis anguria L.	West Indian gherkin	Immature fruit
Cucumis melo L. Cantalupensis group	Cantaloupe, muskmelon	Fruit
Cucumis melo L. Cassaba group	Cassaba melon	
Cucumis melo L. Chito group	Mango	Fruit
Cucumis melo L. Conomon group	Oriental pickling melon	Young fruit
Cucumis melo L. Flexuosus group	Japanese cucumber, snake melon	Immature fruit
Cucumis melo L. Ibericus group	Piel de sapo, amarillo, tendral	Fruit

Cucumis melo L. Inodorus group	Honeydew melon	Fruit
Cucumis metuliferus E. Meyer ex Scrhad.	African horned cucumber	Fruit
Cucumis sativus L.	Cucumber	Immature fruit
Cucurbita argyrosperma Huber	Pumpkin, Cushaw	Young/mature fruit and seed
Cucurbita ficifolia Bouché	Fig-leaf gourd, Malabar gourd	Fruit
Cucurbita maxima Duchesne	Giant pumpkin, winter squash	Mature fruit and seed
Cucurbita moschata Duchesne	Butternut squash, tropical pumpkin	Young and mature fruit
Cucurbita pepo L.	Summer squash, zucchini	Young fruit
Cucurbita pepo L.	Common field pumpkin, winter squash	Mature fruit and seed
Cyclanthera pedata (L.) Schrader var. pedata	Achocha, lady's slipper	Immature fruit
Lagenaria siceraria (Mol.) Standl.	Bottle gourd, calabash gourd	Immature fruit, tender shoot and leaf
Luffa acutangula (L.) Roxb.	Angled loofah, Chinese okra	Immature fruit
Luffa aegyptiaca Miller	Smooth loofah, sponge gourd	Immature fruit and leaf
Momordica charantia L.	Bitter gourd, balsam pear	Immature fruit and young leaf
Benincasa fistulosa (Stocks) Schaef. & Rennner	Squash melon	Fruit
Sechium edule (Jacq.) Swartz.	Chayote, mirliton, vegetable pear	Fruit, tender shoot, leaf
Sicana odorifera (Vell.) Naudin	Casabanana	Immature/mature fruit
Telfairia occidentalis Hook.f	Fluted gourd, fluted pumpkin	Seed, leaf, tender shoot
Telfairia pedata (Sims) Hook.	Oyster nut	Seed

TABLE 1.1. BOTANICAL NAMES, COMMON NAMES, AND EDIBLE PARTS OF PLANTS USED AS VEGETABLES (Continued)

Botanical Name	Common Name	Edible Plant Part
Trichosanthes cucumerina subsp. *anguinea* (L.) Greb	Snake gourd	Immature fruit, leaf, and tender shoot
Trichosanthes cucumeroides (Ser.) Maxim.	Japanese snake gourd	Immature fruit
Trichosanthes dioica Roxb.	Pointed gourd	Immature fruit, tender shoot
Euphorbiaceae	SPURGE FAMILY	
Cnidoscolus aconitifolius (Miller) Johnston	Chaya, Cabbage star	Leaf
Codiaeum variegatum (L.) Rumph. Ex A. Juss.	Croton	Young leaf
Manihot esculenta Crantz	Yuca, cassava, manioc	Root and leaf
Breynia androgyna (L.) Chakrab. & N.P. Blakr.	Common sauropus, Katuk	Leaf
Fabaceae	PEA FAMILY	
Arachis hypogaea L.	Peanut, ground nut	Immature/mature seed
Tylosema esculentum Burch. A. Schreib.	Marama bean	Immature pod and root
Cajanus cajan (L.) Huth.	Congo pea, pigeon pea	Immature pod/leaf
Canavalia ensiformis (L.) DC.	Jack bean, horse bean	Immature seed
Canavalia gladiata (Jacq.) DC.	Sword bean, horse bean	Immature seed
Cicer arietinum L.	Garbanzo, chick pea	Seed
Cyamopsis tetragonoloba (L.) Taub.	Cluster bean, guar	Immature pod and seed
Flemingia vestita Benth. ex Bak.	Flemingia	Tuber

Glycine max (L.) Merr.	Soybean	Immature and sprouted seed
Lablab purpureus (L.) Sweet.	Hyacinth bean	Immature seed
Lathyrus sativus L.	Chickling pea, blue vetchling	Immature pod/seed
Lathyrus tuberosus L.	Groundnut	Tuber
Lens culinaris Medikus	Lentil	Immature pod, sprouted seed
Lupinus spp.	Lupin	Seed
Macrotyloma geocarpum (Harms) Marechal & Baudet	Hausa groundnut	Seed
Macrotyloma uniflorum (Lam.) Verdc.	Horse gram	Seed
Medicago sativa L.	Alfalfa, lucerne	Leaf, young shoot, sprouted seed
Mucuna pruriens (L.) DC.	Buffalo bean, velvet bean	Seed
Neptunia prostrata (Lam.) Baill.	Water mimosa, garden puff	Leaf and tender shoot
Pachyrhizus ahipa (Wedd.) Parodi	Yam bean	Root
Pachyrhizus erosus (L.) Urban	Jicama, Mexican yam bean	Root, immature pod, and seed
Pachyrhizus tuberosus (Lam.) Sprengel	Potato bean	Root and immature pod
Phaseolus acutifolius A. Gray	Tepary bean	Seed, immature pod
Phaseolus coccineus L.	Scarlet runner bean	Immature pod and seed
Phaseolus lunatus L.	Lima bean, broad bean	Immature seed, mature seed
Phaseolus vulgaris L.	Garden bean, snap bean	Immature pod and seed
Pisum sativum var. *sativum*	Pea, garden pea	Immature seed, tender shoot

TABLE 1.1. BOTANICAL NAMES, COMMON NAMES, AND EDIBLE PARTS OF PLANTS USED AS VEGETABLES (Continued)

Botanical Name	Common Name	Edible Plant Part
Pisum sativum var. *macrocarpon*	Snow pea, edible-podded pea	Immature pod
Psophocarpus tetragonolobus (L.) DC.	Goa bean, winged bean	Immature pod, seed, leaf, root
Pueraria montana var. *lobata*	Kudzu, Japanese arrowroot	Root, leaf, tender shoot
Sphenostylis stenocarpa (Hochst. ex. A. Rich.) Harms.	African yam bean	Tuber and seed
Lotus tetragonolobus L.	Asparagus pea, winged pea	Immature pod
Trigonella foenum-graecum L.	Fenugreek	Leaf, tender shoot, immature pod
Vicia faba L.	Fava bean, broad bean, horse bean	Immature seed
Vigna aconitifolia (Jacq.) Maréchal	Moth bean, dew bean	Immature pod and seed
Vigna angularis (Willd.) Ohwi & Ohashi	Adzuki bean	Seed
Vigna mungo (L.) Hepper	Black gram, urd	Immature pod and seed
Vigna radiata (L.) Wilcz.	Mung bean, golden gram	Immature pod, sprouted seed, seed
Vigna subterranea (L.) Verdc.	Bambarra groundnut, Hog peanut	Immature/mature seed
Vigna umbellata (Thunb.) Ohwi & Ohashi	Rice bean, Oriental bean	Seed
Vigna unguiculata subsp. *sesquipedalis* (L.) Verdc.	Asparagus bean, yard-long bean	Immature pod and seed
Vigna unguiculata subsp. *unguiculata*	Southern pea, cowpea, black-eyed pea	Immature pod and seed

Gnetaceae	GNETUM FAMILY	
Gnetum gnemon L.	Bucko, Spanish joint-fir	Leaf, tender shoot, and fruit
Haloragaceae	WATER MILFOIL FAMILY	
Myriophyllum aquaticum (Vellozo) Verdc.	Parrot's feather	Shoot tip
Icacinaceae	ICACINA FAMILY	
Icacina oliviformis (Poir) J. Raynal	False yam	Tuber
Lamiaceae	MINT FAMILY	
Lycopus lucidus Turcz.	Shiny bugleweed	Rhizome
Mentha pulegium L.	Pennyroyal mint	Leaf
Mentha spicata L.	Spearmint	Leaf and inflorescence
Ocimum basilicum L.	Common basil, sweet basil	Leaf
Ocimum americanum L.	American basil	Young leaves
Origanum vulgare L.	Marjoram, oregano	Flowering plant and inflorescence
Perilla frutescens (L.) Britt. var. *Crispa* (Thunb.) Deane	Perilla	Leaf and seed
Plectranthus esculentus N.E. Br.	Kaffir potato	Tuber
Satureja hortensis L.	Savory, summer savory	Leaf and young shoot
Plectranthus rotundifolius (Poir.) Spreng.	Hausa potato	Tuber
Stachys affinis Bunge	Japanese artichoke, Chinese artichoke	Tuber
Malvaceae	MALLOW FAMILY	
Abelmoschus esculentus (L.) Moench	Okra, gumbo	Immature fruit
Abelmoschus manihot (L.) Medikus	Hibiscus root	Leaf and tender shoot
Hibiscus acetosella Wel. ex Ficalho	False roselle	Young leaf and shoot
Hibiscus sabdariffa L.	Jamaican sorrel, Indian sorrel	Calyx and leaf
Malva pusilla L.	Mallow	Leaf and young shoot

21

TABLE 1.1. BOTANICAL NAMES, COMMON NAMES, AND EDIBLE PARTS OF PLANTS USED AS VEGETABLES (Continued)

Botanical Name	Common Name	Edible Plant Part
Cannabaceae	HEMP FAMILY	
Humulus lupulus L.	Hops	Tender shoot
Nelumbonaceae	LOTUS FAMILY	
Nelumbo nucifera Gaertn.	Lotus root	Rhizome, leaf, seed
Nyctaginaceae	FOUR O'CLOCK FAMILY	
Mirabilis expansa (Ruiz & Paron) Standley	Mauka	Tuber
Nymphaeaceae	WATER LILY FAMILY	
Euryale ferox Salisb. ex K.D. Koenig & Sims	Foxnut	Seed, tender shoot, root
Nymphaea nouchali Burm. f.	Water lily	Rhizome, flower stalk, seed
Onagraceae	EVENING PRIMROSE FAMILY	
Oenothera biennis L.	Evening primrose	Leaf and tender shoot
Orobanchaceae	BROOMRAPE FAMILY	
Orobanche crenata Forsskal.	Broomrape	Shoot
Oxalidaceae	OXALIS FAMILY	
Oxalis tuberosa Molina	Oka, oca	Tuber
Passifloraceae	PASSION FLOWER FAMILY	
Passiflora biflora Lam.	Passion flower	Shoot, young leaf, flower
Pedaliaceae	PEDALIUM FAMILY	
Sesamum radiatum Schum. ex Thonn.	Gogoro	Young shoot
Phytolaccaceae	POKEWEED FAMILY	
Phytolacca acinosa Roxb.	Indian poke	Leaf and young shoot
Phytolacca americana L.	Poke	Leaf and young shoot

Phytolacca icosandra L.	Inkweed	Leaf and young shoot
Plantaginaceae	PLANTAIN FAMILY	
Plantago coronopus L.	Buckshorn plantain	Leaf
Polygonaceae	BUCKWHEAT FAMILY	
Rheum rhabarbarum L.	Rhubarb, pieplant	Petiole
Rumex acetosa L.	Sorrel	Leaf
Rumex patientia L.	Dock	Leaf
Rumex scutatus L.	French sorrel	Leaf
Montiaceae	WATER CHICKWEED FAMILY	
Claytonia perfoliata subsp. *perfoliata*	Winter purslane, miner's lettuce	Leaf
Portulacaceae	PURSLANE FAMILY	
Portulaca oleracea L.	Purslane	Leaf and young shoot
Talinaceae		
Talinum paniculatum (Jacq.) Gaertn.	Fameflower	Young shoot
Talinum fruticosum (L.) Juss.	Waterleaf, Ceylon spinach	Leaf
Resedaceae	MIGNONETTE FAMILY	
Reseda odorata L.	Mignonette	Leaf and flower
Rosaceae	ROSE FAMILY	
Fragaria ×*Ananassa* (Weston) Rozier	Strawberry	Fruit
Saururaceae	LIZARD'S-TAIL FAMILY	
Houttuynia cordata Thunb.	Saururis, tsi, chamelion	Leaf
Solanaceae	NIGHTSHADE FAMILY	
Capsicum annuum L.	Bell, serrano, Anaheim, and other peppers	Fruit
Capsicum baccatum L. var. *baccatum*	Small pepper	Fruit

TABLE 1.1. BOTANICAL NAMES, COMMON NAMES, AND EDIBLE PARTS OF PLANTS USED AS VEGETABLES (Continued)

Botanical Name	Common Name	Edible Plant Part
Capsicum chinense Jacq.	Scotch bonnet pepper, habanero pepper	Fruit
Capsicum frutescens L.	Tabasco pepper, bird-chilli	Fruit
Capsicum pubescens Ruiz & Pavon	Rocoto	Fruit
Solanum betaceum Cav.	Tamarillo, tree tomato	Ripe fruit
Lycium chinense Mill.	Boxthorn, Chinese teaplant	Leaf
Solanum lycopersicon L.	Tomato	Ripe fruit
Solanum pimpinellifolium Mill. ex Dunal	Currant tomato	Ripe fruit
Alkekengi officinarum Moench.	Chinese lantern plant	Ripe fruit
Physalis philadelphica subsp. *ixocarpa* Sobr.-Vesp. & Sanz-Elorza	Tomatillo	Unripe fruit
Physalis peruviana L.	Cape gooseberry	Ripe fruit
Solanum americanum Mill.	American black nightshade	Tender shoot, leaf, unripe fruit
Solanum incanum L.	Garden egg, bitter apple	Unripe fruit
Solanum aethiopicum L.	Scarlet eggplant, tomato eggplant, gilo	Immature fruit
Solanum macrocarpon L.	African eggplant	Leaf and fruit
Solanum melongena L.	Eggplant, aubergine	Immature fruit
Solanum muricatum Ait.	Pepino, sweet pepino	Ripe fruit
Solanum nigrum L.	Black nightshade	Mature fruit, leaf, tender shoot
Solanum quitoense Lam.	Naranjillo	Ripe fruit

Scientific name	Common name	Part used
Solanum torvum Swartz	Pea eggplant, devil's fig	Tender shoot, immature fruit
Solanum tuberosum L.	Irish potato	Tuber
Malvaceae	MALLOW FAMILY	
Corchorus olitorius L.	Jew's marrow	Leaf and tender shoot
Lythraceae	LOOSTRIFE FAMILY	
Trapa natans var. bicornis (Osbeck) Makino	Water chestnut	Seed
Trapa natans L.	Water chestnut, Bullnut	Seed
Tropacolaceae	NASTURTIUM FAMILY	
Tropaeolum majus L.	Nasturtium, Indian cress	Leaf, flower
Tropaeolum tuberosum Ruiz & Pavon	Tuberous nasturtium, Anu	Tuber
Urticaceae	NETTLE FAMILY	
Pilea glaberrima (Blume) Blume	Pilea	Leaf
Pilea melastomoides (Poir.) Wedd.	Pilea	Leaf
Urtica dioica L.	Stinging nettle	Leaf
Caprifoliaceae	HONEYSUCKLE FAMILY	
Valerianella eriocarpa Desv.	Italian corn-salad	Leaf
Valerianella locusta subsp. *locusta* (L.) Laterr.	Lewiston corn-salad	Leaf
Violaceae	VIOLET FAMILY	
Viola tricolor L.	Violet, pansy	Flower, leaf
Vitaceae	GRAPE FAMILY	
Cissus discolor Blume	Kangaroo vine, climbing begonia	Leaf, young shoot

Adapted from S.J. Kays and J.C. Silva Dias, *Cultivated Vegetables of the World* (Athens, GA: Exon Press, 1996). Taxonomic data from POWO *Plants of the World Online* (2019), http://www.plantsoftheworldonline.org/ (accessed 17 Feb. 2021); From GBIF: The Global Biodiversity Information Facility (2021), https://www.gbif.org (accessed 17 Feb. 2021).

TABLE 1.2. COMMON NAMES OF COMMON VEGETABLES IN NINE LANGUAGES

English	Danish	Dutch	French	German	Italian	Portuguese	Spanish	Swedish
Artichoke	Artiskok	artisjok	artichaut	Artischocke	carciofo	alcachofra	alcachofa	kronärtskocka
Asparagus	Asparges	asperge	asperge	Spargel	asparago	espargo	espárrago	sparris
Bean, broad	Hestebønne	tuinboon	feve	Puffbohne	fava	fava	haba	bondböna
Bean, snap/green	Bønne	boon	haricot	Bohne	fagiolino	feijão	judia verde	böna
Beet	rødbede	biet	betterave rouge	Rote Rübe	bietola da orta	beterraba	remolacha	rödbeta
Broccoli	broccoli	broccoli	brocoli	Brokkoli	cavolo broccolo	brócolos	bróculi	broccoli
Brussels sprouts	rosenkål	spruitkool	chou de Bruxelles	Rosenkohl	cavolo di Bruxelles	couve de Bruxelas	col de Bruselas	brysselkål
Cabbage	kål	kool	chou	Kohl	cavolo	couve	col	kål
Carrot	gulerød	wortel	carotte	Karotte	carota	cenoura	zanahoria	morot
Cauliflower	blomkål	bloemkool	chou-fleur	Blumenkohl	cavolfiore	couve-flor	coliflor	blomkål
Celery	selleri	selderij	céleri	Schnittselleri	sedano da erbucci	aipo	apio	selleri
Celeriac	knoldselleri	knolselderij	céleri-rave	Knollensellerie	sedano rapa	aipo-rábano	apionabo	rotselleri
Chicory	cikorie	cichorei	chicorée	Zichorienwurzel	cicoria	chicória	achicoria	cikoria
Chinese cabbage	kinakål	Chinese kool	chou chinois	Chinakohl	cavolo cinese	couve chinesa	repollo chino	Salladkål
Corn, sweet	sukkermajs	suikermais	mais sucré	Zuckermais	mais dolce	milho doce	maiz dulce	sockermajs
Cucumber	agurk	komkommer	concombre	Gurke	cetriolino	pepino	pepino	gurka
Eggplant	auberginer	aubergine	aubergine	Aubergine	melanzana	beringela	berenjena	aubergin
Endive	endivie	andijvie	chicorée frisée	Endivie	indivia	endivia	escarola	endiviesallat
Garlic	hvidløg	knoflook	ail	knoblauch	aglio	alho	ajo	vitlök
Horseradish	peberrod	mierikswortel	raifort	Meerrettich	rafano	rábano	rábano	pepparrot

Kale	grønkål	boerenkool	chou frisé	Grunkohl	cavolo a foglia riccia	couve galega	berza enana	grönkål
Kohlrabi	kålrabi	koolrabi	chou-rave	Kohlrabi	cavolo-rapa	couve-rábano	colirrábano	kålrabbi
Leek	porre	prei	poireau	Porree	porro	alho porro	puerro	purjolök
Lettuce	salat	sla	laitue	Salat	lattuga	alface	lechuga	sallad
Melon	melon	meloen	melon	Melone	melone	melão	melón	melon
Onion	løg	ui	oignon	Zwiebel	cipolla	cebola	cebolla	lök
Parsley	persille	peterselie	persil	Petersilie	prezzemola	salsa	perejil	persilja
Parsnip	pastinak	pastinaak	panais	Pastinake	pastinaca	pastinaca	chirivía	palsternacka
Pea	haveaert	erwt	pois	Erbse	pisello	ervilha	guisante	ärt
Pepper	peberfrugt	paprika	poivron	Paprika	peperone dolce	pimento	pimiento	paprika
Potato	kartoffel	aardappel	pomme de terre	Kartoffel	patata	batata	patata	potatis
Pumpkin	centnergraeskar	pompoen	potiron	Kürbis	zucca gigante	abóbora	calabaza grande	pumpa
Radish	radis	radijs	radis	Radi	ravanello	rabanete	rábano	rädisa
Rhubarb	rabarber	rabarber	rhubarbe	Rhabarber	rabarbaro	ruibarbo	ruibarbo	rabarber
Rutabaga	kålroer	koolraap	rutabaga	Kohlrübe	rutabaga	rutabaga	colinabo	kålrot
Spinach	spinat	spinazie	épinard	Spinat	spinaci	espinafre	espinaca	spenat
New Zealand spinach	nyzeelandsk spinat	Nieuwzeelandse spinazie	tetragone	Neuseeländsk Spinat	spinacio di Nuova Zelanda	espinafre da Nova Zelândia	espinaca Nueva Zelandia	nyzeeländsk spenat

TABLE 1.2. COMMON NAMES OF COMMON VEGETABLES IN NINE LANGUAGES (*Continued*)

English	Danish	Dutch	French	German	Italian	Portuguese	Spanish	Swedish
Strawberry	jordbaer	aardbei	fraise	Erdbeere	fragola	morango	fresa	jordgubbe
Summer squash	sommersquash	zomerpompoen	courge d'été	Garten	zucca	abóbora porqueira	calabaza	sommarsquash
Swiss chard	bladbede	snijbiet	bette à carde	Mangold	bietola da costa	acelga	acelga	mangold
Tomato	tomat	tomaat	tomate	Tomate	pomodoro	tomate	tomate	tomat
Turnip	majroe	meiraap	navet	Mairübe	rapa bianca	nabo	nabo-colza	rova
Watermelon	vandmelon	watermeloen	pastèque	Wassermelone	anguria	melancia	sandia	vattenmelon

P. J. Stadhouders (chief editor), *Elsevier's Dictionary of Horticultural and Agricultural Plant Production* (New York: Elsevier Science Publishing Co., Inc., 1990).Used with permission.
Updated using DeepL translation and Linguee dictionaries (GmbH, Maarweg 165, 50825 Cologne, Germany) (accessed Apr. 2021).

TABLE 1.3. BOTANICAL NAMES, COMMON NAMES, FLOWER COLOR, AND TASTE OF SOME EDIBLE FLOWERS

Cautions:

- Proper identification of edible flowers is necessary.
- Edible flowers should be pesticide free.
- Flowers of plants treated with fresh manure should not be used.
- Introduce new flowers into the diet slowly so that possible allergic reactions can be identified.

Botanical Name	Common Name	Flower Color	Taste
Amaryllidaceae	ONION FAMILY		
Allium schoenoprasum L.	Chive	Lavender	Onion, strong
Allium tuberosum Rottl. ex. Sprengel	Chinese chive	White	Onion, strong
Tulbaghia violacea Harv.	Society garlic	Lilac	Onion
Apiaceae	CARROT FAMILY		
Anethum graveolens L.	Dill	Yellow	Stronger than leaves
Anthriscus cerefolium (L.) Hoffm.	Chervil	White, pink, yellow, red, orange	Parsley
Coriandrum sativum L.	Coriander	White	Milder than leaf
Foeniculum vulgare Mill.	Fennel	Pale yellow	Licorice, milder than leaf

TABLE 1.3. BOTANICAL NAMES, COMMON NAMES, FLOWER COLOR, AND TASTE OF SOME EDIBLE FLOWERS (Continued)

Botanical Name	Common Name	Flower Color	Taste
Asparagaceae			
Muscari neglectum Guss. ex	Grape hyacinth	Pink, blue	Grapey
Yucca filamentosa L.	Yucca	Creamy white with purple tinge	Slightly bitter
Asphodelaceae			
Hemerocallis fulva L.	Daylily	Tawny orange	Cooked asparagus/ zucchini
Asteraceae	SUNFLOWER FAMILY		
Bellis perennis L.	English daisy	White to purple petals	Mild to bitter
Calendula officinalis L.	Calendula	Yellow, gold, orange	Tangy and peppery
Carthamus tinctorius L.	Safflower	Yellow to deep red	Bitter
Chamaemelum nobilis All.	English chamomile	White petals, yellow center	Sweet apple
Chrysanthemum L.	Chrysanthemum	Various	Strong to bitter
Cichorium intybus L.	Chicory	Blue to lavender	Similar to endive
Glebionus coronarium L.	Garland chrysanthemum	Yellow to white	Mild
Leucanthemum vulgare Lam.	Oxeye daisy	White, yellow center	Mild
Tagetes erecta L.	African marigold	White, gold	Variable, mild to bitter

Tagetes tenuifolia Cav.	Signet marigold	White, gold, yellow, red	Citrus, milder than *T. erecta*
Taraxicum ceratophorum (Ledeb.) DC.	Dandelion	Yellow	Bitter
Begonaiceae	BEGONIA FAMILY		
Begonia tuberhybrida Voss	Tuberous begonia	Various	Citrus
Boraginaceae	BORAGE FAMILY		
Borago officinalis L.	Borage	Blue, purple, lavender	Cucumber
Brassicaceae	MUSTARD FAMILY		
Brassica spp.	Mustard	Yellow	Tangy to hot
Eruca vesicaria Mill.	Arugala	White	Nutty, smoky
Raphanus sativus L.	Radish	White, pink	Spicy
Caryophyllaceae	PINK FAMILY		
Dianthus spp.	Pinks	Pink, white, red	Spicy, cloves
Cucurbitaceae	GOURD FAMILY		
Cucurbita pepo L.	Summer squash, pumpkin	Yellow	Mild, raw squash
Fabaceae	PEA FAMILY		
Cercis canadensis L.	Redbud	Pink	Bean-like to tart apple
Phaseolus coccineus L.	Scarlet runner bean	Bright orange to scarlet	Mild raw bean
Pisum sativum L.	Garden pea	White, tinged pink	Raw pea
Trifolium pratense L.	Red clover	Pink, lilac	Hay

31

TABLE 1.3. BOTANICAL NAMES, COMMON NAMES, FLOWER COLOR, AND TASTE OF SOME EDIBLE FLOWERS (*Continued*)

Botanical Name	Common Name	Flower Color	Taste
Geraniaceae	GERANIUM FAMILY		
Pelargonium L'Hér.	Scented geraniums	White, red, pink, purple	Various, e.g., apple, lemon, rose, and spice
Iridaceae	IRIS FAMILY		
Gladiolus Gaertn.	Gladiolus	Various	Mediocre
Lamiaceae	MINT FAMILY		
Hyssopus officinalis L.	Hyssop	Blue, pink, white	Bitter, similar to tonic
Lavandula angustifolia Mill.	Lavender	Lavender, purple, pink, white	Highly perfumed
Melissa officinalis L.	Lemon balm	Creamy white	Lemony, sweet
Mentha spp. L.	Mint	Lavender, pink, white	Minty
Monarda didyma L.	Bee balm	Red, pink, white, lavender	Tea-like
Ocimum basilicum L.	Basil	White to pale pink	Spicy
Origanum vulgare L.	Oregano	White	Spicy, pungent
Origanum majorana L.	Marjoram	Pale pink	Spicy, sweet
Salvia elegans Vahl	Pineapple sage	Scarlet	Pineapple/sage overtones

Salvia rosmarinus Schleid.	Rosemary	Blue, pink, white	Mild rosemary
Salvia officinalis L.	Sage	Blue, purple, white, pink	Flowery sage
Satureja hortensis L.	Summer savory	Pink	Mildly peppery, spicy
Satureja montana L.	Winter savory	Pale blue to purple	Mildly peppery, spicy
Thymus spp. L.	Thyme	Pink, purple, white	Milder than leaves
Liliaceae	LILY FAMILY		
Tulipa spp. L.	Tulip	Various	Slightly sweet or bitter
Malvaceae	MALLOW FAMILY		
Abelmoschus esculentus (L.) Moench.	Okra	Yellow, red	Mild, sweet, slightly mucilaginous
Alcea rosea L.	Hollyhock	Various	Slightly bitter
Hibiscus rosa-sinensis L.	Hibiscus	Orange, red, purple	Citrus, cranberry
Hibiscus syriacus L.	Rose-of-Sharon	Red, white, purple, violet	Mild, nutty
Myrtaceae	MYRTLE FAMILY		
Feijoa sellowiana O. Berg	Pineapple guava	White to deep pink	Papaya or exotic melon
Oleaceae	OLIVE FAMILY		
Syringa vulgaris L.	Lilac	White, pink, purple, lilac	Perfume, slightly bitter
Rosaceae	ROSE FAMILY		
Malus spp. Mill.	Apple, crabapple	White to pink	Slightly floral to sour
Poterium sanguisorba subsp. *sanguisorba*	Burnet	Red	Cucumber

TABLE 1.3. BOTANICAL NAMES, COMMON NAMES, FLOWER COLOR, AND TASTE OF SOME EDIBLE FLOWERS (Continued)

Botanical Name	Common Name	Flower Color	Taste
Rubiaceae	MADDER FAMILY		
Galium odoratum (L.) Scop.	Sweet woodruff	White	Sweet, grassy, vanilla
Rutaceae	RUE FAMILY		
Citrus limon (L.) Burm.	Lemon	White	Citrus, slightly bitter
Citrus sinensis (L.) Osbeck.	Orange	White	Citrus, sweet/strong
Tropaeolaceae	NASTURTIUM FAMILY		
Tropaeolum majus L.	Nasturtium	Variable	Watercress, peppery
Violaceae	VIOLET FAMILY		
Viola odorata L.	Violet	Violet, pink, white	Sweet
Viola × wittrockiana Gams.	Pansy	Various, multicolored	Stronger than violets
Viola tricolor L.	Johnny-jump-up	Violet, white, yellow	Stronger than violets

Adapted from S. E. Newman and A. S. O'Connor, *Edible Flowers* (Colorado Cooperative Extension Fact Sheet 7.237, 2020), https://extension.colostate.edu/docs/pubs/garden/07237.pdf (accessed 17 Aug. 2021). Useful reference: J. Mackin, *Cornell Book of Herbs & Edible Flowers* (Ithaca, NY: Cornell Cooperative Extension, 1993).
Taxonomic data from POWO *Plants of the World Online* (2019), http://www.plantsoftheworldonline.org/ (accessed 17 Feb. 2021); From GBIF: The Global Biodiversity Information Facility (2021), https://www.gbif.org (accessed 17 Feb. 2021).

TABLE 1.4. U.S. VEGETABLE PRODUCTION STATISTICS: LEADING FRESH MARKET AND PROCESSING VEGETABLE STATES, 2017[1]

	Fresh Acreage		Processing Acreage	
Rank	State	% of Total	State	% of Total
1	California	33.2	California	17.2
2	Florida	7.9	Washington	14.1
3	Idaho	7.8	Wisconsin	11.1
4	Arizona	5.4	Minnesota	10.5
5	North Carolina	4.7	Idaho	8.6

Adapted from the USDA National Agricultural Statistics Service, 2017 Agricultural Census (2020), https://www.nass.usda.gov/Publications/AgCensus/2017 (accessed 15 Mar. 2021).

[1] Includes data for artichoke, asparagus, lima bean, snap bean, beets, broccoli, brussels sprouts, cabbages, cantaloupe, carrot, cauliflower, celery, chicory, collards, cucumber, daikon, eggplant, escarole/endive, garlic, ginger root, ginseng, fresh herbs, honeydew melon, horseradish, kale, lettuces, mustard greens, okra, onion (dry and green), peas (sugar, snow and green), southern peas, pepper (bell and other), potato, pumpkin, radish, rhubarb, spinach, squash (summer and winter), sweet corn, sweet potato, taro, tomato (field), turnip greens, turnip, watercress, and watermelon.

TABLE 1.5. HARVESTED ACREAGE, PRODUCTION, AND VALUE OF U.S. VEGETABLES FOR FRESH MARKET AND PROCESSING, 2017–2019 AVERAGE

Crop	Acres	Production (1,000 cwt)	Value ($1,000)	Yield per Acre (cwt)
Artichoke[1]	7,100	983	69,059	138
Asparagus[1]	22,900	792	96,057	35
Bean, lima	23,925	677	20,558	28
Bean, snap	205,200	17,236	334,945	84
Broccoli[1]	118,033	18,365	845,624	156
Cabbage	57,967	21,646	452,773	373
Cantaloupe	59,017	14,238	310,301	241
Carrot	78,433	42,227	764,198	527
Cauliflower[1]	45,400	9,546	438,702	210
Celery[1]	30,033	16,444	411,192	549
Cucumber	109,633	16,436	331,269	150
Garlic[1]	29,967	4,844	376,628	162
Honeydew melon	11,700	3,346	76,230	285
Lettuce, head	125,167	43,801	1,651,509	350
Lettuce, leaf	63,400	12,290	679,828	195
Lettuce, romaine	101,633	30,983	1,136,716	305
Onion	136,833	74,907	977,006	548
Pea, green	131,133	5,374	72,725	41
Pepper, bell[1]	40,067	13,140	572,181	328
Pepper, chile[1]	13,467	2,548	90,680	186
Potato	999,000	441,787	4,138,895	442
Pumpkin[1]	67,033	14,926	190,725	223
Spinach	63,993	8,600	471,100	134
Squash[1]	44,733	7,509	216,165	168
Sweet corn	443,133	70,700	802,767	160
Sweet potato	150,133	31,666	626,458	211
Tomato	302,033	254,709	1,714,061	843
Watermelon	107,200	39,059	606,131	365

Adapted from the USDA National Agricultural Statistics Service, Vegetables 2019 Summary (2020), https://www.nass.usda.gov/Publications/Todays_Reports/reports/vegean20.pdf. (accessed 29 Mar. 2021); From USDA National Agricultural Statistics Service, Potatoes 2019 Summary (2020), https://www. nass.usda.gov/Publications/Todays_Reports/reports/pots0920.pdf (accessed 29 Mar. 2021).
[1] Includes fresh market and processing.

TABLE 1.6. HARVESTED ACREAGE AND VALUE FOR U.S. VEGETABLES, HERBS, AND VEGETABLE TRANSPLANTS GROWN IN PROTECTED CULTIVATION, 2017

Crop	Square Feet Cultivated (1,000 sf)	Value ($1,000)	Rank of Top Three States (and Percentage of Total Crop Produced) #1	#2	#3
Tomato	63,929	418,960	CA (22.8%)	TX (8.8%)	NY (6.6%)
Other vegetables and fresh cut herbs	48,634	329,341	CA (42.5%)	FL (4.3%)	TX (3.8%)
Vegetable transplants for farm fields	29,859	250,168	CA (60.0%)	FL (9.2%)	GA (6.9%)

Adapted from the USDA National Agricultural Statistics Service, 2017 Agricultural Census (2020), https://www.nass.usda.gov/Publications/AgCensus/2017 (accessed 15 Mar. 2021).

TABLE 1.7. HARVESTED ACREAGE AND VALUE OF U.S. VEGETABLES GROWN IN PROTECTED CULTIVATION, 2019

Crop	Square Feet Cultivated (1,000 sf)	Value ($1,000)	Production (cwt) Hydroponic	Total
Cucumber	6,337	45,691	340,153	510,300
Herbs	10,685	65,153	73,598	326,309
Lettuce	5,531	71,129	363,885	551,716
Pepper	2,481	8,577	3,454	110,739
Strawberry	659	937	1,837	11,792
Tomato	52,576	345,025	2,533,489	4,165,635
Other vegetables	10,674	166,957	905,334	2,180,310
TOTAL	88,943	703,469	905,334	2,180,310

Adapted from the USDA National Agricultural Statistics Service, 2019 Census of Horticultural Specialties (2020), https://www.nass.usda.gov/Publications/AgCensus/2017/Online_Resources/Census_of_Horticulture_Specialties/HORTIC.pdf (accessed 20 Apr. 2021).

TABLE 1.8. IMPORTANT STATES IN THE PRODUCTION OF U.S. FRESH MARKET AND PROCESSING VEGETABLES BY ACRES GROWN, 2017

Crop	Rank, Acres Harvested for Fresh Market			Rank, Acres Harvested for Processing		
	#1	#2	#3	#1	#2	#3
Artichoke	CA	WA	OR	CA	—	—
Asparagus	MI	CA	WA	MI	WA	—
Bean, green lima	CA	NC	TN	DE	MD	WA
Bean, snap	FL	GA	TN	WI	NC	MI
Beets	CA	AZ	FL	WI	NY	OR
Broccoli[1]	CA	AZ	FL	OR	CA	VA
Brussels sprout	CA	NY	MI	CA	—	—
Cabbage	CA	NY	GA	NY	WI	MI
Cantaloupe	CA	AZ	TX	—	—	—
Carrot	CA	AZ	TX	WA	CA	WI
Cauliflower	CA	AZ	NY	CA	OR	—
Celery	CA	AZ	MI	MI	CA	—
Collard greens	NC	CA	FL	—	—	—
Cucumber	FL	MI	GA	MI	FL	CA
Garlic	CA	NV	NY	CA	—	—
Ginger	HI	FL	CA	—	—	—
Honeydew melon	CA	AZ	TX	—	—	—
Lettuce, head	CA	AZ	NM	—	—	—
Lettuce, leaf	CA	AZ	FL	—	—	—
Lettuce, romaine	CA	AZ	NJ	—	—	—
Mushroom[1]	PA	CA	WA	—	—	—
Okra	TX	FL	AL	—	—	—
Onion, dry	CA	WA	OR	CA	OR	WA
Onion, green	CA	OH	PA	—	—	—
Parsley	CA	NJ	FL	—	—	—
Pea, sugar/snow	CA	MN	ID	WI	ID	CA
Pea, green	WA	FL	GA	MN	WA	WI
Pepper, bell	CA	FL	GA	CA	ID	NJ
Pepper, other	CA	TX	NJ	CA	TX	OH

TABLE 1.8. IMPORTANT STATES IN THE PRODUCTION OF U.S. FRESH MARKET AND PROCESSING VEGETABLES BY ACRES GROWN, 2017 (*Continued*)

Crop	Rank, Acres Harvested for Fresh Market			Rank, Acres Harvested for Processing		
	#1	#2	#3	#1	#2	#3
Potato	ID	CO	WI	ID	WA	ND
Pumpkin[1]	IL	PA	NY	—	—	—
Radish	FL	CA	OH	—	—	—
Rhubarb	MI	NY	MA	—	—	—
Spinach	CA	AZ	TX	CA	TX	MD
Squash, summer	FL	CA	MI	—	—	—
Squash, winter	CA	MI	NC	—	—	—
Sweet corn[1]	MN	WA	WI	—	—	—
Sweet potato	NC	CA	MS	NC	MS	CA
Tomato (field)	FL	CA	TN	CA	IN	OH
Tomato (greenhouse)	CA	TX	NY	—	—	—
Turnips	CA	NY	AL	—	—	—
Watermelon	FL	TX	CA	—	—	—

Adapted from the USDA National Agricultural Statistics Service (2020). 2017 Agricultural Census. Available at: https://www.nass.usda.gov/Publications/AgCensus/2017 (accessed 15 Mar. 2021).
[1] Includes processing as well as fresh use.

TABLE 1.9.   UTILIZATION OF THE U.S. POTATO CROP, 2017–2019 AVERAGE

	Amount	
Item	1,000 cwt	% of Total
A. Sales	412,205	93
1. Table stock	104,717	24
2. Processing	287,893	65
a. Chips and shoestrings	60,363	14
b. All dehydrated (including starch and flour)	45,483	10
c. Frozen french fries	160,444	36
d. Other frozen products	13,869	3
e. Canned potatoes	1,054	<1
f. Other canned products	766	<1
3. Other sales	24,373	6
a. Livestock feed	1,379	<1
b. Seed	22,933	5
B. Nonsales	28,896	6
1. Seed used on farms where grown	3,669	1
2. Shrinkage	25,227	5
Total production	469,997	

Adapted from USDA National Agricultural Statistics Service, Potatoes 2019 Summary (2020), https://www.nass.usda.gov/Publications/Todays_Reports/reports/pots0920.pdf (accessed 29 Mar. 2021).

04

CONSUMPTION OF VEGETABLES IN THE UNITED STATES

TABLE 1.10.   TRENDS IN U.S. PER CAPITA CONSUMPTION OF VEGETABLES

	Amount (lb)[1]		
Year	Fresh	Processed	Total
1970	154	182	336
1975	149	190	339
1980	152	187	339
1985	159	202	361
1990	176	215	392
1995	188	226	414
2000	201	224	425
2005	196	218	415
2010	190	207	397
2015	186	193	379
2018	191	211	401

Adapted from USDA ERS Food Availability Per Capita Data (published 23 Sept. 2020), https://www.ers.usda.gov/data-products/food-availability-per-capita-data-system/ (accessed Apr. 2021).

[1] Fresh-weight equivalent.

TABLE 1.11.   U.S. PER CAPITA CONSUMPTION OF VEGETABLES, 2019[1]

	Amount (lb)			
Vegetable	Fresh	Canned	Frozen	Total
Artichoke, all	1.4	—	—	1.4
Asparagus	1.8	0.1	0.1	1.9
Bean, dry, all	—	—	—	6.9
Bean, lima	0.01	0.04	0.33	0.4
Bean, snap	1.8	2.9	1.9	6.6
Beets	—	0.5	—	0.5

Broccoli	6.2	—	2.6	8.8
Brussels sprout	0.8	—	—	0.8
Cabbage	6.5	0.7	—	7.1
Cantaloupe	7.0	—	—	7.0
Carrot	13.6	1.2	1.8	16.6
Cauliflower	3.0	—	0.7	3.7
Celery	5.3	—	—	5.3
Collard greens	0.8	—	—	0.8
Cucumber	8.0	3.4	—	11.4
Eggplant, all	0.9	—	—	0.9
Escarole/endive	0.2	—	—	0.2
Garlic, all	1.8	—	—	1.9
Honeydew melon	2.0	—	—	2.0
Kale	1.9	—	—	1.9
Lettuce, head	12.7	—	—	12.7
Lettuce, leaf and romaine	12.4	—	—	12.4
Mushroom, all	2.8	1.0	—	3.8
Mustard greens	0.7	—	—	0.7
Okra	0.6	—	—	0.6
Onion	20.4	—	—	22.3[2]
Pea, green	—	0.7	1.3	1.9
Pea and lentil, dry, all	—	—	—	4.2
Pepper, bell	11.3	—	—	11.3
Pepper, Chile	—	7.3	—	7.3
Potato	34.2	33.4[3]	51.5	119.1
Pumpkin	5.8	—	—	5.8
Radish	0.5	—	—	0.5
Spinach, all	2.5	0.13	0.7	3.3
Strawberry	7.1	—	1.8	8.9
Sweet corn	6.8	5.3	6.9	18.9
Sweet potato, all uses	7.9	—	—	7.9
Tomato	20.3	68.1	—	88.5
Turnip greens	0.3	—	—	0.3
Watermelon	14.0	—	—	14.0
Other vegetables, all	—	—	—	12.1

Adapted from USDA ERS, Food Availability Per Capita Data (2020), https://www.ers.usda.gov/data-products/food-availability-per-capita-data-system/ (accessed 20 Apr. 2021).

[1] Data for cantaloupe, honeydew, and strawberry are from 2018.
[2] Includes fresh and dehydrated onion.
[3] Other processed potato.

TABLE 1.12. IMPORTANT VEGETABLE PRODUCING
COUNTRIES, 2017–2019

Crop	First	Second	Third
Artichoke	Italy	Egypt	Spain
Asparagus	China	Peru	Mexico
Bean, string (snap, green)	United States	France	Morocco
Cabbage	China	India	Korea
Carrot and Turnip	China	Uzbekistan	United States
Cauliflower and Broccoli	China	India	United States
Cucumber	China	Turkey	Russian Federation
Eggplant	China	India	Egypt
Garlic	China	India	Bangladesh
Lettuce and Chicory	China	United States	India
Melons	China	Turkey	India
Mushroom	China	Japan	United States
Okra	India	Nigeria	Mali
Onion, dry	China	India	United States
Pea, green	China	India	France
Pepper and chillies, green	China	Mexico	Turkey
Potato	China	India	Russian Federation
Pumpkin, squash and gourds	China	India	Ukraine
Spinach	China	United States	Japan
Strawberry[1]	China	United States	Mexico
Sweet corn	United States	Mexico	Croatia
Sweet potato	China	Malawi	Tanzania
Tomato	China	United States	India
Watermelon	China	Turkey	India

Adapted from Data from the Food and Agriculture Organization of the United Nations, http://www.fao.org/faostat/en/#data/QC (accessed Apr. 2021).

TABLE 1.13. WORLD VEGETABLE PRODUCTION, 2019

Country	Production (million cwt)	(%)
China	13,022	52
India	2,907	12
United States	661	3
Turkey	558	2
Vietnam	374	1
Nigeria	368	1
Others	7,075	28
World	24,969	100

Adapted from Data from the Food and Agriculture Organization of the United Nations, http://www.fao.org/faostat/en/#data/QC/ (accessed Apr. 2021).

NUTRITIONAL COMPOSITION OF VEGETABLES

TABLE 1.14. COMPOSITION OF THE EDIBLE PORTIONS OF FRESH, RAW VEGETABLES

Vegetable	Water (%)	Energy (kcal)	Protein (g)	Fat (g)	Carbohydrate (g)	Fiber (g)	Ca (mg)	P (mg)	Fe (mg)	Na (mg)	K (mg)
					Amount/100 g Edible Portion						
Artichoke	85	47	3.3	0.2	10.5	5.4	44	90	1.3	94	370
Asparagus	93	20	2.2	0.1	3.9	2.1	24	52	2.1	2	202
Bean, green	90	31	1.8	0.1	7.1	3.4	37	38	1.0	6	209
Bean, lima	70	113	6.8	0.9	20.2	4.9	34	136	3.1	8	467
Beet greens	91	22	2.2	0.1	4.3	3.7	117	41	2.6	226	762
Beet roots	88	43	1.6	0.2	9.6	2.8	16	40	0.8	78	325
Broccoli	89	34	2.8	0.4	6.6	2.6	47	66	0.7	33	316
Broccoli raab	93	22	3.2	0.5	2.9	2.7	108	73	2.1	33	196
Brussels sprouts	86	43	3.4	0.3	9.0	3.8	42	69	1.4	25	389
Cabbage, common	92	24	1.4	0.1	5.6	2.3	47	23	0.6	18	246
Cabbage, red	90	31	1.4	0.6	7.4	2.1	45	30	0.8	27	243
Cabbage, savoy	91	27	2.0	0.1	6.1	3.1	35	42	0.4	28	230
Carrot	88	41	0.9	0.2	9.6	2.8	33	35	0.3	69	320
Cauliflower	92	25	2.0	0.1	5.3	2.5	22	44	0.4	30	303
Celeriac	89	42	1.5	0.3	9.2	1.8	43	115	0.7	100	300
Celery	95	14	0.7	0.2	3.0	1.6	40	24	0.2	80	260

Chayote	95	17	0.8	0.1	3.9	1.7	17	18	0.3	2	125
Chicory, witloof	95	17	0.9	0.1	4.0	3.1	28	26	0.2	2	211
Chinese cabbage	95	13	1.5	0.2	2.2	1.0	105	37	0.8	65	252
Collards	91	30	2.5	0.4	5.7	3.6	145	10	0.2	20	169
Cucumber	95	15	0.7	0.1	3.6	0.5	16	24	0.3	2	147
Eggplant	92	24	1.0	0.2	5.7	3.4	9	25	0.2	2	230
Endive	94	17	1.3	0.2	3.4	3.1	52	28	0.8	22	314
Garlic	59	149	6.4	0.5	33.1	2.1	181	153	1.7	17	401
Kale	84	50	3.3	0.7	10.0	2.0	135	56	1.7	43	447
Kohlrabi	91	27	1.7	0.1	6.2	3.6	24	46	0.4	20	350
Leek	83	61	1.5	0.3	14.1	1.8	59	35	2.1	20	180
Lettuce, butterhead	96	13	1.4	0.2	2.3	1.1	35	33	1.2	5	238
Lettuce, crisphead	96	14	0.9	0.1	3.0	1.2	18	20	0.4	10	141
Lettuce, green leaf	94	18	1.3	0.3	3.5	0.7	68	25	1.4	9	264
Lettuce, red leaf	96	16	1.3	0.2	2.3	0.9	33	28	1.2	25	187
Lettuce, romaine	95	17	1.2	0.3	3.3	2.1	33	30	1.0	8	247
Melon, cantaloupe	90	34	0.8	0.2	8.2	0.9	9	15	0.2	16	267
Melon, casaba	92	28	1.1	0.1	6.6	0.9	11	5	0.2	9	182
Melon, honeydew	90	36	0.5	0.1	9.1	0.8	6	11	0.2	18	228
Mushroom	92	22	3.1	0.3	3.2	1.2	3	85	0.5	4	314
Mustard greens	91	26	2.7	0.2	4.9	3.3	103	43	1.5	25	354
Okra	90	31	2.0	0.1	7.0	3.2	81	63	0.8	8	303
Onion, bunching	90	32	1.8	0.2	7.3	2.6	72	37	1.5	16	276
Onion, dry	89	42	0.9	0.1	10.1	1.4	22	27	0.2	3	144
Parsley	88	36	3.0	0.8	6.3	3.3	138	58	6.2	56	554
Parsnip	80	75	1.2	0.3	18.0	4.9	36	71	0.6	10	375

TABLE 1.14. COMPOSITION OF THE EDIBLE PORTIONS OF FRESH, RAW VEGETABLES (*Continued*)

Amount/100 g Edible Portion

Vegetable	Water (%)	Energy (kcal)	Protein (g)	Fat (g)	Carbohydrate (g)	Fiber (g)	Ca (mg)	P (mg)	Fe (mg)	Na (mg)	K (mg)
Pea, edible-podded	89	42	2.8	0.2	7.6	2.6	43	53	2.1	4	200
Pea, green	79	81	5.4	0.4	14.5	5.1	25	108	1.5	5	244
Pepper, hot, chili	88	40	2.0	0.2	9.5	1.5	18	46	1.2	7	340
Pepper, sweet	94	20	0.9	0.2	4.6	1.7	10	20	0.3	3	175
Potato	79	77	2.0	0.1	17.5	2.2	12	57	0.8	6	421
Pumpkin	92	26	1.0	0.1	6.5	0.5	21	44	0.8	1	340
Radicchio	93	23	1.4	0.3	4.5	0.9	19	40	0.6	22	302
Radish	95	16	0.7	0.1	3.4	1.6	25	20	0.3	39	233
Rhubarb	94	21	0.9	0.2	4.5	1.8	86	14	0.2	4	288
Rutabaga	90	36	1.2	0.2	8.1	2.5	47	58	0.5	20	337
Salsify	77	82	3.3	0.2	18.6	3.3	60	75	0.7	20	380
Shallot	80	72	2.5	0.1	16.8	-	37	60	1.2	12	334
Southernpea	77	90	3.0	0.4	18.9	5.0	126	53	1.1	4	431
Spinach	91	23	2.9	0.4	3.6	2.2	99	49	2.7	79	558
Squash, acorn	88	40	0.8	0.1	10.4	1.5	33	36	0.7	3	347
Squash, butternut	86	45	1.0	0.1	11.7	2.0	48	33	0.7	4	352
Squash, hubbard	88	40	2.0	0.5	8.7	-	14	21	0.4	7	320
Squash, scallop	94	18	1.2	0.2	3.8	-	19	36	0.4	1	182
Squash, summer	95	16	1.2	0.2	3.4	1.1	15	38	0.4	2	262

Squash, zucchini	97	16	1.2	0.2	3.4	1.1	15	38	0.4	10	262
Strawberry	91	32	0.7	0.3	7.7	2.0	16	24	0.4	1	153
Sweet corn	76	86	3.2	1.2	19.0	2.7	2	89	0.5	15	270
Sweet potato	77	86	1.6	0.1	20.1	3.0	30	47	0.6	55	337
Swiss chard	93	19	1.8	0.2	3.7	1.6	51	46	1.8	213	379
Taro	71	112	1.5	0.2	26.5	4.1	43	84	0.6	11	591
Tomato, green	93	23	1.2	0.2	5.1	1.1	13	28	0.5	13	204
Tomato, ripe	95	18	0.9	0.2	3.9	1.2	10	24	0.3	5	237
Turnip greens	90	32	1.5	0.3	7.1	3.2	190	42	1.1	40	296
Turnip roots	92	28	0.9	0.1	6.4	1.8	30	27	0.3	67	191
Watermelon	92	30	0.6	0.2	7.6	0.4	7	10	0.2	1	112
Waxgourd	96	13	0.4	0.2	3.0	2.9	19	19	0.4	111	6

Data are from USDA, Nutrient Database for Standard Reference, Release 17 (2005). These legacy and additional food nutrient data are available at: U.S. Department of Agriculture, Agricultural Research Service, FoodData Central [Online] (2019), fdc.nal.usda.gov.

TABLE 1.15. VITAMIN CONTENT OF FRESH RAW VEGETABLES

Vegetable	Amount/100 g Edible Portion					
	Vitamin A (IU)	Thiamine (mg)	Riboflavin (mg)	Niacin (mg)	Ascorbic acid (mg)	Vitamin B$_6$ (mg)
Artichoke	0	0.07	0.07	1.05	11.7	0.12
Asparagus	756	0.14	0.14	0.98	5.6	0.09
Bean, green	690	0.08	0.11	0.75	16.3	0.07
Bean, lima	303	0.22	0.10	1.47	23.4	0.20
Beet greens	6,326	0.10	0.22	0.40	30.0	0.11
Beet roots	33	0.03	0.04	0.33	4.9	0.07
Broccoli	660	0.07	0.12	0.64	89.2	0.18
Broccoli rabb	2,622	0.16	0.13	1.2	20.2	0.17
Brussels sprouts	754	0.14	0.09	0.75	85.0	0.22
Cabbage, common	171	0.05	0.04	0.30	32.2	0.10
Cabbage, red	1,116	0.06	0.07	0.42	57.0	0.21
Cabbage, savoy	1,000	0.07	0.03	0.30	31.0	0.19
Carrot	12,036	0.07	0.06	1.0	5.9	0.14
Cauliflower	13	0.06	0.06	0.53	46.4	0.22
Celeriac	0	0.05	0.06	0.70	8.0	0.17
Celery	449	0.02	0.06	0.32	3.1	0.07
Chayote	0	0.03	0.03	0.47	7.7	0.08
Chicory, witloof	29	0.6	0.03	0.16	2.8	0.04

Chinese cabbage	4,468	0.04	0.07	0.50	45.0	0.19
Collards	6,668	0.05	0.13	0.74	35.3	0.17
Cucumber	105	0.03	0.03	0.10	2.8	0.04
Eggplant	27	0.04	0.04	0.65	2.2	0.08
Endive	2,167	0.08	0.08	0.40	6.5	0.02
Garlic	0	0.20	0.11	0.70	31.2	1.20
Kale	15,376	0.11	0.13	1.00	120.0	0.27
Kohlrabi	36	0.05	0.02	0.40	62.0	0.15
Leek	1,667	0.06	0.03	0.40	12.0	0.23
Lettuce, butterhead	3,312	0.06	0.06	0.40	3.7	0.08
Lettuce, crisphead	502	0.04	0.03	0.12	2.8	0.04
Lettuce, green leaf	7,405	0.07	0.08	0.38	18.0	0.09
Lettuce, red leaf	7,492	0.06	0.08	0.32	3.7	0.10
Lettuce, romaine	5,807	0.10	0.10	0.31	24.0	0.07
Melon, cantaloupe	3,382	0.04	0.02	0.73	36.7	0.07
Melon, casaba	0	0.02	0.03	0.23	21.8	0.16
Melon, honeydew	40	0.08	0.02	0.60	24.8	0.06
Mushroom	0	0.09	0.42	3.85	2.4	0.12
Mustard greens	10,500	0.08	0.11	0.80	70.0	0.18
Okra	375	0.20	0.06	1.00	21.1	0.22

TABLE 1.15. VITAMIN CONTENT OF FRESH RAW VEGETABLES (*Continued*)

Vegetable	Amount/100 g Edible Portion						
	Vitamin A (IU)	Thiamine (mg)	Riboflavin (mg)	Niacin (mg)	Ascorbic acid (mg)	Vitamin B$_6$ (mg)
Onion, bunching	997	0.06	0.08	0.53	18.8	0.06
Onion, dry	2	0.05	0.03	0.08	6.4	0.15
Parsley	8,424	0.09	0.10	1.31	133.0	0.09
Parsnip	0	0.09	0.05	0.70	17.0	0.09
Pea, edible-podded	1,087	0.15	0.08	0.60	60.0	0.16
Pea, green	640	0.27	0.13	2.09	40.0	0.17
Pepper, hot, chili	1,179	0.09	0.09	0.95	242.5	0.28
Pepper, sweet	370	0.06	0.03	0.48	80.4	0.22
Potato	2	0.08	0.03	1.05	19.7	0.30
Pumpkin	7,384	0.05	0.11	0.60	9.0	0.06
Radicchio	27	0.02	0.03	0.26	8.0	0.06
Radish	7	0.01	0.04	0.25	14.8	0.07
Rhubarb	102	0.02	0.03	0.30	8.0	0.02
Rutabaga	2	0.09	0.04	0.70	25.0	0.10
Salsify	0	0.08	0.22	0.50	8.0	0.28
Shallot	12	0.06	0.02	0.2	8.0	0.35
Southernpea	0	0.11	0.15	1.45	2.5	0.07
Spinach	9,377	0.08	0.19	0.72	28.1	0.20
Squash, acorn	367	0.14	0.01	0.70	11.0	0.15

Squash, butternut	10,630	0.10	0.02	1.20	21.0	0.15
Squash, hubbard	1,367	0.07	0.04	0.50	11.0	0.15
Squash, scallop	110	0.07	0.03	0.60	18.0	0.11
Squash, summer	200	0.05	0.14	0.49	17.0	0.22
Squash, zucchini	200	0.05	0.14	0.49	17.0	0.22
Strawberry	12	0.02	0.02	0.39	58.8	0.05
Sweet corn	208	0.20	0.06	1.70	6.8	0.06
Sweet potato	14,187	0.08	0.06	0.56	2.4	0.80
Swiss chard	6,116	0.04	0.09	0.40	30.0	0.10
Taro	76	0.10	0.03	0.60	4.5	0.28
Tomato, green	642	0.06	0.04	0.50	23.4	0.08
Tomato, ripe	833	0.04	0.02	0.60	12.7	0.08
Turnip greens	0	0.07	0.10	0.60	60.0	0.26
Turnip roots	0	0.04	0.03	0.40	21.0	0.09
Watermelon	569	0.03	0.02	0.18	8.1	0.05
Waxgourd	0	0.04	0.11	0.40	13.0	0.04

Data are from USDA, Nutrient Database for Standard Reference, Release 17 (2005). These legacy and additional food nutrient data are available at: U.S. Department of Agriculture, Agricultural Research Service, FoodData Central [Online] (2019), fdc.nal.usda.gov.

Knott's Handbook for Vegetable Growers, Sixth Edition. George J. Hochmuth
and Rebecca G. Sideman.
© 2023 John Wiley & Sons Ltd. Published 2023 by John Wiley & Sons Ltd.
Companion website: https://www.wiley.com/go/hochmuth/vegetablegrowers6e

01
SEED LABELS

Seeds entering into interstate commerce must meet the requirements of the Federal Seed Act. Most state seed laws conform to federal standards. However, the laws of the individual states vary considerably with respect to the kinds and tolerances for noxious weeds. The noxious weed seed regulations and tolerances, if any, may be obtained from the State Seed Laboratory of any state. Vegetable seed in packets or in larger containers shall be labeled in any form that is clearly legible with the following required information:

- *Kind, variety, and hybrid.* The name of the kind and variety and hybrid, if appropriate, must be on the label. Words or terms that create a misleading impression as to the history or characteristics of kind or variety shall not be used.
- *Name of shipper or consignee.* The full name and address of either the shipper or consignee shall appear on the label.
- *Germination.* Vegetable seeds in containers of 1 lb or less with germination equal to or more than the standards need not be labeled to show the percentage germination or date of test. Vegetable seeds in containers of more than 1 lb shall be labeled to show the percentage of germination, the month and year of test, and the percentage of hard seed, if any.
- *Lot number.* The lot number or other lot identification of vegetable seed in containers of more than 1 lb shall be shown on the label and shall be the same as that used in the records pertaining to the same lot of seed.
- *Seed treatment.* Any vegetable seed that has been treated shall be labeled in no smaller than eight-point type to indicate that the seed has been treated and to show the name of any substance used in such treatment.

Adapted from Federal Seed Act Regulations, https://www.ams.usda.gov/rules-regulations/fsa (accessed Apr. 2021).

SEED GERMINATION TESTS

According to the Federal Seed Act §201.4, each person transporting vegetable seed in interstate commerce must keep a complete record of each seed lot, including a sample, for a period of 3 years. The complete record shall include the records of all laboratory tests for germination and hard seed for each seed lot. The record shall show the kind of seed, lot number, date of test, and percentage of germination and hard seeds. The conditions required for vegetable seed germination tests are shown in Table 2.1. In addition to these conditions, the substrate must be moist enough to supply the needed moisture to the seeds at all times.

TABLE 2.1. REQUIREMENTS FOR VEGETABLE SEED GERMINATION TESTS

						Additional Directions
Seed	Substrata[1]	Temperature[2] (°F)	First Count (Days)	Final Count (Days)	Specific Requirements[3]	Fresh and Dormant Seed
Artichoke	B, T	68–86	7	21		
Asparagus	B, T, S	68–86	7	21		
Bean, asparagus	B, T, S	68–86	5	8[4]		
Bean, garden	B, T, S, TC	68–86; 77	None	8		Use 0.3–0.6% Ca(NO₃)₂ to moisten substratum for retesting if hypocotyl collar rot is observed in initial test
Bean, lima	B, T, C, S	68–86	5	9[4]		
Bean, runner	B, T, S	68–86	5	9[4]		
Beet	B, T, S	68–86	3	14	Presoak seeds in water for 2 hrs	
Broadbean	S, C	64	4	14[4]		Prechill at 50°F for 3 days
Broccoli	B, P, T	68–86	3	10		Prechill at 41 or 50°F for 3 days; KNO₃ and light
Brussels sprouts	B, P, T	68–86	3	10		Prechill at 41 or 50°F for 3 days; KNO₃ and light
Burdock, great	B, T	68–86	7	14		
Cabbage	B, P, T	68–86	3	10		Prechill at 41 or 50°F for 3 days; KNO₃ and light

59

TABLE 2.1. REQUIREMENTS FOR VEGETABLE SEED GERMINATION TESTS (Continued)

Seed	Substrata[1]	Temperature[2] (°F)	First Count (Days)	Final Count (Days)	Additional Directions	
					Specific Requirements[3]	Fresh and Dormant Seed
Cabbage, Chinese	B, T	68–86	3	7		
Cabbage, tronchuda	B, P	68–86	3	10		Prechill at 41 or 50°F for 3 days; KNO_3 and light
Cantaloupe	B, T, S	68–86	4	10	Keep substratum on dry side; remove excess moisture	
Cardoon	B, T	68–86	7	21		
Carrot	B, T	68–86	6	14		
Cauliflower	B, P, T	68–86	3	10		Prechill at 41 or 50°F for 3 days; KNO_3 and light
Celeriac	P	59–77; 68	10	21	Light; 750–1,250 lux from cool-white fluorescent source	
Celery	P	59–77; 68	10	21	Light; 750–1,250 lux from cool-white fluorescent source	

Kind	Substrata	Temperature (°F)	First count (days)	Final count (days)	Additional directions	Additional directions
Chard, Swiss	B, T, S	68–86	3	14	Presoak seed in water for 2 hrs	
Chicory	P, TS	68–86	5	14	Light; KNO_3 or soil	
Chives	B, T	68	6	14		
Citron	B, T	68–86	7	14	Soak seeds 6 hrs	Test at 86°F
Collards	B, P, T	68–86	3	10		Prechill at 41 or 50°F for 3 days; KNO_3 and light
Corn, sweet	B, T, S, TC, TCS	68–86; 77	4	7		
Corn salad	B, T	59	7	28		Test at 50°F
Cowpea	B, T, S	68–86	5	8[4]		
Cress:						
Garden	B, P, T	59	4	10	Light; KNO_3	Light
Upland	P	68–86		7		Make first count when necessary or desirable
Water	P	68–86	4	14	Light	
Cucumber	B, T, S	68–86	3	7	Keep substratum on dry side; remove excess moisture	
Dandelion	P, TB	68–86	7	21	Light, 750–1,250 lux	
Dill	B, T	68–86	7	21		
Eggplant	P, TB, RB, T	68–86	7	14		Light; KNO_3
Endive	P, TS	68–86	5	14	Light, KNO_3 or soil	

TABLE 2.1. REQUIREMENTS FOR VEGETABLE SEED GERMINATION TESTS (Continued)

Seed	Substrata[1]	Temperature[2] (°F)	First Count (Days)	Final Count (Days)	Specific Requirements[3]	Fresh and Dormant Seed
Fennel	B, T	68–86	6	14		Prechill at 41 or 50°F for 3 days; KNO_3 and light
Kale	B, P, T	68–86	3	10		Prechill at 41 or 50°F for 3 days; KNO_3 and light
Kale, Chinese	B, P, T	68–86	3	10		
Kale, Siberian	B, P, T	68–86; 68	3	7		
Kohlrabi	B, P, T	68–86	3	10		Prechill at 41 or 50°F for 3 days; KNO_3 and light
Leek	B, T	68	6	14		
Lettuce	P	68	None	7	Light	Prechill at 50°F for 3 days or test at 59°F
Mustard, India	P	68–86	3	7	Light	Prechill at 50°F for 7 days and test for 5 additional days
Mustard, spinach	B, T	68–86	3	7		
Okra	B, T	68–86	4	14[4]		
Onion	B, T	68	6	10		
Alternate method	S	68	6	12		

62

Name	Codes	Temp			Notes
Onion, Welsh	B, T	68	10	6	
Pak-choi	B, T	68–86	7	3	
Parsley	B, T, TS	68–86	28	11	
Parsnip	B, T, TS	68–86	28	6	
Pea	B, T, S	68	8^4	5	
Pepper	TB, RB, T, B, P	68–86	14	6	Light and KNO_3
Pumpkin	B, T, S	68–86	7	4	Keep substratum on dry side; remove excess moisture
Radish	B, T	68	6	4	
Rhubarb	TB, TS	68–86	21	7	Light
Rutabaga	B, T	68–86	14	3	
Sage	B, T, S	68–86	14	5	
Salsify	B, T	59	10	5	Prechill at 50°F for 3 days
Savory, summer	B, T	68–86	21	5	
Sorrel	P, TB, TS	68–86	14	3	Light
Soybean	B, T, S, TC, TCS	68–86; 77	8^4	5	Test at 59°F
Spinach	TB, T	59, 50	21	7	Keep substratum on dry side; remove excess moisture

TABLE 2.1. REQUIREMENTS FOR VEGETABLE SEED GERMINATION TESTS (*Continued*)

Seed	Substrata[1]	Temperature[2] (°F)	First Count (Days)	Final Count (Days)	Specific Requirements[3]	Additional Directions — Fresh and Dormant Seed
Spinach, New Zealand	T	59, 68	5	21	Soak fruits overnight (16 hrs), air dry 7 hrs; plant in very wet towels; do not rewater unless later counts exhibit drying out	On 21st day, scrape fruits and test for 7 additional days
Squash	B, T, S	68–86	4	7	Keep substratum on dry side; remove excess moisture	
Tomato	B, P, RB, T	68–86	5	14		
Tomato, husk	P, TB	68–86	7	28	Light; KNO$_3$	
Turnip	B, T	68–86	3	7		
Watermelon	B, T, S	68–86; 77	4	14	Keep substratum on dry side; remove excess moisture	Light; KNO$_3$ Test at 86°F

64

TABLE 2.1.   REQUIREMENTS FOR VEGETABLE SEED GERMINATION TESTS (*Continued*)

Adapted from Federal Seed Act Regulations, https://www.ams.usda.gov/rules-regulations/fsa (accessed Apr 2021).

[1] B = between blotters

TB = top of blotters

T = paper toweling, used either as folded towel tests or as roll towel tests in horizontal or vertical position

S = sand or soil

TS = top of sand or soil

P = covered petri dishes: with two layers of blotters; with one layer of absorbent cotton; with five layers of paper toweling; with three thicknesses of filter paper; or with sand or soil

C = creped cellulose paper wadding (0.3-in. thick Kimpak or equivalent) covered with a single thickness of blotter through which holes are punched for the seed that are pressed for about one-half their thickness into the paper wadding.

TC = on top of creped cellulose paper without a blotter

RB = blotters with raised covers, prepared by folding up the edges of the blotter to form a good support for the upper fold which serves as a cover, preventing the top from making direct contact with the seeds.

[2] A single number indicates a constant temperature. Two numerals separated by a dash indicate an alternation of temperature; the test is to be held at the first temperature for approximately 16 hrs and at the second temperature for approximately 8 hrs per day. In cases where two temperatures are indicated (separated by a semicolon) the first temperature shall be regarded as the regular method and the second as an alternate method.

[3] *Light* means that seeds shall be illuminated with a cool-white fluorescent light, at an intensity of 75–125 foot-candles (750–1,250 lux), for at least 8 hrs every 24 hrs. *Potassium nitrate* (KNO$_3$) means that a two-tenths (0.2) percent solution of potassium nitrate (KNO$_3$) shall be used to moisten the substratum. Such solution is prepared by dissolving 2 g of KNO$_3$ in 1,000 mL of distilled water.

[4] Hard seeds. Seeds that remain hard at the end of the prescribed test because they have not absorbed water, due to an impermeable seed coat, are to be counted as "hard seed." If at the end of the germination period provided for legume and okra there are still present swollen seeds or seeds that have just started to germinate, all seeds or seedlings except the above-stated shall be removed and the test continued for five additional days and the normal seedlings included in the percentage of germination.

65

SEED GERMINATION STANDARDS

The following germination standards for vegetable seeds in interstate commerce (Table 2.2), which shall be construed to include hard seed, are determined and established under the Federal Seed Act.

TABLE 2.2. GERMINATION STANDARDS FOR VEGETABLE SEEDS IN INTERSTATE COMMERCE

Seed	%
New Zealand spinach, watercress	40
Chives, okra, tomatillo	50
Carrot, celeriac, celery, pepper, summer savory	55
Artichoke, burdock (great), cardoon, upland cress, dandelion, dill, eggplant, leek, parsley, parsnip, rhubarb, sage, spinach	60
Beet, chard (Swiss), chicory, citron, sorrel	65
Asparagus, bean (garden), bean (lima), brussels sprouts, cabbage (tronchuda), corn salad (Valerianella), endive, onion, scallion, watermelon	70
Bean (asparagus), bean (runner), broad bean, broccoli, cabbage, cauliflower, cabbage (Chinese), corn (sweet), cowpea, garden cress, kale, kohlrabi, melon, mustard, pak-choi, pumpkin, radish, rutabaga, salsify, soybean, squash, tomato	75
Collards, cucumber, lettuce, pea, turnip	80

Adapted from Federal Seed Act Regulations, https://www.ams.usda.gov/rules-regulations/fsa (accessed Apr 2021).

SEED PRODUCTION

Self-pollinated vegetable crops. The following crops are primarily self-pollinated and have very little outcrossing: bean (all types), chicory, endive, eggplant, lettuce, okra, pea, pepper, and tomato. Consequently, the only isolation necessary is to have plantings spaced far enough apart to prevent mechanical mixture at planting or harvest. A tall-growing crop is often planted between different varieties. Some of these (fava and runner bean, eggplant, pepper, tomato) outcross at a rate of 1–5%; for absolute genetic control, treat these plants as outcrossers and hand-pollinate, cage, or observe isolation distances (see below in Tables 2.3 and 2.4).

Cross-pollinated vegetables. Cross-pollination of vegetables may occur by wind or insect activity. Therefore, seed producers must isolate plantings of different varieties of the same crop or different crops in the same family that will cross with each other. Some general isolation guidelines are provided; however, the seed grower should follow the recommendations of the seed company for whom the seed is being grown.

TABLE 2.3. ISOLATION DISTANCES BETWEEN PLANTINGS OF VEGETABLES FOR OPEN-POLLINATED SEED PRODUCTION

Primarily Wind-pollinated	Distance (Miles)
Beet	½–2; 5 from sugar beet or Swiss chard
Sweet corn	1
Spinach	¼–3
Swiss chard	¾–5; 5 from sugar beet or Swiss chard

Primarily Insect-pollinated	Distance (Miles)
Asparagus	¼
Broccoli	½–3
Brussels sprouts	½–3
Cabbage	½–3
Cauliflower	½–3
Collards	¾–3; 5 from other cole crops
Kale	¾–2; 5 from other cole crops
Kohlrabi	½–3
Carrot	½–3
Celeriac	1
Celery	1
Chinese cabbage	1
Cucumber	1½ for varieties; ¼ from other cucurbits
Eggplant	¼
Gherkin	¼
Leek	1
Melons	1½–2 for varieties; ¼ from other cucurbits
Mustard	1
Onion	1–3
Parsley	½–1
Pepper	½
Pumpkin	1½–2 for varieties; ¼ from other cucurbits
Radish	¼–2
Rutabaga	¼–2
Spinach	¼–2
Squash	1½–2 for varieties; ¼ from other cucurbits
Turnip	¼–2

Adapted in part from Seed Production in the Pacific Northwest (Pacific Northwest Extension Publications, 1985).

TABLE 2.4. MINIMUM LAND, ISOLATION, FIELD, AND SEED STANDARDS FOR CLASSES OF CERTIFIED VEGETABLE SEED

Vegetable	Foundation				Registered				Certified			
	Land[1]	Isolation[2]	Field[3]	Seed[4]	Land[1]	Isolation[2]	Field[3]	Seed[4]	Land[1]	Isolation[2]	Field[3]	Seed[4]
Bean	1	0	2,000	0.05	1	0	1,000	0.1	1	0	400	0.2
Bean, broad	1	0	2,000	0.05	1	0	1,000	0.1	1	0	500	0.2
Bean, mung	1	0	1,000	0.1	1	0	500	0.2	1	0	200	0.5
Corn, sweet	—	—	—	—	1	—	—	—	0	660	1000	0.5
Okra	1	1,320	0	0	1	1,320	2,500	0.5	1	825	1,250	1.0
Onion	1	5,280	200	0	1	2,640	200	0.5	1	1,320	200	1.0
Pepper	1	200	0	0	1	100	300	0.5	1	30	150	1.0
Southern pea	1	0	2,000	0.1	1	0	1,000	0.2	1	0	500	0.5
Tomato	1	200	0	0	1	100	300	0.5	1	30	150	1.0
Watermelon	1	2,640	0	0	1	2,640	0	0.5	1	1,320	500	1.0

Adapted from Federal Seed Act Regulations, USDA, AMS. Certified Seed. Minimum Land, Isolation, Field, and Seed Standards (2021), https://www.ecfr. gov/cgi-bin/text-idx?SID=52aa08c19f803ff094df2c9c5fa0a7ff&mc=true&node=se7.3.201_1176&rgn=div8 https://www.ecfr.gov/current/title-7/subtitle-B/ chapter-I/subchapter-K/part-201/subject-group-ECFR1d5877781d6f9694/section-201.76

[1] Years that must elapse between destruction of a previous crop and establishment of a new crop of the same kind.
[2] Distance in feet from any contaminating source.
[3] Minimum number of plants in which one off-type plant is permitted.
[4] Maximum percentage of off-type seeds that are permitted in cleaned seed.

GENERAL VEGETABLE SEED PRODUCTION RESOURCES

Onion Seed Production in California, https://anrcatalog.ucanr.edu/pdf/8008.pdf

Cucurbit Seed Production in California, https://ucanr.edu/sites/sbc/files/200655.pdf

Carrot Seed Production, http://www.ars.usda.gov/research/docs.htm?docid=5235

Crop Profile for Table Beet Seed in Washington, https://ipmdata.ipmcenters.org/documents/cropprofiles/WAbeetseed2007.pdf

Crop Profile for Cabbage Seed in Washington, https://mtvernon.wsu.edu/path_team/WA-cabbage-seed-Crop-Profile-Dec-2007.pdf

Crop Profile for Spinach Seed in Washington, https://ipmdata.ipmcenters.org/documents/cropprofiles/WAspinachseed.pdf

05

ORGANIC SEED PRODUCTION

The USDA's National Organic Program (NOP) requires that certified organic growers use organic seed when it is commercially available. The NOP allows certified organic growers to use conventional seed when an equivalent organic variety is not commercially available; for this reason, not all organic farming uses organic seed. Some research, but not all, has shown that varieties bred under organic conditions are likely to be better adapted to organic production conditions than conventionally-bred varieties. Many seed companies now provide organic seeds, and some experienced organic market growers are diversifying into organic seed production, tapping into new market opportunities.

Organic vegetable seed must be grown according to the standards for organic food production, and organic seed producers must be certified by an accredited certifying agent. There are several challenges unique to organic seed production. Crops grown for seed must stay healthy for a much longer period than the same crop grown for direct consumption; this longer cropping season provides more opportunities for pests and pathogens to attack. Organic seed crops cannot be treated with some chemicals used in conventional seed production to keep seed crops disease-free and to enhance pollination. Organic seeds must also remain free of contamination by genetically-engineered organisms, which often requires that seed producers provide extra isolation for their fields and pay for supplemental testing.

ORGANIC VEGETABLE SEED PRODUCTION RESOURCES

Organic Seed Alliance, https://seedalliance.org/all-publications/
Northern Organic Vegetable Improvement Collaborative, https://eorganic.info/group/5751
The Organic Seed Grower: A farmer's guide to vegetable seed production, https://www.sare.org/resources/the-organic-seed-grower/
Principles and Practices of Organic Spinach Seed Production in the Pacific Northwest, https://projects.sare.org/wp-content/uploads/1626spinach_seed_manual.pdf

TABLE 2.5.   VEGETABLE SEED YIELDS[1]

Vegetable	(lb/acre)	
	Average Yield	Yield Range
Asparagus		
o.p.[2]	925	380–2,800
F_1	500	
F_2	750	
Bean, snap	1,800	1,400–2,800
Bean, lima	2,220	1,500–3,000
Beet		
o.p.	1,950	1,800–2,500
F_1	1,150	900–1,400
Broccoli		
o.p.	725	350–1,000
F_1	375	250–500
Brussels sprouts		
o.p.	900	800–1,000
F_1	425	250–600
Cabbage		
o.p.	740	500–1,000
F_1	440	300–600
Cantaloupe		
o.p.	420	350–500
F_1	225	175–300
Carrot		
o.p.	840	500–1,000
F_1	450	200–800
Cauliflower		
o.p.	540	350–1,000
F_1	175	100–250
Celeriac	1,200	800–2,000
Celery	835	500–1,200
Chard, Swiss	1,600	1,000–2,000
Chicory	500	400–600

TABLE 2.5. VEGETABLE SEED YIELDS (*Continued*)

	(lb/acre)	
Vegetable	Average Yield	Yield Range
Chinese cabbage		
o.p.	900	800–1,000
F_1	400	300–500
Cilantro	2,000	1,500–2,500
Corn, sweet		
Su	1,940	1,500–2,500
sh_2	1,100	400–1,700
Cucumber		
Beit alpha	450	350–550
Pickle	650	450–850
Slicer		
o.p.	500	275–600
F_1	290	200–550
Eggplant		
o.p.	640	500–775
F_1	500	400–625
Endive	735	650–800
Florence fennel		
o.p.	1,500	1,000–2,000
F_1	700	600–800
Kale		
o.p.	1,100	1,000–1,200
F_1	650	600–700
Kohlrabi		
o.p.	875	850–900
F_1	450	400–500
Leek		
o.p.	625	500–850
F_1	300	200–400
Lettuce	600	450–800
Mustard	1,325	1,300–1,350
New Zealand spinach	1,750	1,500–2,000

TABLE 2.5. VEGETABLE SEED YIELDS (*Continued*)

Vegetable	Average Yield	Yield Range
	(lb/acre)	
Okra		
o.p.	1,600	1,200–2,000
F_1	650	600–700
Onion		
o.p.	690	575900
F_1	450	350–550
Parsley	900	600–1,200
Parsnip	975	600–1,300
Pea	2,085	1,000–3,000
Pepper		
o.p.	170	100–300
F_1	125	100–150
Pumpkin		
o.p.	575	300–850
F_1	300	235–400
Radish		
o.p.	1,200	600–2,000
F_1	525	200–1,000
Rutabaga	2,200	1,800–2,500
Salisfy	800	600–1,000
Southern pea	1,350	1,200–1,500
Spinach		
o.p.	1,915	1,000–2,500
F_1	1,100	1,000–1,200
Squash, summer		
o.p.	760	400–1,200
F_1	360	250–425
Squash, winter		
o.p.	620	300–1,200
F_1	310	200–400
Tomatillo	600	500–700
Tomato		
F_1	125	75–170

TABLE 2.5. VEGETABLE SEED YIELDS (*Continued*)

| | (lb/acre) | |
Vegetable	Average Yield	Yield Range
Turnip		
o.p.	2,300	2,000–3,000
F$_1$	1,350	1,200–1,500
Watermelon		
o.p.	405	350–600
F$_1$	220	190–245
Triploid (seedless)	40	20–70

[1] Yields are from information provided by representations of several major seed companies. Yields of some hybrids may be very much lower because of difficulties of seed production.

[2] o.p. = open pollinated

F$_1$ = first-generation hybrid

F$_2$ = second-generation hybrid

07

SEED STORAGE

High moisture and temperature cause rapid deterioration of vegetable seeds. A general rule of thumb is that seeds should be stored where the temperature (in degrees Fahrenheit) plus the relative humidity (in percent) are no greater than 100. Storage temperatures near 32°F are not necessary. Between 40 and 50°F is quite satisfactory when the moisture content of the seed is low. If seed moisture content is reduced to 4–5% and the seeds are stored in sealed containers, a storage temperature of 70°F will be satisfactory for more than 1 year.

The moisture content of seeds can be lowered by drying them in moving air at 120°F. For seeds with a high initial moisture content (25–40%), it is preferable to dry them at 110°F. Drying time required will depend on the depth of the layer of seeds, seed size, the volume of air, dryness of air, and original moisture content of seed. When seeds cannot be dried in this way, seal them in airtight containers over, but not touching, a desiccant such as calcium chloride or dried silica gel. Use enough desiccant so that the moisture absorbed from the seeds will produce no visible change in the chemical.

The moisture content of seeds will reach an equilibrium with the atmosphere after a period of time. This takes about 3 weeks for small seeds and 3–6 weeks for large seeds. Bean and okra may develop hard or dormant seeds, which will not germinate satisfactorily, if dried to very low moisture content. White-seeded beans are likely to become hard if the moisture content is reduced to about 10%. Dark-colored beans can be dried to less than 10% moisture before they become hard.

Seed priming increases germination speed and uniformity by beginning the germination process. Seeds are hydrated to provide enough water to initiate germination, and are then dried down, packed, and marketed. Primed seeds usually have a shorter shelf life than nonprimed seeds.

TABLE 2.6. **APPROXIMATE LIFE EXPECTANCY OF VEGETABLE SEEDS STORED UNDER FAVORABLE CONDITIONS**

Vegetable	Years	Vegetable	Years
Asparagus	3	Kohlrabi	3
Bean	3	Leek	2
Beet	4	Lettuce	6
Broccoli	3	Muskmelon	5
Brussels sprout	4	Mustard	4
Cabbage	4	New Zealand spinach	3
Cardoon	5	Okra	2
Carrot	3	Onion	1
Cauliflower	4	Parsley	1
Celeriac	3	Parsnip	1
Celery	3	Pea	3
Chard, Swiss	4	Pepper	2
Chervil	3	Pumpkin	4
Chicory	4	Radish	5
Chinese cabbage	3	Roselle	3
Collards	5	Rutabaga	4
Corn, sweet	2	Salsify	1
Corn salad (mâche)	5	Scorzonera	2
Cress, garden	5	Sea kale	1
Cress, water	5	Sorrel	4
Cucumber	5	Southern pea	3
Dandelion	2	Spinach	3
Eggplant	4	Squash	4
Endive	5	Tomato	4
Fennel	4	Turnip	4
Kale	4	Watermelon	4

Adapted from J. F. Harrington and P. A. Minges, *Vegetable Seed Germination* (University of California Agricultural Extension Leaflet, 1954) unnumbered.

08

SEED PRIMING

Seed priming is a physiology-based seed enhancement technique designed to improve the germination characteristics of seeds. Germination speed, uniformity, and seedling vigor are all improved by priming. These benefits are especially pronounced under adverse temperature and/or moisture conditions.

The commercial applications of seed priming have been expanding rapidly in recent years. Important vegetable crops that are now enhanced through priming include brassicas, carrot, celery, cucurbits, lettuce, onion, pepper, and tomato. More crop species are being added on an ongoing basis.

Priming is accomplished by partially hydrating seed and maintaining it under defined moisture, temperature, and aeration conditions for a prescribed period of time. In this state, the seed is metabolically active. In an optimally hydrated, metabolically active state, important germination steps can be accomplished within the seed. These include repair of membranes and/or genetic material, development of immature embryos, alteration of tissues covering the embryo, and destruction or removal of dormancy blocks.

At the conclusion of the process, the seed is redried to its storage moisture level. The gains made in priming are not lost during storage. Primed seed is physiologically closer to germination than nonprimed seed. When planted at a later date, primed seed starts at this advanced state and moves directly into the final stages of germination and growth.

There are several commercial methods of seed priming. All are based on the basic principles of hydrated seed physiology. They differ in the methods used to control hydration, aeration, temperature, and dehydration. The most important commercial priming methods include:

Liquid osmotic. In this approach, seed is bubbled in a solution of known osmotic concentration (accomplished with various salts or organic osmotic agents). The osmotic properties of the solution control water uptake by the seed. The bubbling is necessary to provide sufficient oxygen to keep the seed alive during the process. The temperature of the solution is controlled throughout the process. After priming is completed, the seeds are removed, washed, and dried.

Membrane and / or flat media osmotic. This method is a variation of liquid osmotic priming. With this method, the seed is placed on a porous membrane suspended on the surface of the osmotic solution. This method addresses some of the aeration concerns associated with liquid osmotic priming, but is limited by practical considerations to smaller seed lots.

Drum hydration. With this method, seeds are placed in a rotating drum and controlled quantities of water are sprayed onto the seed, bringing it to the desired moisture level. Drum rotation provides the necessary aeration to the seeds, and temperature and air flow are controlled throughout the process. After the priming period, the seed is dried by flushing air through the drum. Drum priming is a patented technology.

Solid matrix priming (SMP). With the SMP method, water uptake is controlled by suspending seed in a defined medium (or matrix) of solids (organic and/or inorganic) of known water holding properties. The seed and matrix compete for available water, coming to equilibrium at precisely the right point for priming to occur. Aeration and temperature are precisely controlled throughout the process. After the process is complete, the seed and matrix are separated. The seed is dried back to its original moisture. The SMP method is a patented technology.

In maintaining processing conditions during priming, it is important to prevent the seed from progressing too far through the germination process. If germination is allowed to progress beyond the early stages, it is too late to return to a resting state. The seed is "committed" to growth and cannot be redried without damage and/or reduced shelf life.

Priming alters many basic characteristics of germination and seedling emergence as indicated below:

- *Germination speed.* Primed seed has already accomplished the early stages of germination and begins growing much more rapidly. The total time required is cut approximately in half. This is especially important with slow-germinating species such as celery or carrot.

- *Increased temperature range.* Primed seed will emerge under both cooler and warmer temperatures than unprimed seed. Generally, the temperature range is extended by 5–8°F in both directions.

- *More uniform emergence.* The distribution of germination times within most seed lots is greatly reduced resulting in improved uniformity.

- *Germination at reduced seed water content.* Primed seed will germinate at a lower seed water content than unprimed seed.

- *Control of dormancy mechanisms.* In many cases, priming overcomes dormancy mechanisms that slow germination.

- *Germination percentages.* An increase in the germination percentage occurs in many instances with individual seed lots as a result of the priming process. The increase is generally due to repair of weak or abnormal seeds within the lot.

79

CONSIDERATIONS WITH PRIMED SEED

Shelf life of primed seed. Shelf life is a complicated subject and is influenced by many different factors. The most important factors include crop species, seed lot quality, seed moisture content in storage, transportation and storage conditions (especially temperature), the degree to which a lot has been primed, and subsequent seed treatments (fungicides, film coating, pelleting).

Assuming proper transportation and storage conditions and no other complicating factors (such as coating), deterioration in seed lot performance is rarely experienced during the growing season for which a lot was primed (generally 4 months). In most cases (assuming the same qualifiers listed above), lot performance is maintained for much longer.

As storage time increases, the risk of loss also increases. Most lots are stable, but a percentage deteriorate rapidly. Not only is the priming effect lost, but generally a significant percentage of the lot dies. Screening methods to predict high-risk lots are needed. The results of research in this area are promising, but a usable method of predicting deterioration is not yet available.

Seed only should be primed for planting during the immediate growing season. Priming seed for planting in subsequent years is discouraged. In cases where primed seed must be held for extended periods, the seed should be retested before planting to assess whether deterioration may have occurred.

Treating, coating, and pelleting primed seed. The compatibility of primed seed with any subsequent seed treatment, coating, or pelleting must be determined on a case-by-case basis. The germination characteristics may be influenced. In some cases, priming is performed to improve the vigor of lots that would otherwise not tolerate the stress of coating or pelleting. In other cases, primed seeds may be more sensitive than unprimed seeds and experience deterioration. Combinations must be tested after priming, on a case-by-case basis, before other commercial treatments are performed.

Transport and storage conditions. Exposure to high temperatures, even for brief periods, can induce rapid deterioration of all seeds. The risk is greater with primed seeds. In storage and transport, it is important to maintain seeds that have been enhanced under dry, cool conditions (temperatures of 70°F or less are recommended). Unfavorable conditions may negatively influence shelf life.

Adapted from J. A. Eastin and J. S. Vendeland, *Presented at Florida Seed Association Seminar* (Lincoln, NE: Kamterter Products, Inc., 1996); C. Parera and D. Cantliffe, *Seed Priming*, pp. 109–141, Vol. 16, Horticultural Reviews (1994).

09

SEEDBORNE PATHOGENS AND HOT WATER SEED TREATMENT

Some plant pathogens, including many bacterial pathogens, can penetrate and survive within the seed, out of reach of surface seed treatments. Treating seeds with hot water can prevent establishing or re-introducing seedborne pathogens on the farm. Hot water seed treatment also has the beneficial effect of priming seeds, resulting in faster germination than untreated seed.

To decide whether to use hot water treatment, first determine the likelihood that seedborne pathogens could be present based on the crop (see Table 2.7). Next, ask your seed supplier to find out if the seed has already been treated or primed. Chemically treated, pelleted, or primed seed should not be hot-water treated. Hot water seed treatment can decrease germination rates, especially for older seed (more than 1 year old) or seeds that were grown under stressful environmental conditions. Treat no more seed than you think you will use during one season, as hot water treatments can reduce seed life expectancy.

It is important to use the appropriate protocol for each crop to control pathogens without damaging the seed. Large-seeded crops (beans, cucurbits, peas, corn, etc.) are usually not effectively disinfested with hot water treatment because the temperature required to heat the whole seed would kill the outer seed tissue and reduce germination. While hot water seed treatment can be done effectively on a stovetop in a large pot with an accurate thermometer and careful temperature control, it is much better to use precision water baths that provide an even, stable, and accurate temperature. Home "sous vide" machines are appliances designed to cook food inside vacuum-sealed bags that are immersed in circulating water, and they offer precision temperature control at a lower cost than scientific water baths.

Before you treat seed, you may want to conduct a seed germination test, as different varieties and lots may react differently to hot water treatment. Treat a 50- or 100-seed sample using the procedure below, then test the germination of both the treated seeds and an equal number of untreated seeds, either in the same growing media that you plan to use for transplant production, or in a moist paper towel. If the test gives acceptable germination rates, treat as much seed as you expect to use in the coming season.

1. **Preheat water baths.** Heat one bath to 100°F and another to your treatment temperature (see Table 2.7). The first bath will be used to preheat the seed so that the temperature of the treatment bath does not drop when the seeds are added. Use enough water so that seeds can move around freely. Use an accurate laboratory thermometer. Water must be maintained at a uniform temperature throughout the vessel, the recommended temperature must not be exceeded, and the seed must be treated no longer than the time interval specified. A stirring

TABLE 2.7. HOT WATER SEED TREATMENT CONDITIONS FOR SEEDBORNE PATHOGENS OF VEGETABLE CROPS

Crop	Treatment Temperature (°F)	Treatment Time (min)	Pathogens/Diseases Controlled
Broccoli	122	20	Alternaria leaf spot, bacterial leaf
Brussels sprouts	122	25	spot, blackleg, black rot
Cabbage	122	25	
Collards	122	20	
Kale	122	20	
Brussels sprouts	122	25	
Carrot	122	20	Alternaria leaf blight, bacterial leaf blight, Cercospora leaf spot, crater rot/foliar blight
Celery/ celeriac	118	30	Bacterial leaf spot, Cercospora leaf spot, Septoria leaf spot, Phoma crown, and root rot
Eggplant	122	25	Anthracnose, early blight, Phomopsis, verticillium wilt
Lettuce	118	30	Anthracnose, bacterial leaf spot, lettuce mosaic virus, Septoria leaf spot, verticillium wilt
Onion	122	20	Purple blotch, Stemphylium leaf blight, basal rot, Botrytis blight, smudge, black mold, Downy mildew
Pepper	125	30	Anthracnose, bacterial leaf spot, cucumber mosaic virus, pepper mild mosaic virus, tobacco mosaic virus, tomato mosaic virus
Parsley	122	30	Bacterial leaf blight, alternaria leaf blight, Black rot, Cercospora leaf blight, Septoria blight
Spinach	122	25	Anthracnose, Cladosporium leaf spot, cucumber mosaic virus, downy mildew, fusarium wilt, Stemphylium leaf spot, verticillium wilt

82

TABLE 2.7. HOT WATER SEED TREATMENT CONDITIONS FOR SEEDBORNE PATHOGENS OF VEGETABLE CROPS (*Continued*)

Crop	Treatment Temperature (°F)	Treatment Time (min)	Pathogens/Diseases Controlled
Tomato	122	25	Alfalfa mosaic virus, anthracnose, bacterial canker, bacterial speck, bacterial spot, cucumber mosaic virus, early blight, fusarium wilt, leaf mold, Septoria leaf spot, tomato mosaic virus, verticillium wilt, double virus streak

Adapted from G. Higgins, University of Massachusetts Vegetable Program Fact Sheet (2018), https://ag.umass.edu/vegetable/fact-sheets/hot-water-seed-treatment (accessed 10 Oct. 2021); M. McGrath, A. Wyenandt, and K. Holmstrom, Managing pathogens inside seed with hot water, https://www.vegetables.cornell.edu/pest-management/disease-factsheets/managing-pathogens-inside-seed-with-hot-water/ (accessed 19 Oct. 2021).

hot plate helps to provide continuous agitation and uniform water temperature, though it can be done with continuous manual agitation or an aquarium bubbler. It helps to have a separate container of room temperature water close by to add, if necessary, to prevent overheating.

2. **Prepare the seed.** Make a packet for the seeds out of cheese cloth, screen, or insect netting. Fill each packet no more than halfway with seed, to allow for water movement throughout the packet. Include a metal bolt, coin, or other weight to keep the seed submerged. Label all packets, especially if you're treating more than one variety at once!

3. **Preheat the seed.** Submerge the seed in the preheat bath for 10 min, constantly checking the temperature to ensure that it does not rise above 100°F.

4. **Treat the seed.** Move the seed to the treatment bath and treat for recommended time (see Table 2.7). Check the temperature constantly to ensure that it does not rise above the recommended temperature. Remove the seeds promptly and run them under room temperature tap water to cool them.

5. **Dry the seed.** Pat dry with towels, then air dry at 70–75°F by spreading the seed on dry paper towels.

10
VEGETABLE VARIETIES

This section outlines requirements for naming vegetable seed. It is based on the Federal Seed Act, a truth-in-labeling law intended to protect growers and home gardeners who purchase seed.

WHO NAMES NEW VARIETIES?

The originator or discoverer of a new variety may give that variety a name. If the originator or discoverer chooses not to name a variety, someone else may give that variety a name for marketing purposes. In such a case, the name first used when the seed is introduced into commerce will be the name of the variety.

It is illegal to change a variety name once the name has been legally assigned. In other words, a buyer may not purchase seed labeled as variety "X" and resell it as variety "Y." An exception to this rule occurs when the original name is determined to be illegal. In that case, the variety must be renamed according to the rules mentioned above. Another exception to this rule applies to varieties already being marketed under several names before 1956. (See section on synonyms.)

WHAT'S IN A NAME

"Kind" is the term used for the seed of one or more related plants known by a common name (e.g., carrot, radish, tomato, or watermelon). "Variety" is a subdivision of a kind. A variety has different characteristics from another variety of the same kind of seed. For example, "Bolero" carrot differs from "Danvers 126" carrot and "SuperSeedless 7167" differs from "Jamboree" watermelon.

The rules for naming plants relate to both kinds and varieties of seed:

1. A variety must be given a name that is unique to the kind of seed to which the variety belongs. For instance, there can only be one variety of squash called "Dividend."
2. Varieties of two or more different kinds of seed may have the same name if the kinds are not closely related, or known to intercross. For example, there could be a "Dividend" squash and a "Dividend" tomato because squash and tomato are kinds of seed not closely related. On the other hand, it would not be permissible to have a "Dividend" squash and a "Dividend" pumpkin because the two kinds of seed are closely related.

3. Once assigned to a variety, the name remains exclusive. Even if "Dividend" squash has not been marketed for many years, a newly developed and different squash variety can't be given the name "Dividend" unless the original owner agrees to withdraw "Dividend" squash.

4. A company name may be used in a variety name as long as it is part of the original, legally assigned name. Once part of a legal variety name, the company name must be used by everyone including another company that might market the seed. When a company name is not a part of the variety name, it should not be used in any way that gives the idea that it is part of the variety name. For example, Don's Seed Company can't label or advertise "Dividend" squash variety as "Don's Dividend" since "Don's" may be mistaken to be part of the variety name. The simplest way to avoid confusion is to separate the company and variety names in advertising or labeling.

5. Although USDA discourages it, you may use descriptive terms in variety names as long as such terms are not misleading. "X3R," for instance, is accepted among pepper growers as meaning "resistant to Bacterial Leaf Spot races 1, 2, and 3." It would be misleading to include "X3R" as part of a variety name if that variety were not "X3R" resistant. Similarly, if a cantaloupe variety is named "Burpee Hybrid PMT," the name would be illegal if this variety was not tolerant of powdery mildew.

6. A variety name cannot be misleading, e.g., similar to an existing name but differing only in spelling or punctuation. "Alan" cucumber would not be permissible if an "Allen" cucumber was already on the market.

HYBRIDS

Hybrid designations, whether they are names or numbers, also are variety names. Every rule discussed here applies to hybrid seed as well as to nonhybrid seed. In the case of hybrids, however, the situation is potentially more complex since more than one seed producer or company might use identical parent lines to produce a hybrid variety. One company could then produce a hybrid that was the same as one already introduced by another firm.

When this happens, both firms must use the same name since they are marketing the same variety. If those who developed the parent lines have given the hybrid variety a name, that is the legal name. Otherwise, the proper name is the one given by the company that first introduced the hybrid seed into commerce.

SYNONYMS-VARIETIES WITH SEVERAL NAMES

As noted earlier, the name originally assigned to a variety is the name that must be used forever. It can't be changed unless it is illegal. This does not mean that all varieties must be marketed under a single name. In fact, some old varieties may be marketed legally under more than one name. If several names for a single variety of a vegetable seed were in broad general use before 28 July 1956, those names still may be used.

Here are some examples:

The names "Acorn," "Table Queen," and "Des Moines" have been known for many years to represent a single squash variety. They were in broad general use before 28 July 1956, so seed dealers may continue to use these names interchangeably.

Apart from old varieties with allowable synonym names, all vegetable and agricultural varieties may have *only one* legally recognized name, and that name must be used by anyone who represents the variety name in labeling and advertising. This includes interstate seed shipments and seed advertisements sent in the mail or in interstate or foreign commerce.

IMPORTED SEED

Seed imported into the United States can't be renamed if the original name of the seed is in the Roman alphabet. For example, cabbage seed labeled, "Fredrikshavn" and shipped to the United States from Denmark can't be given a different variety name such as "Bold Blue." Seed increased from imported seed also can't be renamed. If "Fredrikshavn" were increased in the United States, the resulting crop still couldn't be named "Bold Blue." Seed with a name that is not in the Roman alphabet must be given a new name. In such a case, the rules for naming the variety are the same as stated previously.

TRADEMARKS

Variety names may contain trademarks (names registered with the Trademark Division of the U.S. Patent Office), but the trademark status is lost in the sense that anyone marketing seed of that variety must use the entire variety name including the trademark. In addition:

- A trademark symbol or registered trademark symbol cannot be displayed in the variety name.
- A trademark by itself cannot be a variety name and a variety name cannot be trademarked.

SUMMARY

If the naming, labeling, and advertising of a seed variety is truthful, it is probably in compliance with the Federal Seed Act. Keep these simple rules in mind to help eliminate violations and confusion in the marketing of seed:

- Research the proposed variety name before adopting it.
- Make sure the name cannot be confused with company names, brands, trademarks, or names of other varieties of the same kind of seed.
- Never change the variety name, whether marketing seed obtained from another source, or from your own production.

Adapted from Variety Naming Guidelines to Comply with the Federal Seed Act, https://www.ams. usda.gov/rules-regulations/fsa/variety-naming (accessed June 2021); See also the Vegetable Variety Name List, https://www.ams.usda.gov/services/seed-testing/variety-name-list (last updated May 2021).

SELECTION OF VEGETABLE VARIETIES

Selection of the variety or cultivar (e.g., *cultivated variety*) to plant is one of the most important decisions the commercial vegetable grower must make each season. Each year seed companies and experiment stations release dozens of new varieties to compete with those already available. Growers should evaluate some new varieties each year on a trial basis to observe performance on their own farms. A limited number of new varieties should be evaluated so that observations on plant performance and characteristics and yields can be noted and recorded. It is relatively easy to establish a trial but very time-consuming to make all the observations necessary to make a decision on adoption of a new variety.

Variety selection is a very dynamic process. Some varieties are marketed for many years, whereas others might be used only a few seasons. If a variety was released from the public sector (e.g., the USDA or a university), many seed companies may offer it. Varieties developed by a seed company may be available only from that source or may be distributed through many sources. Cooperative Extension Services in most states publishes annual or periodic lists of recommended varieties. In addition, there are regional production guides that feature such lists.

11
GENETICALLY-ENGINEERED VEGETABLE VARIETIES

Genetic engineering refers to processes by which laboratory tools are used to insert, alter, or remove pieces of DNA that contain gene(s) of interest. These methods make it possible to transfer one or a few genes of interest, no matter where the gene originates. The first genetically-engineered (GE) crops were transgenic, incorporating genes from other organisms to incorporate breeding goals that are difficult or impossible to achieve using traditional plant breeding methods. Recent advances in GE technologies include genome editing, which may encourage the production of crops with more subtle changes, like altered expression of existing genes.

The first GE vegetable varieties were commercialized in the 1990s, and to date, the number of transgenic varieties and traits commercially available remains small. Resistance to insects, pathogens, and herbicides is the primary application of this technology in vegetable crops. GE varieties are available to commercial growers with some limitations: growers must sign stewardship agreements and must acknowledge in writing that they know they are growing GE varieties.

Consumer acceptance of GE technologies varies widely, and the technology remains controversial. Genetic engineering is prohibited in certified organic production, and by some other markets including export markets to certain countries. Within the United States, some states and counties prohibit the planting of any GE crops. Thus, growers should take their consumers and markets into account when deciding whether to grow GE varieties.

TABLE 2.8 COMMERCIALIZED VEGETABLE CROPS WITH GENETICALLY-ENGINEERED TRAITS

Crop	Trait	Brand Names
Sweet corn	Bt/caterpillar resistance	Seminis Performance series, Syngenta Attribute I, Attribute II, and Attribute Plus series
Sweet corn	Glyphosate tolerance	Seminis Performance series, Syngenta Attribute II series
Sweet corn	Glufosinate tolerance	Syngenta Attribute I, Attribute II, and Attribute Plus series
Summer squash	Virus resistance	Conqueror III, Destiny III, XPT 1832 III, and others.
Potato	Late blight resistance	Three varieties from J.R. Simplot, available to licensed growers
Eggplant	Bt/fruit and shoot borer	Not currently available in the United States
Zucchini	Virus resistance	Justice III, SV0914YG, and others

B. Phillips and P. Porder, Sweet corn production Handy Bt Trait Table, https://www.texasinsects.org/bt-corn-trait-table.html (accessed 18 Oct. 2021); M. Lobato-Gómez, S. Hewitt, T. Capell et al., Transgenic and genome-edited fruits: background, constraints, benefits, and commercial opportunities. *Horticultural Research* 8:166 (2021). https://doi.org/10.1038/s41438-021-00601-3.

-A-

Adaptive Seeds
25079 Brush Creek Road
Sweet Home, OR 97386
Ph (541) 367-1105
http://www.adaptiveseeds.com

American Takii, Inc.
301 Natividad Road
Salinas, CA 93906
Ph (831) 443-4901
http://www.takii.com

Arkansas Valley Seed Solutions
4300 Monaco Street
Denver, CO 80216
Ph (877) 907-3337
http://avseeds.com

-B-

Baker Creek Heirloom Seeds
2278 Baker Creek Road
Mansfield, MO 65704
Ph (417) 924-8917
Fax (417) 924-8887
http://www.rareseeds.com

Ball Horticultural Company
622 Town Road
West Chicago, IL 60185-2698
Ph (800) 879-2255
http://www.fredgloeckner.com

Bejo Seeds, Inc.
1972 Silver Spur Place
Oceano, CA 93445
Ph (805) 473-2199
http://www.bejoseeds.com

BHN Seed
P.O. Box 3267
Immokalee, FL 34142
Ph (239) 352-1100
http://www.bhnseed.com

Bonanza Seeds International, Inc.
3818 Railroad Avenue
Yuba City, CA 95991
Ph (530) 673-7253
http://www.bonanzaseeds.com

W. Brotherton Seed Co., Inc.
P.O. Box 1136
Moses Lake, WA 98837
Ph (509) 765-1816
http://brothertonseed.com/

Burgess Seed & Plant Co.
905 Four Seasons Road
Bloomington, IL 61701
Ph (309) 622-7761
http://www.eburgess.com

W. Atlee Burpee & Co.
300 Park Avenue
Warminster, PA 18974
Ph (800) 888-1447
http://www.burpee.com

-C-

California Asparagus Seed
2815 Anza Avenue
Davis, CA 95616
Ph (530) 753-2437
http://californiaasparagusseed.com/

Clifton Seed Company
P.O. Box 206
Faison, NC 28341
Ph (800) 231-9359
http://www.cliftonseed.com

The Cook's Garden
P.O. Box 6530
Warminster, PA 18974
Ph (800) 457-9703
http://www.cooksgarden.com

Corona Seeds Worldwide
590-F Constitution Avenue
Camarillo, CA 93012
Ph (805) 388-2555
http://www.coronaseeds.com

Crookham Co.
P.O. Box 520
Caldwell, ID 83606-0520
Ph (208) 459-7451
http://www.crookham.com

Crop King Inc.
134 West Drive
Lodi, OH 44254
Ph (330) 302-4203
Fax (330) 769-2616
http://www.cropking.com

Cutter Asparagus Seed
516 Young Avenue
Arbuckle, CA 95912
Ph (530) 475-3647
http://www.asparagusseed.com

-E-

Elsoms Seeds Ltd.
Pinchbeck Road
Spalding, Lincolnshire PE11 1QG
England, UK
Ph 0 1775 715000
Fax 0 1775 715001
http://www.elsoms.com

Enza Zaden
7 Harris Place
Salinas, CA 93901
Ph (831) 754-2300
http://www.enzazaden.com/

-F-

Farmer Seed & Nursery Co.
1706 Morrissey Drive
Bloomington, IL 61704
Ph (507) 334-1623
http://www.farmerseed.com

Fedco Seeds
P.O. Box 520
Clinton, ME 04927
Ph (207) 426-9900
http://www.fedcoseeds.com

Field and Forest Products
501 Hart Dr,
Peshtigo, WI 54157
Ph (800) 792-6220
http://fieldforest.net

-G-

Germania Seed Co.
P.O. Box 31787
Chicago, IL 60631
Ph (800) 380-4721
http://www.germaniaseed.com

Golden Valley Seed
P.O. Box 1600
El Centro, CA 92243
Ph (760) 337-3100
Fax (760) 337-3135
http://www.goldenvalleyseed.com

Gurney's Seed & Nursery Co.
P.O. Box 4178
Greendale, IN 47025-4178
Ph (513) 354-1492
http://www.gurneys.com

-H-

Harris Seeds
355 Paul Road
Rochester, NY 14624
Ph (800) 544-7938
http://www.harrisseeds.com

The Chas. C. Hart Seed Co.
304 Main Street
Wethersfield, CT 06109
Ph (860) 529-2537
http://www.hartseed.com

Hazera Seeds USA Inc.
3155 SW 10th Street Suite 6L
Deerfield Beach, FL 33442
Ph (954) 429-9445
https://www.hazera.us.com/

High Mowing Organic Seeds
76 Quarry Road
Wolcott, VT 05680
Ph (866) 735-4454
http://www.highmowingseeds.com

Hollar Seeds
P.O. Box 106
Rocky Ford, CO 81067
Ph (719) 254-7411
http://www.hollarseeds.com

Hydro-Gardens, Inc.
8765 Vollmer Road
Colorado Springs, CO 80932
Ph (888) 693-0578
Fax (888) 693-0578
http://www.hydro-gardens.com

-I-

Illinois Foundation Seeds, Inc.
1083 County Road 900N
Tolono, IL 61880
Ph (217) 485-6260
http://www.ifsi.com

-J-

Johnny's Selected Seeds
955 Benton Avenue
Winslow, ME 04901
Ph (877) 564-6697
http://www.johnnyseeds.com

Jordan Seeds, Inc.
6400 Upper Aston Road
Woodbury, MN 55125
Ph (651) 738-3422
http://www.jordanseeds.com

J. W. Jung Seed Co.
335 S. High Street
Randolph, WI 53956
Ph (800) 247-5864
http://www.jungseed.com

-K-

Keithly-Williams Seeds
420 Palm Avenue
Holtville, CA 92250
Ph (800) 533-3465
http://www.keithlywilliams.com

Kitazawa Seed Company
201 4th Street, #206
Oakland, CA 94607
Ph (510) 595-1188
http://kitazawaseed.com

Known-You Seed Co., Ltd.
26360 Reuther Avenue
Santa Clarita, CA 91350
Ph (661) 360-3985
http://www.knownyou.com

-L-

Livingston Seed Co.
202 S. Washington Street
Norton, MA 02755
Ph (508) 285-5800
http://www.livingstonseed.com

-M-

Earl May Seed & Nursery Co.
208 N. Elm Street
Shenandoah, IA 51603
Ph (712) 246-1020
http://www.earlmay.com

Henry F. Michell Co.
P.O. Box 60160
King of Prussia, PA 19406-0160
Ph (800) 422-4678
http://www.michells.com

-N-

Native Seeds/SEARCH
3584 E. River Road
Tucson, AZ 85718
Ph (520) 622-0830
http://www.nativeseeds.org

New England Seed Co.
122 Park Ave Building H.
Hartford, CT 06108
Ph (877) 229-5477
http://www.neseed.com

Nichol's Garden Nursery
P.O. Box 1288
Philomath, IR 97370
Ph (800) 422-3985
http://www.nicholsgardennursery.com

Nirit Seeds Ltd.
Moshar Hadar-Am
42935 Israel
Ph (972) 9 832 24 35
Fax (972) 9 832 24 38
http://www.niritseeds.com

Nourse Farms, Inc.
41 River Road
South Deerfield, MA 01373
Ph (413) 665-2658
http://www.noursefarms.com

Nunhems USA, Inc.
1200 Anderson Corner Road
Parma, ID 83660
Ph (208) 673-4000
http://www.nunhems.com

-O-

Osborne Quality Seeds
2428 Old Hwy 99 South Road
Mount Vernon, WA 98273
Ph (360) 424-7333
http://www.osborneseed.com

OSC Seeds
P.O. Box 7
Waterloo, Ontario
Canada N2J 3Z9
Ph (519) 886-0557
http://www.oscseeds.com

Ornamental Edibles
5723 Trowbidge Way
San Jose, CA 95138
Ph (408) 528-7333
http://www.ornamentaledibles.com

Orsetti Seed Co., Inc.
2300 Technology Parkway, Ste 1
P.O. Box 2350
Hollister, CA 95023
Ph (831) 636-4822
http://www.orsettiseed.com

Outstanding Seed Company
P.O. Box 1584
Beaver Falls, PA 15010
Ph (800) 385-9254
http://www.outstandingseed.com

-P-

DP Seeds LLC
8269 E. US Hwy 95
Yuma, AZ 85365
Ph (928) 341-8494
http://www.dpseeds.com

Paramount Seeds, Inc.
7998 SW Jack James Drive
Stuart, FL 34997
Ph (772) 221-0653
http://www.paramountseeds.com

Park Seed Co.
3507 Cokesbury Road
Hodges, SC 29653
Ph (800) 845-3369
http://www.parkseed.com

Penn State Seed Co.
RR1 Box 390, Hwy 309
Dallas, PA 18612-9781
Ph (800) 847-7333
http://www.pennstateseed.com

Pepper Gal
400 N W 20th Street
Ft. Lauderdale, FL 33311
Ph (954) 537-5540
http://www.peppergal.com

W.H. Perron
P.O. Box 2500
Georgetown, Ontario
Canada L7G 4A2
Ph (905) 873-3037
http://www.dominion-seed-house.com

Pinetree Garden Seeds
P.O. Box 300
New Gloucester, ME 04260
Ph (207) 926-3400
http://www.superseeds.com

-R-

Redwood City Seed Co.
P.O. Box 361
Redwood City, CA 94064
Ph (650) 325-7333
http://www.ecoseeds.com

Renee's Garden Seeds
6060 Graham Hill Road
Felton, CA 95018
Ph (888) 880-7228
http://www.reneesgarden.com

Rijk Zwaan
701 La Guardia Street
Salinas, CA 93905
Ph (831) 455-3000
http://www.rijkzwaanusa.com

Rupp Seeds, Inc.
17919 County Road B
Wauseon, OH 43567
Ph (419) 337-1841
http://www.ruppseeds.com

Rispens Seeds, Inc.
P.O. Box 310
Beecher, IL 60401
Ph (888) 874-0241
http://www.rispenseeds.com

-S-

Sakata Seed America, Inc.
18095 Serene Drive
Morgan Hill, CA 95037
Ph (408) 778-7758
http://www.sakatavegetables.com

Seeds of Change
Mars, Inc. 6885 Elm Street
McLean, VA 22101
Ph (703) 821-4900
http://www.seedsofchange.com

Seedway, Inc.
P.O. Box 250, 1734 Railroad Place
Hall, NY 14463
Ph (800) 836-3710
http://www.seedway.com

Seminis Vegetable Seeds
8090 North Lindbergh Road
St. Louis, MO 63167
Ph (866) 334-1056
http://www.seminis-us.com

R.H. Shumway
334 W. Stroud Street
Randolph, WI 53956-1274
Ph (800) 342-9461
http://www.rhshumway.com

Snow Seed Co.
1147 Madison Lane
Salinas, CA 93907
Ph (831) 758-9869
http://www.snowseedcompany.com

Southern Exposure Seed Exchange
P.O. Box 460
Mineral, VA 23117
Ph (540) 894-9480
http://www.southernexposure.com

Stokes Seeds, Inc.
13031 Reflections Dr.
Holland, MI 49424
Ph (616) 786-4999
http://www.stokesseeds.com

Syngenta Seeds, Inc.
P.O. Box CH-4002
Basel, Switzerland
Ph (208) 322-7272
http://www.syngenta-us.com/

-T-

The Territorial Seed Co.
434 E. Main Street
Cottage Grove, OR 97424
Ph (800) 626-0866
http://www.territorialseed.com

Tomato Grower's Supply Co.
P.O. Box 60015
Fort Myers, FL 33906
Ph (239) 768-1119
http://www.tomatogrowers.com

Otis S. Twilley Seed Co., Inc.
121 Gary Road
Hodges, SC 29653
Ph (800) 622-7333
http://www.twilleyseed.com

-U-

United Genetics
8000 Fairview Road
Hollister, CA 95023
Ph (831) 636-4882
http://www.unitedgenetics.com

-V-

Vermont Bean Seed Co., Inc.
334 West Stroud Street
Randolph, WI 53956-1274
Ph (800) 349-1071
http://www.vermontbean.com

Vesey's Seeds Ltd.
P.O. Box 9000
Charlottetown, PE
Canada C1A 8K6
Ph (800) 363-7333
http://www.veseys.com

Victory Seed Co.
P.O. Box 192
Molalla, OR 97038
Ph (503) 829-3126
http://www.victoryseeds.com

Vilmorin North America
3 Harris Place
Salinas, CA 93901-4856
Ph (831) 771-1500
http://www.vilmorinnorthamerica.com

Vitalis Organic Seeds
Enza Coastal Seeds Inc.
Salinas, CA 93901
info@vitalisorganic.com
http://usa.vitalisorganic.com
https://eu.biovitalis.eu/contact

-W-

Walker Brothers, Inc.
105 Porchtown Road
Pittsgrove, NJ 08318
Ph (856) 358-2548
http://walkerplants.com

West Coast Seeds
3925 64th Street
Delta, BC
Canada V4K 3N2
Ph (888) 804-8820
http://www.westcoastseeds.com

Wild Garden Seed
P.O. Box 1509
Philomath, OR 97370
Ph (541) 929-4068
http://www.wildgardenseed.com

Willhite Seed Co.
Box 23
Poolville, TX 76487-0023
Ph (817) 599-8656
http://www.willhiteseed.com

Wyatt-Quarles Seed Co.
P.O. Box 739
Garner, NC 27529
Ph (877) 767-7333
http://www.wqseeds.com

Information in this section was correct at the time of preparation. However, there are frequent changes in phone area codes, postal codes, addresses, and even company names.

PART **3**

SEEDLING AND TRANSPLANT PRODUCTION

Knott's Handbook for Vegetable Growers, Sixth Edition. George J. Hochmuth and Rebecca G. Sideman.
© 2023 John Wiley & Sons Ltd. Published 2023 by John Wiley & Sons Ltd.
Companion website: https://www.wiley.com/go/hochmuth/vegetablegrowers6e

TRANSPLANT PRODUCTION

Vegetable crops are established in the field by direct seeding or by use of vegetative propagules (see Part 5) or transplants. While many vegetable crops may be transplanted, some are more easily transplanted than others (Table 3.1). Transplants are produced in containers of various sorts in greenhouses, protected beds, and open fields. Either greenhouse-grown containerized or field-grown bare-root transplants can be used successfully. Generally containerized transplants get off to a faster start, but are more expensive. Containerized transplants, sometimes called "plug" transplants, have become the norm for melons, pepper, tomato, and eggplant. Because transplant production demands suitable facilities and careful attention to detail, many vegetable growers choose to purchase transplants from production specialists rather than grow them themselves.

TABLE 3.1. RELATIVE EASE OF TRANSPLANTING VEGETABLES[1]

Easy	Moderate	Require Special Care[2]
Beet	Celery	Sweet corn
Broccoli	Eggplant	Cantaloupe
Brussels sprouts	Onion	Cucumber
Cabbage	Pepper	Summer squash
Cauliflower		Watermelon
Chard		
Lettuce		
Tomato		

[1] Although containerized transplant production is the norm for most vegetables, there is information on bare-root transplants available at University of Georgia Extension Bulletin 1144, Commercial Production of Vegetable Transplants (revised 2017), https://extension.uga.edu/publications/detail. html?number=B1144&title=Commercial%20Production%20of%20Vegetable%20Transplants.

[2] Containerized transplants are recommended.

PLANT GROWING CONTAINERS

TABLE 3.2   ADVANTAGES AND DISADVANTAGES OF VARIOUS PLANT GROWING CONTAINERS

Container	Advantages	Disadvantages
Single peat pellet	No media preparation, low storage requirement	Requires individual handling in setup, limited sizes
Prespaced peat pellet	No media preparation, can be handled as a unit of 50	Limited to rather small sizes
Single peat pot	Good root penetration, easy to handle in field, available in large sizes	Difficult to separate, master container is required, dries out easily, may act as a wick in the field if not properly covered
Strip peat pots	Good root penetration, easy to handle in field, available in large sizes, saves setup and filling time	May be slow to separate in the field, dries out easily
Paper pot systems	Transplanting system saves labor	Designed to use with specialized transplanting tool, pots are single use
Soil blocks	Avoid waste, expense of containers, minimal root disturbance	Requires adjustment of mixes so they hold together properly, labor required to make the soil blocks, difficult to handle
Plastic flat with unit	Easily handled, reusable, good root penetration	Requires storage during off season, may be limited in sizes
Plastic pack	Easily handled	Roots may grow out of container causing handling problems, limited in sizes, requires some setup labor
Plastic pot	Reusable, good root penetration	Requires handling as single plant

TABLE 3.2   ADVANTAGES AND DISADVANTAGES OF VARIOUS PLANT GROWING CONTAINERS (Continued)

Container	Advantages	Disadvantages
Polyurethane foam flat	Easily handled, requires less media than similar sizes of other containers, comes in many sizes, reusable	Requires regular fertilization, plants grow slowly at first because cultural systems use low levels of nitrogen
Expanded polystyrene tray	Lightweight, easy to handle, variable cell sizes and shapes, reusable, automation compatible	Need sterilization between uses, moderate investment, as trays age, roots can penetrate sidewalls of cells
Injection-molded trays	Variable cell sizes, reusable, long life, compatible for automation	Large investment, need sterilization between uses
Vacuum-formed tray	Low capital investment, automation incompatible due to damage to tray	Short life span, needs sterilization between uses

Adapted in part from D. C. Sanders and G. R. Hughes (eds.), *Production of Commercial Vegetable Transplants* (North Carolina Agricultural Extension Service -337, 1984).

03

SEEDS AND SEEDING

SEEDING SUGGESTIONS FOR GROWING TRANSPLANTS

1. *Media.* A desirable seeding mix should provide good drainage but retains moisture well enough to prevent rapid fluctuations, has good aeration, is low in soluble salts, and is free from insects, diseases, and weed seeds. Field soil alone usually is not a desirable seeding medium because it may crust or drain poorly under greenhouse conditions. Adding sand or a sand and peat mix may produce a very good seeding mixture. Many growers use soilless mixes (see Plant Growing Mixes) because of the difficulty of obtaining field soil that is free from pests and contaminating chemicals.

2. *Seeding.* Approximate seed requirements for growing transplants are provided in Table 3.3. Adjust seeding rates to account for the stated germination percentages and variations in soil temperatures. Excessively thick stands result in spindly seedlings, and poor stands are wasteful of valuable bench or bed space. Seeding into containerized trays can be done mechanically, using pelletized seeds if necessary. Pelletized seeds are seeds that have been coated with a clay material to facilitate planting by machine and easier singulation (one seed per cell in the tray). Seeding depth should be carefully controlled; most seeds should be planted from 1/4 to 1/2 in. deep. Exceptions are celery, which should only be 1/8 in. deep, and the vine crops, sweet corn, and beans, which can be seeded 1 in. or deeper.

3. *Moisture.* Maintain soil moisture in the desirable range by thorough watering after seeding and careful periodic watering as necessary. A combination of "spot watering" of dry areas and overall watering is usually necessary. Do not overwater.

4. *Temperature.* Be certain to maintain the desired temperature. Cooler than optimum temperatures may encourage disease, and warmer temperatures result in spindly seedlings. Germination rate and uniformity are enhanced by placing seeded containerized trays in a germination room where temperature and humidity are controlled. Once germination has initiated, the trays are moved to the greenhouse.

5. *Disease control.* Use disease-free or treated seed to prevent early disease problems. Containers should be new or cleaned and sanitized between uses. A disease-free seeding medium is essential. Maintain a strict sanitation program to prevent introduction of diseases. Carefully control watering and relative humidity. Use approved fungicides as drenches or sprays when necessary. Keep greenhouse environment as dry as possible with air-circulation fans and anticondensate plastic greenhouse covers.

6. *Transplanting.* Start transplanting when seedlings show the first true leaves so that transplanting can be completed before the seedlings become large and overcrowded. Seedlings in containerized trays do not require transplanting to a final transplant growing container.

7. *Fertilization.* Developing transplants need light, water, and fertilization with nitrogen, phosphorus, and potassium to develop a stocky, vigorous transplant, ready for the field. Excessive fertilization, especially with nitrogen, leads to spindly, weak transplants that are difficult to establish in the field. Excessive fertilization of tomato transplants with nitrogen can lead to reduced fruit yield in the field. Only 40–60 ppm nitrogen is needed in the irrigation solution for tomato. Many commercial soilless transplant mixes have a starter nutrient "charge," but this charge needs to be supplemented with a nutrient solution after seedlings emerge.

TABLE 3.3. APPROXIMATE SEED REQUIREMENTS FOR PLANT GROWING

Vegetable	Plants/Ounce of Seed	Seed Required to Produce 10,000 Transplants
Asparagus	550	1¼ lb
Broccoli	5,000	2 oz
Brussels sprouts	5,000	2 oz
Cabbage	5,000	2 oz
Cantaloupe	500	1¼ lb
Cauliflower	5,000	2 oz
Celery	15,000	1 oz
Sweet corn	100	6¼ lb
Cucumber	500	1¼ lb
Eggplant	2,500	4 oz
Lettuce	10,000	1 oz
Onion	4,000	3 oz
Pepper	1,500	7 oz
Summer squash	200	3¼ lb
Tomato	4,000	3 oz
Watermelon	200	3¼ lb

To determine seed requirements per acre:

$$\frac{\text{Desired plant population}}{10,000} \times \text{Seed required for } 10,000 \text{ plants}$$

Example 1: To grow enough broccoli for a population of 20,000 plants/acre:

$$\frac{20,000}{10,000} \times 2 = 4 \text{ oz seed}$$

Example 2: To grow enough summer squash for a population of 3,600 plants/acre:

$$\frac{3,600}{10,000} \times 3\frac{1}{4} = 1\frac{1}{4} \text{ lb approximately}$$

TEMPERATURE AND TIME REQUIREMENTS

TABLE 3.4. RECOMMENDATIONS FOR TRANSPLANT PRODUCTION USING CONTAINERIZED TRAYS

Crop[1]	Cell Size (in.)	Seed Required for 10,000 Transplants	Seeding Depth (in.)	Optimum Germination Temperature (°F)	Days to Germination[2]	pH Tolerance[3]	Time Required (Weeks)[4]
Broccoli	0.8–1.0	2 oz	¼	85	4	6.0–6.8	5–7
Brussels sprouts	0.8–1.0	2 oz	¼	80	5	5.5–6.8	5–7
Cabbage	0.8–1.0	2 oz	¼	85	4	6.0–6.8	5–7
Cantaloupe	1.0	1¼ lb	½	90	3	6.0–6.8	4–5
Cauliflower	0.8–1.0	2 oz	¼	80	5	6.0–6.8	5–7
Celery	0.5–0.8	1 oz	⅛–¼	70	7	6.0–6.8	5–7
Collards	0.8–1.0	2 oz	¼	85	5	5.5–6.8	5–7
Cucumber	1.0	1¼ lb	½	90	3	5.5–6.8	2–3
Eggplant	1.0	4 oz	¼	85	5	6.0–6.8	5–7
Lettuce	0.5–0.8	1 oz	⅛	75	2	6.0–6.8	4
Onion	0.5–0.8	3 oz	¼	75	4	6.0–6.8	10–12
Pepper	0.5–0.8	7 oz	¼	85	8	5.5–6.8	5–7
Squash	0.5–0.8	3¼ lb	½	90	3	5.5–6.8	3–4
Tomato	1.0	3 oz	¼	85	5	5.5–6.8	5–7
Watermelon	1.0	3¼ lb	½	90	3	5.0–6.8	3–4

Adapted from C. S. Vavrina, *An Introduction to the Production of Containerized Transplants* (Florida Cooperative Extension Service Fact Sheet HS 849, 2002), http://edis.ifas.ufl.edu/HS126.

[1] Other crops can be grown as transplants by matching seed types and growing according to the above specifications (example: endive = lettuce). Sweet corn can be transplanted, but tap root is susceptible to breakage.

[2] Under optimum germination temperatures.

[3] Plug pH will increase over time with alkaline irrigation water.

[4] Depends on growing conditions, primarily temperature. Expect the high end of this range when growing at cooler temperatures.

PLANT GROWING MIXES

SOILLESS MIXES FOR TRANSPLANT PRODUCTION

Most commercial transplant producers use some type of soilless media for growing vegetable transplants. Most media employ various mixtures of sphagnum peat and vermiculite or perlite, and growers might incorporate some fertilizer materials as the final media are blended. For small growers or on-farm use, similar types of media can be purchased premixed and bagged. Whether making or purchasing media, avoid storing mixes that contain fertilizer for long periods of time and keep mixes that contain a significant amount of peat from drying out.

Most of the currently used media mixes are based on variations of the Cornell mix recipe (Table 3.5):

TABLE 3.5. CORNELL PEAT-LITE MIXES

Component	Amount (cu yd)
Sphagnum peat	0.5
Horticultural vermiculite	0.5

Additions for Specific Uses (amount/cu yd)

		Greenhouse Tomatoes	
Addition	Seedling or Bedding Plants	Liquid Feed	Slow-Release Feed
Ground limestone (lb)	5	10	10
20% superphosphate (lb)	1–2	2.5	2.5
Calcium or potassium nitrate (lb)	1	1.5	1.5
Trace element mix (oz)	2	2	2
Osmocote (lb)	0	0	10
Mag Amp (lb)	0	0	5
Wetting agent (oz)	3	3	3

Adapted from J. W. Boodley and R. Sheldrake, Jr., *Cornell Peat-lite Mixes for Commercial Plant Growing* (New York State Agricultural Experiment Station, Station Agriculture Information Bulletin 43, 1982).

OTHER ADDITIONS/COMPONENTS

Availability and expense of peat sometimes limit its use. For this reason, other substrates have been explored and are occasionally used to replace a portion of the peat used in a peat-lite mix. *Coconut coir* is another renewable substrate that has good air- and water-holding properties with a higher pH and, often, a higher electrical conductivity (EC, or soluble salts) level than peat moss. *Wood fiber* is made from heat- and mechanically-treated wood, and while it has been shown to tie up some nitrogen, a blend of low rates of wood fiber (10–40% by volume) with peat can have favorable physical characteristics. *Composts* are highly variable, and while they can provide some plant nutrients with good physical properties, consistency of the product, presence of weed seeds and pathogens, and herbicide residues are important considerations. *Perlite*, a heat-treated mined rock that is chemically inert and that does not hold water, is included in mixes primarily to improve drainage.

TABLE 3.6. STERILIZATION OF PLANT GROWING SOILS

Agent	Method	Recommendation
Heat	Steam	30 min at 180°F
	Aerated steam	30 min at 160°F
	Electric	30 min at 180°F
Chemical	Chloropicrin	3–5 cc/cu ft of soil. Cover for 1–3 days. Aerate for 14 days or until no odor is detected before using
	Vapam	1 qt/100 sq ft. Allow 7–14 days before use

Caution: Chemical fumigants are highly toxic. Follow manufacturer's recommendations on the label.

Soluble salts, manganese, and ammonium usually increase after heat sterilization. Delay using heat-sterilized soil for at least 2 weeks to avoid problems with these toxic materials.

Adapted from K. F. Baker (ed.), *The UC System for Producing Healthy Container Grown Plants*, (California Agricultural Experiment Station Manual 23, 1972).

TABLE 3.7. TEMPERATURES REQUIRED TO DESTROY PESTS IN COMPOSTS AND SOIL

Pests	30-min Temperature (°F)
Nematodes	120
Damping-off organisms	130
Most pathogenic bacteria and fungi	150
Soil insects and most viruses	160
Most weed seeds	175
Resistant weeds and resistant viruses	212

Adapted from K. F. Baker (ed.), *The UC System for Producing Healthy Container Grown Plants* (California Agricultural Experiment Station Manual 23, 1972).

07

FERTILIZING AND IRRIGATING TRANSPLANTS

MONITORING NUTRIENT CONTENT IN TRANSPLANT PRODUCTION

Electrical conductivity (EC) measures the concentration of salts in a soil, and it is correlated with the concentration of plant nutrient ions. Several methods are commonly used to measure EC of soils and soilless media. The method used greatly impacts results and interpretation. The various methods are described briefly below. If testing media of potted plants, samples are taken from the lower two-thirds of the pot.

Pour-Through. This method is commonly used for potted plants containing slow-release fertilizers. The growing medium is irrigated thoroughly and allowed to drain for 1 hr. At testing, a small volume of distilled water is poured on the medium at the top of the pot. A small sample of the resulting leachate is collected at the base for testing.

1 part soil : 2 parts water dilution (1 : 2). One volume of soil is mixed with two volumes of water, and this slurry is used for testing.

1 part soil : 1 part water dilution (1 : 1). One volume of soil is mixed with one volume of water, and this slurry is used for testing.

Saturated medium extract (SME) or saturated paste. At testing, just enough water is added to a soil or media sample to saturate air spaces (see Table 3.8).

111

TABLE 3.8. ELECTRICAL CONDUCTIVITY (EC) IN SOIL AND PEAT-LITE MIXES

Method	Test Substrate	EC—Degree of Salinity (dS/m)[1]				
		Nonsaline	Slightly Saline	Moderately Saline	Strongly Saline	Excessive
SME	Soilless media	<0.75	0.75–2.0	2.1–5.0	5.1–6.0	>6.0
	Soil (all types)	<2.0	2.0–4.0	4.1–8.0	8.1–16.0	>16.0
1 : 2	Soilless media	<0.25	0.25–0.75	0.76–1.75	1.76–2.25	>2.25
	Soil (all types)	<0.4	0.4–0.8	0.9–1.6	1.7–3.2	>3.2
1 : 1	Coarse to loamy sand	0–1.1	1.2–2.4	2.5–4.4	4.5–8.9	>9.0
	Loamy fine sand to loam	0–1.2	1.3–2.4	2.5–4.7	4.8–9.4	>9.5
	Silt loam to clay loam	0–1.3	1.4–2.5	2.6–5.0	5.1–10.0	>10.1
	Silty clay loam to clay	0–1.4	1.5–2.8	2.9–5.7	5.8–11.4	>11.5
Pour-Through	Soilless media	<1.0	1.0–2.6	2.7–6.5	6.6–7.8	>7.8

Adapted from Recommended Soil Testing Procedures for the Northeastern United States (Northeastern Regional Publication No. 493, 3rd Edition, Revised 2011), https://sites.udel.edu/canr-nmeq/northeastern-soil-testing-procedures/ (accessed July 2021); On-site Testing of Growing Media and Irrigation Water (British Columbia Ministry of Agriculture, July 2015), https://www2.gov.bc.ca/assets/gov/farming-natural-resources-and-industry/agriculture-and-seafood/animal-and-crops/crop-production/on-site_testing_of_growing_media_and_irrigation_water_2015.pdf (accessed July 2021).

[1] Nonsaline soil or media has very low concentrations of nutrients. Slightly saline soils are appropriate for general plant growth. Moderately saline soils are satisfactory for established plants, but may be too high for seedling establishment or for some salt-sensitive species. Strongly saline soils may limit plant growth, especially if the soil dries out. Plants can be seriously injured if grown in excessively saline soils.

IRRIGATION OF TRANSPLANTS

There are two systems for application of water and fertilizer solutions to transplants produced in commercial operations: overhead sprinklers and subirrigation. Sprinkler systems apply water or nutrient solution by overhead water sprays from various types of sprinkler or emitter applicators. Advantages of sprinklers include the ability to apply chemicals to foliage and the ability to leach excessive salts from media. Disadvantages include high investment cost and maintenance requirements. Chemical and water application can be variable in poorly maintained systems, and nutrients can be leached if excess amounts of water are applied. One type of subirrigation uses a trough of nutrient solution in which the transplant trays are periodically floated, sometimes called ebb and flow or the "float system." Water and soluble nutrients are absorbed by the media and move upward into the media. Advantages of this system include uniform application of water and nutrient solution to all flats in a trough or basin. Subirrigation with recirculation of the nutrient solution minimizes the potential for pollution because all nutrients are kept in an enclosed system. Challenges with subirrigation include the need for care to avoid contamination of the entire trough with a disease organism. In addition, subirrigation systems restrict the potential to vary nutrient needs of different crops or developmental stages of transplants within a specific subirrigation trough.

With either production system, transplant growers need to exercise care in application of water and nutrients to the crop. Some standard fertilizer formulations used for transplant fertilization are provided in Table 3.9, and maximum acceptable water quality indices are presented in Table 3.10. Irrigation and fertilization programs are linked. Changes in one program can affect the efficiency of the other program. Excessive fertilization can lead to soluble salt injury, and excessive nitrogen application can lead to overly vegetative transplants. Excessive irrigation can leach nutrients.

TABLE 3.9. FERTILIZER FORMULATIONS FOR TRANSPLANT FERTILIZATION BASED ON NITROGEN AND POTASSIUM CONCENTRATIONS

	N and K_2O Concentrations (ppm)			
FERTILIZER	50	100	200	400
	oz/100 gal[1]			
20-20-20	3.3	6.7	13.3	26.7
15-0-15	4.5	8.9	17.8	35.6
20-10-20	3.3	6.7	13.3	26.7

TABLE 3.9. FERTILIZER FORMULATIONS FOR TRANSPLANT FERTILIZATION BASED ON NITROGEN AND POTASSIUM CONCENTRATIONS (Continued)

	N and K_2O Concentrations (ppm)			
FERTILIZER	50	100	200	400
Ammonium nitrate	1.4	2.9	5.7	11.4
+ potassium nitrate	1.5	3.0	6.1	12.1
Calcium nitrate	3.0	6.0	12.0	24.0
+ potassium nitrate	1.5	3.0	6.0	12.0
Ammonium nitrate	1.2	2.5	4.9	9.9
+ potassium nitrate	1.5	3.0	6.0	12.0
+ monoammonium phosphate	0.5	1.1	2.2	4.3

Adapted from P. V. Nelson, "Fertilization," in: E. J. Holcomb (ed.), *Bedding Plants IV. A Manual on the Culture of Bedding Plants as a Greenhouse Crop* (Batavia, IL: Ball Publishing, pp. 151–176, 1994).

[1] 1.0 oz in 100 gal is equal to 7.5 g in 100 L.

TABLE 3.10. MAXIMUM ACCEPTABLE WATER QUALITY INDICES FOR BEDDING PLANTS

Variable	Plug Production	Finish Flats and Pots
pH[1] (acceptable range)	5.5–7.5	5.5–7.5
Alkalinity[1]	1.5 me/L (75 ppm)	2.0 me/L (100 ppm)
Hardness[2]	3.0 me/L (150 ppm)	3.0 me/L (150 ppm)
EC	1.0 mS	1.2 mS
Ammonium-N	20 ppm	40 ppm
Boron	0.5 ppm	0.5 ppm

Adapted from P. V. Nelson, "Fertilization," in: E. J. Holcomb (ed.), *Bedding Plants IV. A Manual on the Culture of Bedding Plants as a Greenhouse Crop* (Batavia, IL: Ball Publishing, pp. 151–176, 1994).

[1] Moderately higher alkalinity levels are acceptable when lower amounts of limestone are incorporated into the substrate during its formulation. Very high alkalinity levels require acid injection into water source. Units: me/L = milliequivalent per liter; mg/L = equivalent weight.
[2] High hardness values are not a problem if calcium and magnesium concentrations are adequate and soluble salt level is tolerable.

08

PLANT GROWING PROBLEMS

TABLE 3.11. DIAGNOSIS AND CORRECTION OF TRANSPLANT DISORDERS Various problems may occur during production of transplants. Table 3.11 provides some potential causes and corrective measures for commonly observed transplant problems.

Symptoms	Possible Causes[1]	Corrective Measures
1. Spindly growth	Shade, cloudy weather, excessive watering, excessive temperature	Provide full sun, reduce temperature, restrict watering, ventilate or reduce night temperature, fertilize less frequently, provide adequate space
2. Budless plants	Many possible causes; no conclusive cause	Maintain optimum temperature and fertilization programs
3. Stunted plants	Low fertility	Apply fertilizer frequently in low concentration
A. Purple leaves	Phosphorus deficiency	Apply a soluble, phosphorus-rich fertilizer at 50 ppm P every irrigation for up to 1 week
B. Yellow leaves	Nitrogen deficiency	Apply N fertilizer solution at 50–75 ppm each irrigation for 1 week. Wash the foliage with water after application
C. Wilted shoots	*Pythium* root rot, flooding damage, soluble salt damage to roots	Check for *Pythium* or other disease organism. Reduce irrigation amounts and reduce fertilization
D. Discolored roots	High soluble salts from overfertilization, high soluble salts from poor soil sterilization	Leach the soil by excess watering. Do not sterilize at temperatures above 160°F. Leach soils before planting when soil tests indicate high amounts of soluble salts

TABLE 3.11. DIAGNOSIS AND CORRECTION OF TRANSPLANT DISORDERS (Continued)

Symptoms	Possible Causes[1]	Corrective Measures
4. Tough, woody plants	Overhardening	Apply starter solution (10-55-10 or 15-30-15 at 1 oz/gal to each 6–12 sq ft of bench area) 3–4 days before transplanting
5. Water-soaked and decayed stems near the soil surface	Damping-off	Use a sterile, well-drained medium. Adjust watering and ventilation practices to provide a less moist environment. Use approved fungicidal drenches
6. Poor root growth	Poor soil aeration, poor soil drainage, low soil fertility, excess soluble salts. Low temperature, residue from chemical sterilization, herbicide residue	Determine the cause and take corrective measures
7. Green algae or mosses growing on soil surface	High soil moisture, especially in shade or during cloudy periods	Adjust watering and ventilation practices to provide a less moist environment. Use a better-drained medium

[1] Possible causes are listed here; however, more than one factor may lead to the same symptom. Plant producers should thoroughly evaluate all possible causes of a specific disorder. Most states have a plant diagnostic lab (usually in affiliation with the state land grant university) that can assist with diagnosis of plant problems.

116

SUGGESTIONS FOR MINIMIZING DISEASES IN VEGETABLE TRANSPLANTS

Successful vegetable transplant production depends on attention to disease control. With the lack of labeled chemical pesticides, growers must focus on cultural and greenhouse management strategies to minimize opportunities for disease organisms to attack the transplant crop.

Greenhouse environment: Transplant production houses should be located at least several miles from any vegetable production field to avoid the entry of disease-causing agents in the house. Weeds around the greenhouse should be removed and the area outside the greenhouse should be free of weeds, volunteer vegetable plants, and discarded transplants.

Media and water: All media and irrigation water should be pathogen free. If media is to be blended on site, all mixing equipment and surfaces must be routinely sanitized. Irrigation water should be drawn from pathogen-free sources. Water from ponds or recycling reservoirs should be avoided.

Planting material: Only pathogen-free seed or plant plugs should be brought into the greenhouse to initiate a new transplant crop. Transplant producers should not accept seeds of unknown quality for use in transplant production. This can be a problem especially when producing small batches of transplants from small packages of seed, e.g., for a variety trial.

Cultural practices: Attention must be given to transplant production practices such as fertilization, irrigation, and temperature so that plant vigor is optimum. Free moisture, from sprinkler irrigation or condensation, on plants should be avoided. Ventilation of houses by exhaust fans and horizontal airflow fans will help reduce free moisture on plants. Growers should follow a strict sanitation program that will prevent introduction of disease organisms into the house. Weeds under benches must be removed. Outside visitors to the greenhouse should be strictly minimized, and all visitors and workers should walk through a disinfecting foot bath. All plant material and soil mix remaining between transplant crops should be removed from the house.

CONTROLLING TRANSPLANT HEIGHT

One aspect of transplant quality involves transplants of size and height that are optimum for efficient handling in the field during transplantation and for rapid establishment. Traditional means for controlling plant height included withholding water and nutrients and/or application of growth regulator chemicals. Today, few growth regulator chemicals are labeled for vegetable transplant production. Plant height control research is focusing on nutrient management, temperature manipulation, light quality, and mechanical conditioning of plants.

Nutrient management: Nitrogen applied in excess often causes transplants to grow tall rapidly. Using low-N solutions with 30–50 parts per million (ppm) nitrogen will help control plant height when frequent (daily) irrigations are needed. Higher concentrations of N might be needed when irrigations are infrequent (every 3–4 days). Choosing an intermediate N concentration (e.g. 80 ppm) for the entire transplant life cycle can cause an excessive growth rate. Irrigation frequency should guide the N concentration. Research has shown that excessive N applied to the transplant can lead to reduced fruit yield in the field.

Moisture management: Withholding water is a time-tested method of reducing plant height, but transplants can be damaged by drought. Sometimes transplants growing in Styrofoam trays along the edge of a greenhouse walkway will dry out faster than the rest of the transplants in the greenhouse. These "dry" plants are always shorter compared to the other transplants. Overwatering transplants should therefore be avoided and careful attention should be given to irrigation timing.

Light intensity: Transplants grown under reduced light intensity will stretch; therefore, growers must give attention to maximizing light intensity in the greenhouse. Aged polyethylene greenhouse covers should be replaced and greenhouse roofs and sides should be cleaned periodically, especially in winter. Supplementing light intensity for some transplant crops with lights might be justified.

Temperature management: Transplants grown under cooler temperatures (e.g., 50°F) will be shorter than plants grown under warmer temperatures. Where possible, greenhouse temperatures can be reduced or plants can be moved outdoors. Under cool temperatures, the transplant production cycle will be longer by several days and increased crop turnaround time might be unacceptable. For some crops, such as tomato, growing transplants under cool temperatures might lead to fruit quality problems, e.g., catfacing of fruits.

Mechanical conditioning: Shaking or brushing transplants frequently results in shorter transplants. Transplants can be brushed by several physical methods, for example, by brushing a plastic rod over the tops of the plants. This technique obviously should be practiced on dry plants only to avoid spreading disease organisms.

Day/night temperature management: The difference between the day and night temperatures (DIF) can be employed to help control plant height.

With a negative DIF, the day temperature is cooler than night temperature. Plants grown with a positive DIF are taller than plants grown with a zero or negative DIF. The critical period during the day for controlling height using DIF is from 30 min before sunrise until 2–3 hrs after sunrise. Lowering the temperature during this 3-hour period by a few degrees can help manage plant height. This system is not used during germination but rather is initiated when the first true leaves appear. The DIF system requires the capability to control the greenhouse temperature and would be most applicable to temperate regions in winter and spring seasons when day temperatures are cool and greenhouses can be heated. Some vegetable crops are more responsive to DIF than others (see Table 3.12).

TABLE 3.12. VEGETABLE TRANSPLANT RESPONSE TO THE DIFFERENCE IN DAY AND NIGHT TEMPERATURE (DIF)

Common Name	Response to DIF[1]
Broccoli	3
Brussels sprouts	3
Cabbage	3
Cantaloupe	3
Cucumber	1–2
Eggplant	3
Pepper	0–1
Squash	2
Tomato	2
Watermelon	3

Adapted from E. J. Holcomb (ed.), *Bedding Plants IV* (Batavia, IL: Ball Publishing, 1994); Original source: J. E. Erwin and R. D. Heins, "Temperature effects on bedding plant growth," *Bulletin* 42:1–18, Minnesota Commercial Flower Growers Association (1993). Used with permission.

[1] 0 = no response; 3 = strong response.

09

CONDITIONING TRANSPLANTS

Plants can be conditioned to prepare them to withstand stress conditions in the field, including low temperatures, high temperatures, drying winds, low soil moisture, or injury to the roots in transplanting. Growth rates decrease during conditioning, and the energy otherwise used in growth is stored in the plant to aid in resumption of growth after transplanting. Conditioning is used as an alternative to the older term, "hardening."

Any treatment that restricts growth will increase hardiness. Cool-season crops generally develop hardiness in proportion to the severity of the treatment and length of exposure and will, when well-conditioned, withstand subfreezing temperatures. Warm-season crops, even when highly conditioned, will not withstand temperatures much below freezing.

1. *Water supply*. Gradually reduce water by watering lightly at less frequent intervals. Do not allow the plants to dry out suddenly with severe wilting.

2. *Temperature*. Expose plants to lower temperatures (5–10°F) than those used for optimum growth. High day temperatures may reverse the effects of cool nights, making temperature management difficult. Do not expose biennials to prolonged cool temperatures, for this induces bolting.

3. *Fertility*. Do not fertilize, particularly with nitrogen, immediately before or during the initial stages of conditioning. Apply a starter solution or liquid fertilizer 1 or 2 days before field setting and/or with the transplanting water.

4. *Combinations*. Restricting water and lowering temperatures and fertility, used in combination, are perhaps more effective than any single approach.

Duration: Seven to ten days are usually sufficient to complete the conditioning process. Do not impose conditions so severe that plants will be overconditioned in case of delayed planting because of poor weather. Overconditioned plants require too much time to resume growth, and early yields may be lower.

PRETRANSPLANT HANDLING OF CONTAINERIZED TRANSPLANTS

Field performance of transplants is related not only to production techniques in the greenhouse but also to handling techniques before field planting. In the containerized tray production system, plants can be delivered to the field in the trays if the transplant house is near the production fields. For long-distance transport, the plants are usually pulled from the trays and packed in boxes. Tomato plants left in trays until field planting tend to have more rapid growth rates and larger fruit yields than when transplants were pulled from the trays and packed in boxes. Storage of pulled and packed tomato plants also reduces yields of large fruits compared to plants kept in the trays. If pulled plants must be stored prior to planting, storage temperatures should be selected to avoid chilling or overheating of transplants. Transplants that must be stored for short periods can be kept successfully at 50–55°F. Field performance of transplants is often improved if they are irrigated with a starter solution, or dilute fertilizer solution, at transplant. See Table 13 for some examples of commonly used starter solutions for vegetable transplants.

TABLE 3.13. STARTER SOLUTIONS FOR FIELD TRANSPLANTING[1]

Materials	Quantity to Use in Transplanter Tank
Readily Soluble Commercial Mixtures	
8-24-8, 11-48-0	(Follow manufacturer's directions)
23-21-17, 13-26-13	Usually 3 lb/50 gal of water
6-25-15, 10-52-17	
Straight Nitrogen Chemicals	
Ammonium sulfate, calcium nitrate, or sodium nitrate	2½ lb/50 gal of water
Ammonium nitrate	1½ lb/50 gal of water
Commercial Solutions	
30% nitrogen solution	1½ pt/50 gal of water
8-24-0 solution (N and P_2O_5)	2 qt/50 gal of water
Regular Commercial Fertilizer Grades	
4-8-12, 5-10-5, 5-10-10, etc.	
1 lb/gal for stock solution; stir well and let settle	5 gal of stock solution with 45 gal of water

[1] Apply at a rate of about ½ pt/plant.

121

10
SHIPPING AND TRANSPORTATION OF TRANSPLANTS

The majority of large-scale transplant producers in the United States are located in the southern states, though transplant producers of various scales are located throughout the country. Transplants that are transported a relatively short distance (<100 miles) are often transported in trays. Transplants that travel by truck are often pulled from their trays or from the field, packed in boxes, and shipped over a period of several days to their final destination.

In dark and warm conditions, transplant quality can deteriorate quickly as plants deplete their starch reserves and elongate. Studies have shown that shipping plants at lower temperatures (43 or 55°F) and with dim lighting (as opposed to darkness) improves plant quality. Conditioning plants by withholding water (see above) prior to transportation may also help slow down deterioration during shipping.

11
GRAFTED TRANSPLANTS

Grafting scion varieties onto rootstocks that provide certain attributes is increasingly used for field or high tunnel vegetable production in the United States. This is most widely done with tomato and watermelon, and increasingly with pepper and cucumber, and to a lesser degree with eggplant and muskmelon. Rootstocks provide plants with soilborne disease resistance, resistance or tolerance to biotic or abiotic stresses, increased yields, or enhanced vigor. In many situations, grafted vegetable plants have been shown to outperform ungrafted plants in terms of vigor, stress resistance, or yield.

Grafted transplants are more expensive to produce than ungrafted plants because they have higher seed costs and require specialized facilities and knowledge to perform the grafting process. While some growers prefer to graft their own plants, others prefer to purchase grafted plants from producers they trust who have perfected their grafting skills.

Grafting techniques vary, but it is usually performed on young seedlings using scion and rootstock varieties of matching sizes. This can be challenging due to variable growth rates and germination percentages. A clean and cool work area with good sanitization protocols is required. After grafting, plants are placed in a "healing chamber," which has high humidity and low light. There are many designs for healing chambers at various scales, but in all cases, plants must be gradually exposed to stressors as they heal, and healing is often considered the most critical step in grafting.

Additional resources relating to vegetable grafting:

1. Vegetable Grafting Research-Based Information Portal, published by M. Kleinhenz, Ohio State University, available at: http://www. vegetablegrafting.org/
2. Grafting Manual: How to produce grafted vegetable plants. Editors: C. Kubota, C. Miles and X. Zhao. 2017. Available at: http://www. vegetablegrafting.org/resources/grafting-manual/

12
ORGANIC TRANSPLANT PRODUCTION

Considerations unique to the production of certified organic transplants include (1) choice of media, (2) fertility management, and (3) pest and disease management. Choice of greenhouse construction materials and phytosanitation methods are also impacted by organic regulations, and extensive recordkeeping is required.

Organic transplants must be grown in media that is free of all prohibited synthetic inputs, including synthetic fertilizers and wetting agents. Ingredients commonly used in organic potting mixes are similar to those used in conventional mixes: peat moss, coir, vermiculite, and perlite. Often, organic fertilizers are added to provide slow-release nutrients; e.g., bone meal, alfalfa meal, fish meal. Composts are also often used, comprising 20–30% of the overall potting mix. If using composts, special attention must be paid to ensure that the compost is consistently high quality with a neutral pH and relatively low ammonia and ammonium, is thoroughly composted, and lacks plant pathogens and herbicide residues.

Throughout the duration of transplant growth, the transplants must receive only organic sources of nutrients. Peat-based media has low nutrient content on its own, and producers must take care to provide nutrients to young seedlings once they germinate. Organic transplant producers typically use one of two strategies (or combine them) to meet transplant nutrient needs:

1. *Potting up*. If an organic potting mix contains slow-release sources of plant nutrients (e.g., high-quality compost, blood meal, seed or plant meals, and bone meal), continuing to repot plants into large pots as they grow provides them with new media and therefore additional fertilization. Potting up is labor-intensive and can require large amounts of media, but can produce very high-quality transplants without the need for supplemental fertilization.
2. *Fertigation*. Similar to the strategy most often used for conventional transplants, organic producers can provide plant nutrients in liquid forms. There are many liquid sources of organic nitrogen; fish and seaweed emulsions are the most common. These are low in nutrient content, and therefore must be applied at relatively high rates on a regular basis.

Pest and disease management approaches are similar for organic as for conventional transplant production, but there are fewer curative tools available; the emphasis must be on sanitation and preventative approaches.

Organic potting mixes are not sterile if they are compost based; but the presence of beneficial biological control organisms in composts can reduce losses to plant pathogens like Pythium and Rhizoctonia.

Additional resources relating to organic transplant production:

1. Greenhouse Organic Transplant Production, by J. Biernbaum, 2006, Michigan State University. https://www.canr.msu.edu/hrt/uploads/535/78622/Organic-Transplants-2013-13pgs.pdf
2. Potting Mixes for Certified Organic Production, by G. Kuepper and K. Everett, 2004, ATTRA Horticultural Technical Note. https://attra.ncat.org/product/potting-mixes-for-certified-organic-production/

13

ADDITIONAL TRANSPLANT PRODUCTION RESOURCES

1. https://extension.uga.edu/publications/detail.html?number=
   B1144&title=Commercial%20Production%20of%20Vegetable%
   20Transplants.

2. Growing Vegetable Transplants in Plug Trays, by J. Bodnar and R.
   Garton, Ontario Ministry Agric. Food and Rural Affairs. http://www.
   omafra.gov.on.ca/english/crops/facts/transplants-plugtrays.htm (1996).

3. Organic Plug and Transplant Production, by L. Greer and K. Adam.
   https://attra.ncat.org/product/plug-and-transplant-production-for-
   organic-systems/ (2005).

4. Plug & Transplant Production. A Growers Guide, by R. Styer and
   D. Koranski. Ball Publishing, Batavia, IL (1997).

5. Commercial Transplant Production in Florida. By G. McAvoy and
   M. Ozores-Hampton. University of Florida Extension Bulletin HS714.

6. https://trace.tennessee.edu/cgi/viewcontent.cgi?article=1044&context=
   utk_agexcrop (1999).

7. The Physiology of Vegetable Crops, 2nd Edition. Edited by H.C. Wien
   and H. Stützel. Published by CABI, May, 2020.

8. Commercial Production of Vegetable Transplants. University of Georgia
   Bulletin 1144. Accessed Oct 2021 at https://extension.uga.edu/
   publications/detail.html?number=B1144&title=Commercial%
   20Production%20of%20Vegetable%20Transplants#Water (Updated 2017).

9. Growing Vegetable Transplants in Tennessee. University of Tennessee
   Knoxville Publication No. PB819. https://extension.tennessee.edu/
   publications/Documents/PB819.pdf (1999).

PART 4

GREENHOUSE AND PROTECTED AGRICULTURE PRODUCTION

Knott's Handbook for Vegetable Growers, Sixth Edition. George J. Hochmuth and Rebecca G. Sideman.
© 2023 John Wiley & Sons Ltd. Published 2023 by John Wiley & Sons Ltd.
Companion website: https://www.wiley.com/go/hochmuth/vegetablegrowers6e

01
STRUCTURES

Growing crops in structures that offer some degree of environmental control as compared with field production is collectively referred to as "protected agriculture" or "protected cultivation." Structures vary in the degree to which the environment within can be controlled. The most advanced greenhouses are permanent structures that have systems that enable precise control of temperature, relative humidity, light, water, and CO_2 enrichment. On the other end of the spectrum, simple frames that support coverings to protect field-grown crops from sunlight, rain, wind, or insects (e.g., screenhouses, shade houses, low tunnels, or caterpillar tunnels) offer less-expensive protection and can increase quality and yields and extend the harvest season of some vegetable crops. "High tunnels" or "hoophouses" represent middle ground. In these simple and relatively inexpensive greenhouse structures, simple hoops or bows with minimal framing are covered with plastic films. All sorts of systems are used: ventilation can be active or passive, supplemental heating systems may or may not be used, and crops can be grown in-ground or in soilless systems. A brief introduction to the common structures and their uses follows.

Low tunnels consist of hoops (typically 2–3 ft tall) installed directly over crops, covered with netting or plastic film. They are often used to extend the growing season (protecting young crops, or prolonging the harvest season in the fall months) or to protect crops from rain or insects. They are ventilated either by using specialized perforated or slitted coverings, or by manually raising the sides with a system to hold the coverings in place.

Shade houses or screenhouses comprise relatively simple frames supporting porous coverings; shade cloth or insect screen. They are used to protect crops from insects, sunlight, and wind and are most commonly used in warmer regions.

High tunnels are temporary structures that can be moved, though they may not be. They vary in size from roughly 12 ft wide to over 30 ft wide and can range from 20 to 200 ft long or even longer. The greater the dimensions of the high tunnel, the greater the potential temperature gains. High tunnels are usually covered with polyethylene, either two layers with air inflation between the layers or a single layer. Construction methods vary: tunnels built in regions with significant snow and wind must be constructed with enough strength to withstand those loads.

Greenhouses are designed to be permanent and may be single standalone structures or multibay (gutter-connected) houses. Greenhouses such as Venlo houses are very tall, gutter-connected, gabled structures with roof vents; these are often used in larger acreage installations.

Roof heights, coverings/glazings, and ventilations strategies vary. Roof shapes may be gabled or curved; in general, gabled houses are more suitable for heavy and rigid coverings like glass, and curved structures are covered with lighter materials like polyethylene.

Coverings must strike a balance between allowing light transmission, being durable and economical. Glass is among the most expensive coverings, but it is considered the gold standard in terms of light transmission. Fiberglass is durable and lightweight, but light transmission decreases as the fibers break down when exposed to UV light over time. Polyethylene film is lightweight with moderate light transmission and can provide good insulation (especially when two layers are installed with inflation fans creating an air layer between them), but depending on the grade of plastic purchased, it will need to be replaced every 2–6 years. Polycarbonate is a rigid lightweight material that allows good light transmission and, depending on the number of layers, can offer good insulation properties, but is more expensive than polyethylene or fiberglass. Commercial shade cloth materials range from about 30 to 80% light blocking, each having different applications. Similarly, insect screening materials used in screenhouse construction vary in mesh size, which determines the size of insects that will be excluded. In general, there is a tradeoff, with the smaller mesh size offering better exclusion properties but greatly reducing airflow.

There are additional newer covering materials for various protected agriculture applications, such as ethylene tetrafluoroethylene (ETFE) and bubble film such as SolaWrap. The advantages and drawbacks of these materials, and others in development, are currently being explored.

For more information, see:

S. Agehara, G. E. Vallad and E. A. Torres-Quezada, *Protected Culture for Vegetable and Small Fruit Crops: Types of Structures* (University of Florida/IFAS Extension Publication HS 1224, Revised Sept 2020), https://edis.ifas.ufl.edu/pdf/HS/HS122400.pdf (accessed 2 Nov. 2021).

02

CULTURAL MANAGEMENT

Although most vegetables can be grown successfully in greenhouses, only a few are grown there commercially on a large scale. Tomato, cucumber, and lettuce are the three most commonly grown vegetables in commercial greenhouses. Some general cultural management principles are discussed here. Growers should consult with their local Extension Service for local recommendations.

GREENHOUSE DESIGN

Successful greenhouse vegetable production depends on careful greenhouse design and construction. Important factors include the types of crops to be grown, prevailing climate and seasonal weather conditions, available space, and budget. Consideration must be provided for environmental controls, durability of components, and ease of operations, among other factors. There are many specialized companies that offer turn-key greenhouses for small-to-large operations. The publications listed at the end of this chapter offer helpful advice for construction.

SANITATION

There is no substitute for good sanitation for preventing insect and disease outbreaks in greenhouse crops. Successful sanitation depends on diligently observing the greenhouse for openings that would allow entry of insects and closely looking at the plants for evidence of insect infestations or diseases.

To keep greenhouses clean, remove all dead plants, unnecessary mulch material, seedling flats, weeds, etc. Burn or bury all plant refuse. Do not contaminate streams or water supplies with plant refuse. Weeds growing in and near the greenhouse during and after the cropping period should be destroyed. Do not attempt to overwinter garden or house plants in the greenhouses. Pests can be maintained on these plants and ready for an early invasion of vegetable crops. To prevent disease organisms from carrying over on the structure of the greenhouse and on the heating pipes and walks, treat with approved materials following all safety guidelines.

A 15–20-ft strip of carefully maintained lawn or bare ground around the greenhouse helps decrease trouble from two-spotted mites and other pests. To reduce entry of whiteflies, leafhoppers, and aphids from weeds and other plants near the greenhouses, treat the area with a labeled insecticide and control weeds around the greenhouse. Some pests can be excluded with properly designed screens. Do not locate outdoor field crops or gardens near the greenhouse.

MONITORING PESTS

Insects such as greenhouse and silverleaf whiteflies, thrips, and leaf miners are attracted to shades of yellow and fly toward that color. Thus, insect traps can be made by painting pieces of board with the correct shade of yellow pigment and then covering the paint with a sticky substance. Similar traps are available commercially from several greenhouse supply sources. By placing a number of traps within the greenhouse range, it is possible to check infestations daily and be aware of early infestations. Control programs can then be commenced while populations are low. Two-spotted mites cannot be trapped in this way, but infestations usually begin in localized areas. Check leaves daily and begin control measures as soon as the first infested areas are noted.

SPACING

Good-quality container-grown transplants should be set in arrangements to allow about 4 sq ft/plant for tomato, 5 sq ft/plant for American-type cucumber, and 7–9 sq ft/plant for European-type cucumber. Lettuce requires 36–81 sq in./plant.

TEMPERATURE

Greenhouse tomato varieties may vary in their temperature requirements, but most varieties perform well at a day minimum temperature of 70–75°F and a night minimum temperature of 62–64°F. Temperatures for cucumber seedlings should be 72–76°F day and 68°F night. In a few weeks, night temperature can be gradually lowered to 62–64°F. Night temperatures for lettuce can be somewhat lower than for tomato and cucumber.

In northern areas, provisions should be made to heat irrigation water to be used in greenhouses to about 70°F.

PRUNING AND TYING

Greenhouse tomatoes and cucumbers are usually pruned to a single stem by frequent removal of axillary shoots or suckers. Other pruning systems are possible and sometimes used. Various tying methods are used; one common method is to train the pruned plant around a string suspended from an overhead wire.

POLLINATION

Greenhouse tomatoes must be pollinated by hand or with bumblebees to assure a good set of fruit. This involves tapping or vibrating each flower cluster to transfer the pollen grains from the anther to the stigma. This should be done daily as long as there are open blossoms on the flower cluster. The pollen is transferred most readily during sunny periods and with the most difficulty on dark, cloudy days. The electric or battery-operated hand vibrator is the most widely accepted tool for vibrating tomato flower clusters. Most red-fruited varieties pollinate more easily than pink-fruited varieties and can often be pollinated satisfactorily by tapping the overhead support wires or by shaking flowers in the airstream of a motor-driven backpack air-blower (be sure to wear ear protection with any loud machines). Modern growers now use bumblebees for pollinating tomato. Specially reared hives of bumblebees are purchased by the grower for this purpose.

Pollination of European seedless cucumbers causes off-shape fruit, so outdoor bees must be excluded from the greenhouse. To help overcome this, gynoecious cultivars have been developed that bear almost 100% female flowers. Only completely gynoecious and parthenocarpic (set fruits without pollination) cucumber cultivars are now recommended for commercial production.

American-type (seeded) cucumbers require bees for pollination. One colony of honeybees per house should be provided. It is advisable to shade colonies from the afternoon sun and to avoid excessively high temperatures and humidities. Honeybees fly well in glass and polyethylene plastic houses but fail to work under certain other types of plastic. Under these conditions, crop failures may occur through lack of pollination.

ORGANIC PRODUCTION

The same requirements for organic production in open fields are generally required for indoor production. There may be some special certification needs for switching to organic production systems from a conventional system. Aspiring greenhouse organic growers should check with their certifying agency about meeting requirements for organic production.

Some specific considerations include the following:

1. Greenhouse structural components. Growers may find some treated wood materials are not permitted for structural components in an organic greenhouse system.

2. Organic producers rely on organic sources of plant nutrients including crop residues, manures, organic wastes, and approved minerals. When used in fertigation, many liquid organic fertilizers can clog emitters and present a risk of causing ammonium toxicity, but some plant- or fish-based materials can overcome these challenges.

3. Composts are a key component of the media used in many organic greenhouse production systems. Attention to the quality of any compost used in greenhouse production (C : N ratio, levels of soluble salts and ammonium, pH, and lack of pathogens) is crucial.

4. "Foliar feeding" is popular with many organic growers, especially with those using some sort of soilless system. While foliar feeding is not a cost-effective way to supply most of the nutrients needed by plants, it can be used to treat certain nutrient deficiencies. Special attention must be paid to food safety considerations if using foliar amendments.

5. Pest management will be different in an organic greenhouse and growers will rely more on mechanical control, biological and other natural systems. Solarization or steaming of the soil in ground-bed greenhouses may help reduce soil pests if it can be practiced efficiently within the particular cropping system. Organic growers will needed to be particularly attuned to environmental controls needed to minimize disease infestation, and certain crops will need to be chosen to fit particular times of the year when conditions are most favorable for the crop.

Many of the details and principles outlined in Extension publications dealing with conventional greenhouse production apply equally to organic systems. The main differences revolve around media, fertilizers, and pest control. Crop selection mostly depends on the market. Crop management, e.g., tomato pruning, pollination, irrigation, and more would be similar for conventional and organic crops and more information is provided in the references below.

Adapted from Ontario Ministry of Agriculture Publication Growing Greenhouse Vegetables in Ontario. http://www.omafra.gov.on.ca/english/crops/pub836/p836order.htm (accessed 13 Nov. 2021) and from G. Hochmuth, "Production of Greenhouse Tomatoes," in *Florida Greenhouse Vegetable Production Handbook*, vol. 3, (2018), https://edis.ifas.ufl.edu/publication/CV266 (accessed 13 Nov. 2021).

For more information and greenhouse vegetable production, see:

G. Hochmuth and R. Hochmuth, *Keys to Successful Tomato and Cucumber Production in Perlite Media* (2016), https://edis.ifas.ufl.edu/publication/HS169 (accessed 13 Nov. 2021).

G. Hochmuth and R. Hochmuth, Nutrient solution formulation for hydroponic (perlite, rockwool, NFT) tomatoes in Florida (2018), https://edis.ifas.ufl.edu/publication/cv216 (accessed 13 Nov. 2021).

L. Greer and S. Diver, Organic Greenhouse vegetable production – Horticulture systems guide. ATTRA Appropriate Technology for rural areas (2000), https://attra.ncat.org/product/organic-greenhouse-vegetable-production/ (accessed 13 Nov. 2021).

G. Hochmuth and R. Hochmuth, *Design Suggestions and Greenhouse Management for Vegetable Production in Perlite and Rockwool Media in Florida* (2019), https://edis.ifas.ufl.edu/publication/CV195 (accessed 13 Nov. 2021).

R. Bucklin, "Physical Greenhouse Design Considerations," in *Florida Greenhouse Vegetable Production Handbook,* vol. 2 (2020), https://edis.ifas.ufl.edu/publication/CV254 (accessed 13 Nov. 2021).

S. Walker and I. Joukhadar, Greenhouse vegetable production. New Mexico State University, Circular 556 (2019), https://aces.nmsu.edu/pubs/_circulars/CR556/ (accessed 13 Nov., 2021).

D. Cox, Organic Fertilizers – Thoughts on Using Liquid Organic Fertilizers for Greenhouse Plants. University of Massachusetts Extension (2014), https://ag.umass.edu/greenhouse-floriculture/fact-sheets/organic-fertilizers-thoughts-on-using-liquid-organic-fertilizers (accessed 13 Nov. 2021).

D. Trinklein, Foliar Feeding Revisited. University of Missouri (2019), https://ipm.missouri.edu/MPG/2019/4/foliarFeeding/ (accessed 13 Nov. 2021).

K. Adam, Herbs: organic greenhouse production (2018), https://attra.ncat.org/product/herbs-organic-greenhouse-production/ (accessed 13 Nov. 2021).

Growing hydroponic tomatoes. University of Arizona, (accessed 13 Nov. 2021).

CARBON DIOXIDE ENRICHMENT OF GREENHOUSE ATMOSPHERES

The beneficial effects of adding carbon dioxide (CO_2) to the northern greenhouse environment are well established. The crops that respond most consistently to supplemental CO_2 are cucumber, lettuce, and tomato, although almost other greenhouse crops will benefit. CO_2 enrichment of southern greenhouses probably has little benefit due to frequent ventilation requirements under the warmer temperatures.

Outside air contains about 400 parts per million (ppm) CO_2 by volume. Most plants grow well at this level, but if levels are higher, the plants respond by producing more sugars. During the day, in a closed greenhouse, the plants use the CO_2 in the air and reduce the level below the normal 400 ppm. This is the point at which CO_2 addition is most important. Most crops respond to CO_2 additions up to about 1,300 ppm. Somewhat lower concentrations are adequate for seedlings or when growing conditions are less than ideal.

Carbon dioxide enrichment can be achieved by burning natural gas, propane, or kerosene and directly from containers of pure CO_2. Each source has potential advantages and disadvantages. When natural gas, propane, or kerosene is burned, not only is CO_2 produced but also heat, which can supplement the normal heating system. Incomplete combustion or contaminated fuels may cause plant damage. Most sources of natural gas and propane have sufficiently low levels of impurities, but you should notify your supplier of your intention to use the fuel for CO_2 addition in a greenhouse. Sulfur levels in the fuel should not exceed 0.02% by weight. Caution must be exercised when burning fossil fuels inside a closed environment, and workers should be made aware about burning fossil fuels in a closed environment.

A number of commercial companies have burners available for natural gas, propane, and liquid fuels. The most important feature of a burner is that it burns the fuel completely, leaving no dangerous fumes.

Because photosynthesis occurs only during daylight hours, CO_2 addition is not required at night, but supplementation is recommended on dull days. Supplementation should start approximately 1 hr before sunrise, and the system should be shut off 1 hr before sunset. If supplemental lighting is used at night, intermittent addition of CO_2 or the use of a CO_2 controller may be helpful.

When ventilators are opened, it is not possible to maintain high CO_2 levels. However, it is often during these hours (high light intensity and temperature) that CO_2 supplementation is beneficial. Because maintaining optimal levels is impossible under certain situations, maintaining at least ambient levels is suggested. A CO_2 controller, whereby the CO_2 concentration can be maintained at any level above ambient, is therefore useful.

One important factor is an adequate air distribution system. The distribution of CO_2 mainly depends on the air movement in the greenhouse(s), for CO_2 does not travel far by diffusion. This means that if a single source of CO_2 is used for a large surface area or several connecting greenhouses, a distribution system must be installed. Air circulation (horizontal fans or a fanjet system) that moves a large volume of air provides uniform distribution within the greenhouse.

Adapted from Ontario Ministry of Agriculture and Food AGDEX 290/27 (1984); From G. Hochmuth and R. Hochmuth (eds.), "Greenhouse Vegetable Crop Production Guide," in *Florida Greenhouse Vegetable Production Handbook,* vol. 3, Florida Cooperative Extension Fact Sheet HS784 (2019), https://edis.ifas.ufl.edu/entity/topic/book_florida_greenhouse_v3. (accessed 13 Nov. 2021).

Carbon dioxide in Greenhouses (Aug. 2009), http://www.omafra.gov.on.ca/english/crops/facts/00-077.htm (accessed 13 Nov. 2021).

Greenhouse carbon dioxide supplementation, https://extension.okstate.edu/fact-sheets/greenhouse-carbon-dioxide-supplementation.html (accessed 13 Nov. 2021).

SOILLESS CULTURE OF GREENHOUSE VEGETABLES

Well-managed field soils supply crops with sufficient water and appropriate concentrations of the 14 essential inorganic elements. A combination of desirable soil chemical, physical, and biotic characteristics provides conditions for extensive rooting, which results in anchorage, the third general quality provided to crops by soil.

When field soils are used in the greenhouse for repeated intensive crop culture, desirable soil characteristics can deteriorate rapidly. Diminishing concentrations of essential elements and impaired physical properties must be restored as in the field by applications of lime, fertilizer, and organic matter. Amendment with high levels of organic matter and other amendments, in an environment that does not receive leaching rains, can result in very high levels of soluble salts, which can impede plant growth. Pathogenic microorganisms, weed seeds, and nematode populations can also become problems.

In some regions of the country and the world, soils have been used for greenhouse production over many decades. Strategies used to maintain soil health in protected agriculture include regular leaching (either using natural rainfall every few years as the greenhouse covering is replaced, or by irrigating), the use of fertilizers with comparatively low salt indices, regular inputs of amendments high in organic matter, and in some cases, judicious use of steam sterilization to mitigate pest problems.

Even with the best management, soils may deteriorate in quality over time. In addition, the costs—particularly of steam sterilization—of maintaining greenhouse soils in good condition have escalated so that soilless culture methods are competitive with, or perhaps more economically favorable than soil culture. Accordingly, recent decades have seen a considerable shift from soil culture to soilless culture in greenhouses. Liquid and solid media systems are used.

Liquid Soilless Culture

The nutrient film technique (NFT) is the most commonly used liquid production system.

NFT growing systems consist of a series of narrow channels through which nutrient solution is recirculated from a supply tank (Figure 4.1). A plumbing system of plastic tubing and a submersible pump in the tank are the basic components. The growing channels are generally constructed of opaque plastic film or plastic pipe; asphalt-coated wood and fiberglass are also used. The basic characteristic of all NFT systems is the shallow depth of solution maintained in the channels. Flow is usually continuous, but sometimes systems are operated intermittently by supplying solution a few minutes every hour. The purpose of intermittent flow is to assure adequate aeration of the root systems. This also reduces the energy required, but under rapid growth conditions, plants may experience water stress if the flow period is too short or infrequent. Therefore, intermittent flow management seems better adapted to mild temperature periods or to plantings during the early stages of development. Capillary matting is sometimes used in the bottom of NFT channels, principally to avoid the side-to-side meandering of the solution stream around young root systems; it also acts as a reservoir by retaining nutrients and water during periods when flow ceases.

NFT channels are frequently designed for a single row of plants with a channel width of 6–8 in. Wider channels of 12–15 in. are used to accommodate two rows of plants, but meandering of the shallow solution stream becomes a greater problem with greater width. To minimize this problem, small dams can be created at intervals down the channel by placing thin wooden sticks across the stream, or the channel may be lined with capillary matting. The channels should be sloped 4–6 in./100 ft to maintain gravity flow of the solution. Flow rate into the channels should be in the range of 1–2 qt/min (1,000–2,000 mL/min).

Channel length should be limited to a maximum of 100 ft in order to minimize increased solution temperature on bright days. The ideal solution temperature for tomato is 68–77°F. Temperatures below 59 or above 86°F decrease plant growth and tomato yield. Channels of black plastic film increase solution temperature on sunny days. During cloudy weather, it may be necessary to heat the solution to the recommended temperature. Solution temperatures in black plastic channels can be decreased by shading or painting the outer surfaces white or silver. As an alternative to channels lined with black polyethylene, 4–6-in. PVC pipe may be used. Plant holes are spaced appropriately along the pipe. The PVC system is permanent once it is constructed; the lining in polyethylene-lined channels must be replaced for each crop. Initial costs are higher for the PVC, and sanitation between crops may be more difficult. In addition, PVC pipe systems are subject to root flooding if root masses clog pipes.

Solid Soilless Culture

Lightweight media in upright containers or upright or lay-flat bags and rockwool mats are the most commonly used media culture systems.

Media Culture

Soilless culture in bags, pots, or troughs with a lightweight medium is the simplest, most economical, and easiest to manage of all soilless systems.

The most common media used in containerized systems of soilless culture are perlite, rockwool, peat-lite, coco-fiber (coir), or a mixture of pine bark and wood chips. Container types range from long wooden troughs in which one or two rows of plants are grown to polyethylene bags or rigid plastic pots containing one to three plants. Research is on-going with various mixtures of media sources, and containers. Bag or pot systems using bark chips or peat-lite are in common use throughout the United States and offer major advantages over other types of soilless culture:

1. These materials have excellent retention qualities for nutrients and water.
2. Containers of medium are readily moved in or out of the greenhouse whenever necessary or desirable.
3. Media is lightweight and easily handled.
4. The medium is useful for several successive crops.
5. The containers are significantly less expensive and less time-consuming to install.
6. In comparison with recirculated hydroponic systems, the nutrient-solution system is less complicated and less expensive to manage.

From a plant nutrition standpoint, the latter advantage is of significant importance. In a recirculated system, the solution is continuously changing in its nutrient concentration because of differential plant uptake. In the bag or pot system, the solution is not recirculated. Nutrient solution is supplied from a fertilizer proportioner or large supply tank to the surface of the medium in a sufficient quantity to wet the medium. Any excess is drained away from the system through drain holes in the base of the containers. The drainage or leachate should be collected and reused outside the greenhouse on an outdoor crop or lawn. The leachate should not be allowed to drain into the groundwater or nearby water bodies causing pollution. Thus, the concentration of nutrients in solution supplied to the plants is the same at each application. This eliminates the need to sample and analyze the solution periodically to determine necessary adjustments and avoids the possibility of solution excess or deficiencies.

In the bag or pot system, the volume of medium per container varies from about 1/2 cu ft in vertical polyethylene bags or pots to 2 cu ft in lay-flat bags. In the vertical bag system, 4-mil black polyethylene bags with prepunched drain holes at the bottom are common. One, but sometimes two, tomato or cucumber plants are grown in each bag. Lay-flat bags accommodate two or three plants. In either case, the bags are aligned in rows with spacing appropriate to the type of crop. It is good practice to place vertical bags or pots on a narrow sheet of plastic film to prevent root contact or penetration into the underlying soil. Plants in lay-flat bags, which have drainage slits (or overflow ports) cut along the sides an inch or so above the base, also benefit from a protective plastic sheet beneath them.

Nutrient solution is delivered to the containers by supply lines of black polyethylene tubing, spaghetti tubing, spray sticks, or ring drippers in the containers. The choice of application system is important in order to provide proper wetting of the medium at each irrigation. Texture and porosity of the growing medium and the surface area to be wetted are important considerations in making the choice. Spaghetti tubing provides a point-source wetting pattern, which might be appropriate for fine-textured media that allow water to be conducted laterally with ease. In lay-flat bags, single spaghetti tubes at individual plant holes provide good wetting of peat-lite or rockwool. In a vertical bag containing a porous medium, a spray stick with a 90-degree spray pattern does a good job of irrigation if it is located to wet the majority of the surface. Ring drippers are also a good choice for vertical bags, although somewhat more expensive. When choosing an application system for bag or container culture, remember that the objective of irrigation is to distribute nutrient solution uniformly so that all of the medium is wet.

Rockwool and Perlite Culture

Rockwool is made by melting various types of rocks at very high temperatures. The resulting fibrous particles are formed into growing blocks or mats that are sterile and free of organic matter (Figures 4.2 to 4.8). The growing mats have a high water-holding capacity, no buffering capacity, and an initial pH of 7–8.5, which is lowered quickly with application of slightly acidic nutrient solutions. Uncovered mats, which are covered with polyethylene during setup, or polyethylene enclosed mats can be purchased. The mats are 8–12 in. wide, 36 in. long, and 3 in. thick. Perlite, a volcanic mineral, is heated and expanded into small, granular particles. Perlite has a high water-holding capacity but provides good aeration. Perlite-filled bags can be purchased or a grower may elect to purchase bulk perlite and fill plastic sleeves at the farm.

The greenhouse floor should be carefully leveled and covered with 3-miles black/white polyethylene, which restricts weed growth and acts as a light reflector with the white side up. Alternatively, the greenhouse may be constructed with a concrete floor. The lay-flat bags are placed end to end to form a row; single or double rows are spaced for the crop and greenhouse configuration. Upright containers or bags are spaced appropriately for the crop.

A complete nutrient solution made with good-quality water is used for initial soaking of the growing containers. Large volumes are necessary because of the high water-holding capacity of the mats. Drip irrigation tubing or spaghetti tubing arranged along the plant row are used for initial soaking and, later, for fertigation.

141

Cross-slits, corresponding in size to the propagating blocks, are made in the polyethylene bags/ cover at desired in-row plant spacings; usually two plants are grown in each 30-in.-long mat. The propagating blocks containing the transplant are placed on the mat, and the excess polyethylene from the cross-slit is arranged around the block (see Figure 4.2). Frequent irrigation is required until plant roots are established in the mat; thereafter, fertigation is applied 4–10 times a day depending on the growing conditions and stage of crop growth. The mats are leached with good-quality water when samples taken from the mats with a syringe have excessive conductivity readings.

Adapted in part from H. Johnson Jr., G. J. Hochmuth, and D. N. Maynard, "Soilless Culture of Greenhouse Vegetables," Florida Cooperative Extension Bulletin 218 (1985); From M. Sweat and G. Hochmuth, "Production Systems," in *Florida Greenhouse Vegetable Production Handbook*, vol. 3, Fact Sheet HS785, https://edis.ifas.ufl.edu/publication/CV263 (accessed 13 Nov. 2021).

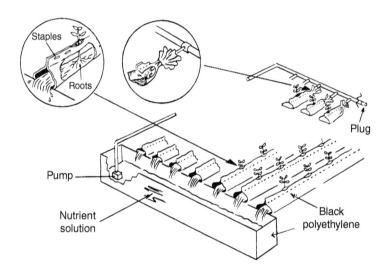

Figure 4.1. NFT culture system using polyethylene film to hold plants and supply nutrient solution through a recirculation system (From Florida Cooperative Extension Bulletin 218).

Figure 4.2. Arranged mats are covered with white/black polyethylene (Adapted from GRODAN® Instructions for cultivation—cucumbers, Grodania A/S, Denmark and used with permission).

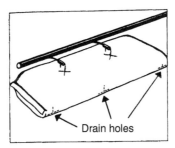

Drain holes

Figure 4.3. Irrigation system and drainage holes for rockwool mats enclosed in a polyethylene bag (Adapted from GRODAN Instructions for cultivation—cucumbers, Grodania A/ S, Denmark and used with permission).

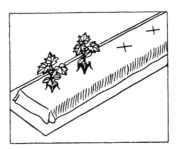

Figure 4.4. Cross-slits are made to accommodate transplants in propagation blocks (Adapted from GRODAN Instructions for cultivation—cucumbers, Grodania A/S, Denmark and used with permission).

143

Figure 4.5. Ordinarily, two plants are placed in each 30-in.-long mat (Adapted from GRODAN Instructions for cultivation—cucumbers, Grodania A/S, Denmark and used with permission).

Figure 4.6. Fertigation supplied by spaghetti tubing to each plant (Adapted from GRODAN Instructions for cultivation—cucumbers, Grodania A/S, Denmark and used with permission).

Figure 4.7. Fertigation supplied by drip irrigation tubing (Adapted from GRODAN Instructions for cultivation—cucumbers, Grodania A/S, Denmark and used with permission).

144

Figure 4.8. Removal of sample from rockwool mat with a syringe for conductivity determination (Adapted from GRODAN Instructions for cultivation—cucumbers, Grodania A/S, Denmark and used with permission).

NUTRIENT SOLUTIONS FOR SOILLESS CULTURE

Because the water and/or media used for soilless culture of greenhouse vegetables is devoid of essential elements, these must be supplied in a nutrient solution.

Commercially available fertilizer mixtures may be used, or nutrient solutions can be prepared from individual chemical salts. Historically the most widely used and generally successful nutrient solution is one developed by D. R. Hoagland and D. I. Arnon at the University of California decades ago. Many commercial mixtures are based on their formula.

Detailed directions for preparation of nutrient solutions based on Hoagland's solution, which are suitable for experimental or commercial use, are presented in (Table 4.1 to 4.5). The "Florida Solutions" below were developed by Dr. George Hochmuth based on greenhouse hydroponic research and on-farm tests in Florida. Information on plant tissue testing is presented in Tables 4.6 and 4.7.

TABLE 4.1.   HOAGLAND'S NUTRIENT SOLUTIONS

Salt	Stock Solution (g to make 1L)	Final Solution (mL to make 1L)
Solution 1		
$Ca(NO_3)_2 \cdot 4H_2O$	236.2	5
KNO_3	101.1	5
KH_2PO_4	136.1	1
$MgSO_4 \cdot 7H_2O$	246.5	2
Solution 2		
$Ca(NO_3)_2 \cdot 4H_2O$	236.2	4
KNO_3	101.1	6
$NH_4H_2PO_4$	115.0	1
$MgSO_4 \cdot 7H_2O$	246.5	2

TABLE 4.1. HOAGLAND'S NUTRIENT SOLUTIONS (Continued)

Micronutrient Solution

Compound	Amount (g) Dissolved in 1 L Water
H_3BO_3	2.86
$MnCl_2 \cdot 4H_2O$	1.81
$ZnSO_4 \cdot 7H_2O$	0.22
$CuSO_4 \cdot 5H_2O$	0.08
$H_2MoO_4 \cdot H_2O$	0.02

Iron Solution

Iron chelate, such as Sequestrene 330, made to stock solution containing 1 g actual iron/L. Sequestrene 330 is 10% iron; thus, 10 g/L are required. The amounts of other chelates must be adjusted on the basis of their iron content.

Procedure: To make 1 L of Solution 1, add 5 mL $Ca(NO_3)_2 \cdot 4H_2O$ stock solution, 5 mL KNO_3, 1 mL KH_2PO_4, 2 mL $MgSO_4 \cdot 7H_2O$, 1 mL micronutrient solution, and 1 mL iron solution to 800 mL distilled water. Make up to 1 L. Some plants grow better on Solution 2, which is prepared in the same way.

Adapted from D. R. Hoagland and D. I. Arnon, "The Water-culture Method for Growing Plants Without Soil," California Agricultural Experiment Station Circular 347 (1950).

SOME NUTRIENT SOLUTIONS FOR COMMERCIAL GREENHOUSE VEGETABLE PRODUCTION

These solutions are designed to be supplied directly to greenhouse vegetable crops.

TABLE 4.2. JOHNSON'S SOLUTION

Compound	Amount (g/100 gal water)
Potassium nitrate	95
Monopotassium phosphate	54
Magnesium sulfate	95
Calcium nitrate	173
Chelated iron (FeDTPA)	9
Boric acid	0.5
Manganese sulfate	0.3
Zinc sulfate	0.04
Copper sulfate	0.01
Molybdic acid	0.005

	N	P	K	Ca	Mg	S	Fe	B	Mn	Zn	Cu	Mo
ppm	105	33	138	85	25	33	2.3	0.23	0.26	0.024	0.01	0.007

TABLE 4.3. JENSEN'S SOLUTION

Compound	Amount (g/100 gal water)
Magnesium sulfate	187
Monopotassium phosphate	103
Potassium nitrate	77
Calcium nitrate	189
Chelated iron (FeDTPA)	9.6
Boric acid	1.0
Manganese chloride	0.9
Cupric chloride	0.05
Molybdic acid	0.02
Zinc sulfate	0.15

	N	P	K	Ca	Mg	S	Fe	B	Mn	Zn	Cu	Mo
ppm	106	62	156	93	48	64	3.8	0.46	0.81	0.09	0.05	0.03

Adapted from H. Johnson, Jr., G. J. Hochmuth, and D. N. Maynard, "Soilless Culture of Greenhouse Vegetables," Florida Cooperative Extension Service Bulletin 218 (1985).

TABLE 4.4. NUTRIENT SOLUTION FORMULATION FOR TOMATO GROWN IN PERLITE, COCO-FIBER, OR ROCKWOOL IN FLORIDA (THE FLORIDA SOLUTION)

	Stage of Growth				
	1 Transplant to First Cluster	2 First Cluster to Second	3 Second Cluster to Third	4 Third Cluster to Fifth	5 Fifth Cluster to Termination
Stock A	3.3 pt Phosphorus[1] 6 lb KC1 10 lb MgSO$_4$ 10 g CuSO$_4$ 35 g MnSO$_4$ 10 g ZnSO$_4$ 40 g Solubor 3 mL Molybdenum[2]	3.3 pt Phosphorus 6 lb KC1 10 lb MgSO$_4$ 10 g CuSO$_4$ 35 g MnSO$_4$ 10 g ZnSO$_4$ 40 g Solubor 3 mL Molybdenum	3.3 pt Phosphorus 6 lb KC1 10 lb MgSO$_4$ 2 lb KNO$_3$ 10 g CuSO$_4$ 35 g MnSO$_4$ 10 g ZnSO$_4$ 40 g Solubor 3 mL Molybdenum	3.3 pt Phosphorus 6 lb KC1 12 lb MgSO$_4$ 2 lb KNO$_3$ 10 g CuSO$_4$ 35 g MnSO$_4$ 10 g ZnSO$_4$ 40 g Solubor 3 mL Molybdenum	3.3 pt Phosphorus 6 lb KC1 12 lb MgSO$_4$ 6 lb KNO$_3$ 10 g CuSO$_4$ 35 g MnSO$_4$ 10 g ZnSO$_4$ 40 g Solubor 3 mL Molybdenum

Stock B	2.1 gal liquid	2.4 gal liquid	2.7 gal liquid	3.3 gal liquid	3.3 gal liquid
Calcium nitrate	$Ca(NO_3)_2$[3] or 11.5 lb dry $Ca(NO_3)_2$ 0.7 lb Fe 330[4]	$Ca(NO_3)_2$ or 13.1 lb dry $Ca(NO_3)_2$ 0.7 lb Fe 330	$Ca(NO_3)_2$ or 14.8 lb dry $Ca(NO_3)_2$ 0.7 lb Fe 330	$Ca(NO_3)_2$ or 18.0 lb dry $Ca(NO_3)_2$ 0.7 lb Fe 330	$Ca(NO_3)_2$ or 18.0 lb dry $Ca(NO_3)_2$ 0.7 lb Fe 330

Adapted from G. Hochmuth (ed.), *Florida Greenhouse Vegetable Production Handbook*, vol. 3, Florida Cooperative Extension Service (2019), https://edis.ifas.ufl.edu/entity/topic/book_florida_greenhouse_v3 (accessed 13 Nov. 2021).

[1] Phosphorus from phosphoric acid (13 lb/gal specific wt., 23% P).

[2] Molybdenum from liquid sodium molybdate (11.4 lb/gal specific wt., 17% Mo).

[3] Liquid $Ca(NO_3)_2$ from a 7-0-11 (N-P_2O_5-K_2O-Ca) solution. When using dry $CaNO_3$ be sure to use greenhouse grade (not field grade) fertilizer.

[4] Iron as Sequestrene 330 (10% Fe).

TABLE 4.5. RECOMMENDED NUTRIENT SOLUTION CONCENTRATIONS FOR TOMATO GROWN IN ROCKWOOL, COCO-FIBER, OR PERLITE IN FLORIDA (THE FLORIDA SOLUTION)

	Stage of Growth				
Nutrient	1 Transplant to First Cluster	2 First Cluster to Second	3 Second Cluster to Third	4 Third Cluster to Fifth	5 Fifth Cluster to Termination
	Final delivered nutrient solution concentration (ppm)				
N	70	80	100	120	150
P	50	50	50	50	50
K	120	120	150	150	200
Ca	150	150	150	150	150
Mg	40	40	40	50	50
S	50	50	50	60	60
Fe	2.8	2.8	2.8	2.8	2.8
Cu	0.2	0.2	0.2	0.2	0.2
Mn	0.8	0.8	0.8	0.8	0.8
Zn	0.3	0.3	0.3	0.3	0.3
B	0.7	0.7	0.7	0.7	0.7
Mo	0.05	0.05	0.05	0.05	0.05

Adapted from G. Hochmuth, "Fertilization Management for Greenhouse Vegetables," in *Florida Greenhouse Vegetable Production Handbook*, vol. 3, Florida Cooperative Extension Service Fact Sheet HS787, https://edis.ifas.ufl.edu/publication/CV265 (accessed 13 Nov. 2021); G. Hochmuth and R. Hochmuth, *Nutrient Solution Formulation for Hydroponic (Perlite, Rockwool, NFT) Tomatoes in Florida* (2018), https://edis.ifas.ufl.edu/publication/cv216 (accessed 13 Nov. 2021).

06
TISSUE COMPOSITION

TABLE 4.6. APPROXIMATE NORMAL TISSUE COMPOSITION OF HYDROPONICALLY GROWN GREENHOUSE VEGETABLES[1]

Element	Tomato	Cucumber
K	5–8%	8–15%
Ca	2–3%	1–3%
Mg	0.4–1.0%	0.3–0.7%
NO_3–N	14,000–20,000 ppm	10,000–20,000 ppm
PO_4-P	6,000–8,000 ppm	8,000–10,000 ppm
Fe	40–100 ppm	90–120 ppm
Zn	15–25 ppm	40–50 ppm
Cu	4–6 ppm	5–10 ppm
Mn	25–50 ppm	50–150 ppm
Mo	1–3 ppm	1–3 ppm
B	20–60 ppm	40–60 ppm

Adapted from H. Johnson, *Hydroponics: A Guide to Soilless Culture Systems* (University of California Division of Agricultural Science Leaflet 2947, 1977).

[1] Values are for recently expanded leaves, fifth or sixth from the growing tip, petiole analysis for macronutrients, leaf blade analysis for micronutrients. Expressed on a dry weight basis.

TABLE 4.7. **SUFFICIENCY NUTRIENT RANGES FOR SELECTED GREENHOUSE VEGETABLE CROPS USING DRIED MOST RECENTLY MATURED WHOLE LEAVES**

	Beginning of Harvest Season		Just Before Harvest
Element	Tomato	Cucumber	Lettuce
		Percent	
N	3.5–4.0	2.5–5.0	2.1–5.6
P	0.4–0.6	0.5–1.0	0.5–0.9
K	2.8–4.0	3.0–6.0	4.0–8.0
Ca	0.5–2.0	0.8–6.0	0.9–2.0
Mg	0.4–1.0	0.4–0.8	0.4–0.8
S	0.4–0.8	0.4–0.8	0.2–0.5
		Parts per million	
B	35–60	40–100	25–65
Cu	8–20	4–10	5–18
Fe	50–200	90–150	50–200
Mn	50–125	50–300	25–200
Mo	1–5	1–3	0.5–3.0
Zn	25–60	50–150	30–200

Adapted from G. Hochmuth, "Fertilization Management for Greenhouse Vegetables," in *Florida Greenhouse Vegetable Production Handbook,* vol. 3, Florida Cooperative Extension Service Fact Sheet HS787, https://edis.ifas.ufl.edu/publication/CV265 (accessed 13 Nov. 2021).

PART 5

FIELD PLANTING

Knott's Handbook for Vegetable Growers, Sixth Edition. George J. Hochmuth and Rebecca G. Sideman.
© 2023 John Wiley & Sons Ltd. Published 2023 by John Wiley & Sons Ltd.
Companion website: https://www.wiley.com/go/hochmuth/vegetablegrowers6e

01

TEMPERATURES FOR VEGETABLES

COOL-SEASON AND WARM-SEASON VEGETABLES CROPS

Vegetables generally can be divided into two broad groups. In general, vegetable crops whose edible portions are vegetative parts (roots, stems, leaves, and buds or immature flower parts) are cool-season crops, and those that develop edible immature and mature fruit are warm-season crops. There are, however, exceptions. For example, sweet potato and New Zealand spinach are warm-season crops, and pea and broad bean are cool-season crops.

Cool-season crops generally differ from warm-season crops in the following respects:

1. The plants are hardy or frost tolerant.
2. Seeds germinate at cooler soil temperatures.
3. Root systems are shallower.
4. Plant size is smaller.
5. Some, the biennials, are susceptible to premature seed stalk development from exposure to prolonged cool weather.
6. They are stored near 32°F, because the harvested product is not subject to chilling injury at temperatures between 32 and 50°F. There are exceptions: the white potato is stored at warmer temperatures, and sweet corn, a warm-season crop, is held at 32°F after harvest.

TABLE 5.1. CLASSIFICATION OF VEGETABLE CROPS ACCORDING TO THEIR ADAPTATION TO FIELD TEMPERATURES

Cool-season Crops

Hardy[1]		Half-hardy[1]
Broad bean	Kohlrabi	Asparagus
Broccoli	Leek	Beet
Brussels sprouts	Mustard	Carrot
Cabbage	Onion	Cauliflower
Chive	Parsley	Celery
Collards	Pea	Chard
Garlic	Radish	Chicory
Horseradish	Rhubarb	Chinese cabbage
Kale	Spinach	Globe artichoke
	Turnip	Endive
		Lettuce
		Parsnip
		Potato

Warm-season Crops

Tender[1]	Very Tender[1]
Cowpea	Cantaloupe
New Zealand spinach	Cucumber
Snap bean	Eggplant
Soybean	Lima bean
Sweet corn	Okra
Tomato	Pepper, hot
	Pepper, sweet
	Pumpkin
	Squash
	Sweet potato
	Watermelon

Adapted from A. A. Kader, J. M. Lyons, and L. L. Morris, Postharvest responses of vegetables to preharvest field temperatures, *HortScience* 9:523–529 (1974) and Arora et al., Frost hardiness of *Asparagus officinalis* L., *HortScience* 27(7):823–824 (1992).

[1] Relative resistance to frost and light freezes.

158

TABLE 5.2. GROWING DEGREE DAY BASE TEMPERATURES

Base Temperature (°F)[1]	Crop (s)
60	Eggplant, okra, sweet potato
55	Cucumber, watermelon
51	Tomato
50	Bean, muskmelon, pepper, corn
45	Squash
40	Asparagus, beet, broccoli, collards, lettuce, pea, potato
39	Strawberry
35	Onion
38	Carrot

Adapted from D. C. Sanders, H. J. Kirk, and C. Van Den Brink, *Growing Degree Days in North Carolina* (North Carolina Agricultural Extension Service AG-236, 1980).

[1] Temperature below which growth is negligible.

TABLE 5.3. APPROXIMATE MONTHLY TEMPERATURES FOR BEST GROWTH AND QUALITY OF VEGETABLE CROPS

Some crops can be planted as temperatures approach the proper range. Cool-season crops grown in the spring must have time to mature before warm weather. Fall crops can be started in hot weather to ensure a sufficient period of cool temperature to reach maturity. Within a crop, varieties may differ in temperature requirements; hence this listing provides general rather than specific guidelines.

Temperatures (°F)

Optimum	Minimum	Maximum	Vegetable
55–75	45	85	Chicory, chive, garlic, leek, onion, salsify, scolymus, scorzonera, shallot
60–65	40	75	Beet, broad bean, broccoli, Brussels sprouts, cabbage, chard, collards, horseradish, kale, kohlrabi, parsnip, radish, rutabaga, sorrel, spinach, turnip
60–65	45	75	Artichoke, cardoon, carrot, cauliflower, celeriac, celery, Chinese cabbage, endive, Florence fennel, lettuce, mustard, parsley, pea, potato
60–70	50	80	Lima bean, snap bean
60–75	50	95	Sweet corn, southern pea, New Zealand spinach
65–75	50	90	Chayote, pumpkin, squash, tomato
65–75	60	90	Cucumber, cantaloupe
70–75	65	80	Sweet pepper
70–85	65	95	Eggplant, hot pepper, martynia, okra, roselle, sweet potato, watermelon

TABLE 5.4.   SOIL TEMPERATURE CONDITIONS FOR VEGETABLE SEED GERMINATION[1]

Vegetable	Minimum (°F)	Optimum Range (°F)	Optimum (°F)	Maximum (°F)
Asparagus	50	60–85	75	95
Bean	60	60–85	80	95
Bean, lima	60	65–85	85	85
Beet	40	50–85	85	95
Cabbage	40	45–95	85	100
Cantaloupe	60	75–95	90	100
Carrot	40	45–85	80	95
Cauliflower	40	45–85	80	100
Celery	40	60–70	70[2]	85[2]
Chard, Swiss	40	50–85	85	95
Corn	50	60–95	95	105
Cucumber	60	60–95	95	105
Eggplant	60	75–90	85	95
Lettuce	35	40–80	75	85
Okra	60	70–95	95	105
Onion	35	50–95	75	95
Parsley	40	50–85	75	90
Parsnip	35	50–70	65	85
Pea	40	40–75	75	85
Pepper	60	65–95	85	95
Pumpkin	60	70–90	90	100
Radish	40	45–90	85	95
Spinach	35	45–75	70	85
Squash	60	70–95	95	100
Tomato	50	60–85	85	95
Turnip	40	60–105	85	105
Watermelon	60	70–95	95	105

[1] Compiled by J. F. Harrington, Department of Vegetable Crops, University of California, Davis.
[2] Daily fluctuation to 60°F or lower at night is essential.

02
SCHEDULING SUCCESSIVE PLANTINGS

Successive plantings are necessary to ensure a continuous supply of produce. This seemingly easy goal is in fact extremely difficult to achieve because of interrupted planting schedules, poor stands, and variable weather conditions.

Maturity can be predicted in part by using "days to harvest" or "heat units." Production for fresh market entails the use of days to harvest, while some processing crops may be scheduled using the heat unit concept. Additional flexibility is provided by using varieties that differ in time and heat units to reach maturity.

Fresh Market Crops

Sweet corn is used as an example, since it is an important fresh-market crop in many parts of the country and requires several plantings to obtain a season-long supply.

Step 1. Select varieties suitable for your area that have a range of days-to-maturity. We will illustrate with five fictitious varieties maturing in 68–84 days from planting with 4-day intervals between varieties.

Step 2. Make the first planting as early as possible in your area.

Step 3. Construct a table like that shown (Table 5.5) and calculate the time of the next planting, so that the earliest variety used matures 4 days after "Late" in the first planting. We chose to use "Mainseason" as the earliest variety in the second planting; thus 88 days – 80 days = 8 days elapsed time before the second and subsequent plantings.

Step 4. As sometimes happens, the third planting was delayed 4 days by rain. To compensate for this delay, "Midseason" is selected as the earliest variety in the third planting to provide corn 96 days after the first planting.

TABLE 5.5. EXAMPLES OF SWEET CORN PLANTINGS

| | | Time (Days) | | |
| | | To | From | To Next |
Planting	Variety	Maturity	First Planting	Planting
First	Early	68	68	
	Second Early	72	72	
	Midseason	76	76	
	Mainseason	80	80	
	Late	84	84	
				8
Second	Mainseason	80	88	
	Late	84	92	
				12
Third	Midseason	76	96	
	Mainseason	80	100	
	Late	84	104	

Adapted from H. Tiessen, *Scheduled Planting of Vegetable Crops* (Ontario Ministry of Agriculture and Food AGDEX 250/22, 1980).

Processing Crops

The heat unit system is used to schedule plantings and harvests for some processing crops, most notably pea and sweet corn. The use of this system implies that accumulated temperatures over a selected base temperature are a more accurate means of measuring growth than a time unit such as days.

In its simplest form, heat units are calculated as follows:

$$\frac{\text{Maximum} + \text{minimum daily temperature}}{2} - \text{Base temperature}$$

The base temperature is 40°F for pea and 50°F for sweet corn. Several variations to this basic formula have been proposed to further extend its usefulness.

Heat unit requirements to reach maturity have been determined for most processing pea and sweet corn varieties and many snap bean varieties.

Processors using the heat unit system assist growers in scheduling plantings to coincide with processing plant operating capacity.

Winter Growing in Cold Climates

The advent of high tunnels (minimally heated or unheated greenhouses), low tunnels, and row covers increases the potential growing and harvest season for vegetable crops. In some regions of the country, it is common practice to seed very cold-hardy greens crops (spinach, brassica greens, arugula, mâche, lettuce, and others) under cover in the autumn for winter and early spring harvest, expecting that their growth will slow and/or stop during low light and low temperature conditions. This practice is sometimes referred to as "stockpiling," because crops are grown during the warm fall months and are then held in the cool growing environment and harvested as they are needed for market. This presents special scheduling challenges: plantings must be sown early enough to permit sufficient growth during the warm and light fall months so that they are ready to harvest by the time growth stops, and late enough so that they don't become overmature before the onset of winter. Year-to-year and site-to-site variability in fall weather makes this a particularly difficult task that requires extensive recordkeeping and often staggering of production to manage this variability and risk.

03

TIME REQUIRED FOR SEEDLING EMERGENCE

TABLE 5.6. DAYS REQUIRED FOR SEEDLING EMERGENCE AT VARIOUS SOIL TEMPERATURES FROM SEED PLANTED ½ IN. DEEP

The days from planting to emergence constitute the time interval when a preemergence weed control treatment can be used safely and effectively. More days are required with deeper seeding because of cooler temperatures and the greater distance of growth.

Vegetable	Soil Temperature (°F)								
	32	41	50	59	68	77	86	95	104
Asparagus	NG	NG	53	24	15	10	12	20	28
Bean, lima	—	—	NG	31	18	7	7	NG	—
Bean snap	NG	NG	NG	16	11	8	6	6	NG
Beet	—	42	17	10	6	5	5	5	—
Cabbage	—	—	15	9	6	5	4	—	—
Cantaloupe	—	—	—	—	8	4	3	—	—
Carrot	NG	51	17	10	7	6	6	9	NG
Cauliflower	—	—	20	10	6	5	5	—	—
Celery	NG	41	16	12	7	NG	NG	NG	—
Corn, sweet	NG	NG	22	12	7	4	4	3	NG
Cucumber	NG	NG	NG	13	6	4	3	3	—
Eggplant	—	—	—	—	13	8	5	—	—

TABLE 5.6. DAYS REQUIRED FOR SEEDLING EMERGENCE AT VARIOUS SOIL TEMPERATURES FROM SEED PLANTED ½ IN. DEEP (Continued)

Vegetable	Soil Temperature (°F)								
	32	41	50	59	68	77	86	95	104
Lettuce	49	15	7	4	3	2	3	NG	NG
Okra	NG	NG	NG	27	17	13	7	6	7
Onion	136	31	13	7	5	4	4	13	NG
Parsley	—	—	29	17	14	13	12	—	—
Parsnip	172	57	27	19	14	15	32	NG	NG
Pea	—	36	14	9	8	6	6	—	—
Pepper	NG	NG	NG	25	13	8	8	9	NG
Radish	NG	29	11	6	4	4	3	—	—
Spinach	63	23	12	7	6	5	6	NG	NG
Tomato	NG	NG	43	14	8	6	6	9	NG
Turnip	NG	NG	5	3	2	1	1	1	NG
Watermelon	—	NG	—	—	12	5	4	3	3

Adapted from J. F. Harrington and P. A. Minges, *Vegetable Seed Germination* (California Agricultural Extension Mimeo Leaflet, 1954).
NG = No germination; — = not tested.

04

SEED REQUIREMENTS

TABLE 5.7. APPROXIMATE NUMBER OF SEEDS PER UNIT WEIGHT AND FIELD SEEDING RATES FOR TRADITIONAL PLANT DENSITIES

Vegetable	Seeds (No.)	Unit Weight	Field Seeding[1] (lb/acre)
Asparagus[2]	14,000–20,000	lb	2–3
Bean, baby lima	1,200–1,500	lb	60
Bean, fordhook lima	400–600	lb	85
Bean, bush snap	1,600–2,000	lb	75–90
Bean, pole snap	1,600–2,000	lb	20–45
Beet	24,000–26,000	lb	10–15
Broad bean	300–800	lb	60–80
Broccoli[3]	9,000	oz	½–1½
Brussels sprouts[3]	9,000	oz	½–1½
Cabbage[3]	9,000	oz	½–1½
Cantaloupe[3]	16,000–20,000	lb	2
Cardoon	11,000	lb	4–5
Carrot	300,000–400,000	lb	2–4
Cauliflower[3]	9,000	oz	½–1½
Celeriac	72,000	oz	1–2
Celery[3]	72,000	oz	1–2
Chicory	27,000	oz	3–5
Chinese cabbage	9,000	oz	1–2
Collards	9,000	oz	2–4
Corn salad	13,000	oz	10
Cucumber	15,000–16,000	lb	3–5
Dandelion	35,000	oz	2
Eggplant[3]	6,500	oz	2
Endive	25,000	oz	3–4
Florence fennel	7,000	oz	3
Kale	9,000	oz	2–4
Kohlrabi	9,000	oz	3–5
Leek[3]	200,000	lb	4
Lettuce, head[3]	20,000–25,000	oz	1–3
Lettuce, leaf	25,000–30,000	oz	1–3
Mustard	15,000	oz	3–5

TABLE 5.7. APPROXIMATE NUMBER OF SEEDS PER UNIT WEIGHT AND FIELD SEEDING RATES FOR TRADITIONAL PLANT DENSITIES (*Continued*)

Vegetable	Seeds (No.)	Unit Weight	Field Seeding[1] (lb/acre)
New Zealand spinach	5,600	lb	15
Okra	8,000	lb	6–8
Onion, bulb[3]	130,000	lb	3–4
Onion, bunching	180,000–200,000	lb	3–4
Parsley	250,000	lb	20–40
Parsnip	192,000	lb	3–5
Pea	1,500–2,500	lb	80–250
Pepper[3]	4,200–4,600	oz	24
Pumpkin	1,500–4,000	lb	2–4
Radish	40,000–50,000	lb	10–20
Roselle	900–1,000	oz	3–5
Rutabaga	150,000–190,000	lb	1–2
Salsify	1,900	oz	8–10
Sorrel	30,000	oz	2–3
Southern pea	3,600	lb	20–40
Soybean	4,000	lb	20–40
Spinach	45,000	lb	10–15
Squash, summer	3,500–4,500	lb	4–6
Squash, winter	1,600–4,000	lb	2–4
Swiss chard	25,000	lb	6–8
Sweet corn, su or se	1,800–2,500	lb	12–15
Sweet corn, sh_2	3,000–5,000	lb	12–15
Tomato[3]	10,000–12,000	oz	½–1
Turnip	150,000–200,000	lb	1–2
Watermelon, small seed[3]	8,000–10,000	lb	1–3
Watermelon, large seed[3]	3,000–5,000	lb	2–4

[1]Actual seeding rates are adjusted to desired plant populations, germination percentage of the seed lot, and weather conditions that influence germination.
[2]6–8 lb/acre for crown production.
[3]Transplants are used frequently, instead of direct field seeding. See Table 3.3 for seeding rates for transplants.

05

PLANTING RATES FOR LARGE SEEDS

Weigh out a 1-oz sample of the seed lot and count the number of seeds. Given the number of seeds per ounce, the following table gives the approximate pounds of seed per acre for certain between-row and in-row spacings of lima bean, pea, snap bean, and sweet corn. These are based on 100% germination. If the seed germinates only 90%, for example, then divide the pounds of seed by 0.90 to get the planting rate. Do the same with other germination percentages.

Example: 30 seeds/oz to be planted in 22-in. rows at 1-in. spacing between seeds.

$$\frac{595}{0.90} = 661 \text{ lb/acre}$$

Only precision planting equipment would begin to approach as exact a job of spacing as this table indicates. Moreover, field conditions such as soil structure, temperature, and moisture will affect germination and final stand.

TABLE 5.8. PLANTING RATES FOR LARGE SEEDS

Spacing Between Rows (in.)

No. of Seeds/ oz	18						20						22					
	1	2	3	4	5	6	1	2	3	4	5	6	1	2	3	4	5	6
	Spacing Between Seeds in Row (in.)																	
	Seed Needed (lb/acre)																	
30	726	364	242	182	146	121	655	328	218	164	131	109	595	298	198	149	119	98
40	545	273	182	136	110	90	491	246	163	123	99	82	446	223	148	112	90	74
50	440	220	146	110	88	74	396	198	132	99	79	66	361	180	120	90	72	60
60	354	178	118	90	76	59	318	159	106	80	64	53	289	145	97	73	58	48
70	312	156	104	78	62	56	281	140	94	70	56	47	256	128	85	64	51	43
80	272	136	90	68	54	46	245	123	82	62	49	41	223	112	74	56	45	37
90	242	120	82	60	48	40	218	109	73	55	44	37	198	99	66	50	40	33
100	216	108	72	54	48	38	198	99	66	50	39	33	181	90	60	45	35	30
110	198	99	66	50	42	34	173	89	59	44	35	30	161	80	54	40	32	27
120	180	90	60	45	40	30	162	81	54	40	33	27	148	74	49	37	30	25
130	168	84	56	42	36	28	152	76	51	38	31	25	138	69	46	34	28	23
140	156	78	52	38	34	26	141	70	47	35	28	24	128	64	43	32	25	22
150	146	73	49	36	30	24	131	66	44	33	26	22	119	60	40	30	24	20

No. of Seeds/oz	24						30						36					
	Spacing Between Seeds in Row (in.)																	
	1	2	3	4	5	6	1	2	3	4	5	6	1	2	3	4	5	6
	Seed Needed (lb/acre)																	
30	545	273	182	136	109	91	437	219	146	109	88	73	363	182	121	91	73	61
40	408	204	136	102	82	68	328	164	106	82	66	54	272	136	91	68	55	45
50	330	165	110	82	66	55	265	132	88	66	53	44	220	110	73	55	44	37
60	265	133	88	67	57	44	212	106	71	59	43	35	177	89	59	45	38	29
70	234	117	78	59	47	39	188	94	63	47	38	31	156	78	52	39	31	26
80	204	102	68	51	41	34	164	82	53	41	33	27	136	68	45	34	27	23
90	181	90	61	45	36	30	146	73	49	37	29	25	121	60	41	30	24	20
100	162	81	55	40	32	28	131	67	44	33	27	22	108	54	37	27	21	19
110	148	74	49	37	30	25	119	60	40	30	24	20	99	49	33	25	20	17
120	135	68	45	34	27	23	108	54	36	27	22	18	90	45	30	23	18	15
130	126	63	42	32	25	21	101	51	34	25	20	17	84	42	28	21	17	14
140	117	58	39	29	23	20	94	47	32	23	19	16	78	39	26	19	15	13
150	109	55	38	27	22	18	88	44	29	22	18	15	73	37	24	18	14	12

SPACING OF VEGETABLE CROPS AND PLANT POPULATIONS

Spacing for vegetables is determined by the equipment used to plant, maintain, and harvest the crop as well as by the area required for growth of the plant without undue competition from neighboring plants. Previously row spacings were dictated almost entirely by the space requirement of cultivating equipment. Many of the traditional row spacings can be traced back to the horse cultivator.

Modern herbicides have largely eliminated the need for extensive cultivation in many crops; thus, row spacings need not be related to cultivation equipment. Instead, the plant's space requirement can be used as the determining factor, which generally increases plant populations. A more uniform product with a higher proportion of marketable vegetables as well as higher total yields result from the closer plant spacings. The term "high-density production" has been developed to describe vegetable spacings designed to satisfy the plant's space requirement.

Table 5.9 provides guidelines for high-density spacing of several vegetable crops, and Table 5.10 provides traditional plant and row spacings for several vegetable crops.

TABLE 5.9.   HIGH-DENSITY SPACING OF VEGETABLES

Vegetable	Spacing (in.)	Plant Population (Plants/Acre)
Snap bean	3 × 12	174,000
Beet	2 × 12	261,000
Carrot	1½ × 12	349,000
Cauliflower	12 × 18	29,000
Cabbage	12 × 18	29,000
Cucumber (processing)	3 × 20	104,000
Lettuce	12 × 18	29,000
Onion	4 × 26	261,000

TABLE 5.10. TRADITIONAL PLANT AND ROW SPACINGS FOR VEGETABLES

Vegetable	Between Plants in Row (in.)	Between Rows (in.)
Artichoke	48–72	84–96
Asparagus	9–15	48–72
Bean, broad	8–10	20–48
Bean, bush	2–4	18–36
Bean, lima, bush	3–6	18–36
Bean, lima, pole	8–12	36–48
Bean, pole	6–9	36–48
Beet	2–4	12–30
Broccoli[1]	12–24	18–36
Broccoli raab	3–4	24–36
Brussels sprouts	18–24	24–40
Cabbage[1]	12–24	24–36
Cantaloupe and other melons	12	60–84
Cardoon	12–18	30–42
Carrot	1–3	16–30
Cauliflower[1]	14–24	24–36
Celeriac	4–6	24–36
Celery	6–12	18–40
Chard, Swiss	12–15	24–36
Chervil	6–10	12–18
Chicory	4–10	18–24
Chinese cabbage	10–18	18–36
Chive	12–18	24–36
Collards	12–24	24–36
Corn	8–12	30–42
Cress	2–4	12–18
Cucumber[1]	8–12	36–72
Dandelion	3–6	14–24
Dasheen (taro)	24–30	42–48
Eggplant	18–30	24–48
Endive	8–12	18–24
Florence fennel	4–12	24–42

TABLE 5.10. TRADITIONAL PLANT AND ROW SPACINGS FOR VEGETABLES (*Continued*)

Vegetable	Between Plants in Row (in.)	Between Rows (in.)
Garlic	1–3	12–24
Horseradish	12–18	30–36
Jerusalem artichoke	15–18	42–48
Kale	18–24	24–36
Kohlrabi	3–6	12–36
Leek	2–6	12–36
Lettuce, cos	10–14	16–24
Lettuce, head[1]	10–15	16–24
Lettuce, leaf	8–12	12–24
Mustard	5–10	12–36
New Zealand spinach	10–20	36–60
Okra	8–24	42–60
Onion	1–4	16–24
Parsley	4–12	12–36
Parsley, Hamburg	1–3	18–36
Parsnip	2–4	18–36
Pea	1–3	24–48
Pepper[1]	12–24	18–36
Potato	6–12	30–42
Pumpkin	36–60	72–96
Radish	½–1	8–18
Radish, storage type	4–6	18–36
Rhubarb	24–48	36–60
Roselle	24–46	60–72
Rutabaga	5–8	18–36
Salsify	2–4	18–36
Scolymus	2–4	18–36
Scorzonera	2–4	18–36
Shallot	4–8	36–48
Sorrel	½–1	12–18
Southern pea	3–6	18–42
Spinach	2–6	12–36
Squash, bush[1]	24–48	36–60
Squash, vining	36–96	72–96
Strawberry[1]	10–24	24–64

TABLE 5.10. TRADITIONAL PLANT AND ROW SPACINGS FOR VEGETABLES (*Continued*)

Vegetable	Between Plants in Row (in.)	Between Rows (in.)
Sweet potato	10–18	36–48
Tomato, flat	18–48	36–60
Tomato, staked	12–24	36–48
Tomato, processing	2–10	42–60
Turnip	2–6	12–36
Turnip greens	1–4	6–12
Watercress	1–3	6–12
Watermelon	24–36	72–96

[1] Some crops can be grown in double rows on polyethylene mulched beds with 10–20 in. between rows.

TABLE 5.11. LENGTH OF ROW PER ACRE AT VARIOUS ROW SPACINGS

Distance Between Rows (in.)	Row Length (ft/acre)	Distance Between Rows (in.)	Row Length (ft/acre)
6	87,120	40	13,068
12	43,560	42	12,445
15	34,848	48	10,890
18	29,040	60	8,712
20	26,136	72	7,260
21	24,891	84	6,223
24	21,780	96	5,445
30	17,424	108	4,840
36	14,520	120	4,356

TABLE 5.12. NUMBER OF PLANTS PER ACRE AT VARIOUS SPACINGS

In order to obtain other spacings, divide 43,560 (the number of square feet per acre) by the product of the between-rows and in-row spacings, each expressed as feet; that is, 43,560 divided by 0.75 (3×0.25 ft) = 58,080 for a spacing of 18 by 6 in.

Spacing (in.)	Plants	Spacing (in.)	Plants	Spacing (ft)	Plants
12×1	522,720	30×3	69,696	6×1	7,260
12×3	174,240	30×6	34,848	6×2	3,630
12×6	87,120	30×12	17,424	6×3	2,420
12×12	43,560	30×15	13,939	6×4	1,815
		30×18	11,616	6×5	1,452
$15^1 \times 1$	418,176	30×24	8,712	6×6	1,210
15×3	139,392				
15×6	69,696	36×3	58,080	7×1	6,223
15×12	34,848	36×6	29,040	7×2	3,111
		36×12	14,520	7×3	2,074
$18^1 \times 3$	116,160	36×18	9,680	7×4	1,556
18×6	58,080	36×24	7,260	7×5	1,244
18×12	29,040	36×36	4,840	7×6	1,037
18×14	24,891			7×7	889
18×18	19,360	40×6	26,136		
		40×12	13,068	8×1	5,445
$20^1 \times 3$	104,544	40×18	8,712	8×2	2,722
20×6	52,272	40×24	6,534	8×3	1,815
20×12	26,136			8×4	1,361
20×14	22,402	42×6	24,891	8×5	1,089
20×18	17,424	42×12	12,445	8×6	907
		42×18	8,297	8×8	680
$21^1 \times 3$	99,564	42×24	6,223		
21×6	49,782	42×36	4,148	10×2	2,178
21×12	24,891			10×4	1,089
21×14	21,336	48×6	21,780	10×6	726
21×18	16,594	48×12	10,890	10×8	544
		48×18	7,260	10×10	435

TABLE 5.12. NUMBER OF PLANTS PER ACRE AT VARIOUS SPACINGS (*Continued*)

Spacing (in.)	Plants	Spacing (in.)	Plants	Spacing (ft)	Plants
24 × 3	87,120	48 × 24	5,445		
24 × 6	43,560	48 × 36	3,630		
24 × 12	21,780	48 × 48	2,722		
24 × 18	14,520				
24 × 24	10,890	60 × 12	8,712		
		60 × 18	5,808		
		60 × 24	4,356		
		60 × 36	2,904		
		60 × 48	2,178		
		60 × 60	1,742		

[1] Equivalent to double rows on beds at 30-, 36-, 40-, and 42-in. centers, respectively.

PRECISION SEEDING

High-density plantings, high costs of hand thinning, and erratic performance
of mechanical thinners have resulted in the development of precision seeding
techniques. The success of precision seeding depends on having seeds with
nearly 100% germination and on exact placement of each seed.
Some of the advantages of precision seeding are:

- Reduced seed costs. Only the seed that is needed is sown.
- Greater crop uniformity. Each seed is spaced equally, fewer harvests
  are necessary, and/or greater yield is obtained at harvest.
- Improved yields. Each plant has an equal chance to mature; yields can
  increase 20–50%.
- Improved plant stands. Seeds are dropped shorter distances, resulting
  in less scatter and a uniform depth of planting.
- Thinning can be reduced or eliminated.

Some precautions must be taken to ensure the proper performance of
precision seeding equipment:

1. A fine, smooth seedbed is required for uniform seeding depth.
2. Seed must have high germination.
3. Seed must be uniform in size; this can be achieved by seed sizing or
   seed coating.
4. Seed must be of regular shape; elongated or angular seeds such as
   carrot, lettuce, cucumber, spinach, and onion must be coated for
   satisfactory precision seeding. Seed size is increased two to five times
   with clay or proprietary coatings.

Several types of equipment are available for precision seeding of vegetables.

Belt type—represented by the StanHay seeder. Circular holes punched in
a belt accommodate the seed size. Holes are spaced along the belt at
specified intervals. Coated seed usually improves the uniformity obtained
with this type of seeder.
Roller type—represented by the Jang JP or Johnny's six-row seeder. Circular
holes in a roller at the bottom of the seed hopper accommodate the seed
size. Sprocket selection on the front drive belt and seed roller determines
final spacing.
Plate type—represented by the John Deere 33, Planet Jr., Earth Way, Sutton,
or Jang TD type seeder. Seeds drop into a notch in a horizontal plate and

are transported to the drop point. The plate is vertical in the Earth Way and catches seed in a pocket in a plastic plate. Most spacing is achieved by gearing the rate of turn of the plate.

Vacuum type—represented by the Gaspardo, Heath, Monosem, StanHay, and several other seeders. Seed is drawn against holes in a vertical plate and is agitated to remove excess seed. Various spacings are achieved through a combination of gears and number of holes per plate. Coated seed should not be used in these planters.

Spoon type—represented by the Nibex. Seed is scooped up out of a reservoir by small spoons (sized for the seed) and then carried to a drop shoot where the spoon turns and drops the seed. Spacing is achieved by spoon number and gearing.

Pneumatic type—represented by the International Harvester cyclo planter. Seed is held in place against a drum until the air pressure is broken. Then it drops in tubes and is blown to the soil. This planter is recommended only for larger vegetable seed.

Grooved cylinder type—represented by the Gramor seeder. This seeder requires round seed or seed that is made round by coating. Seven seeds fall from a supply tube into a slot at the top of a metal case into a metal cylinder. The cylinder turns slowly. As it reaches the bottom of the case, the seeds drop out of a diagonal slot. The seed is placed in desired increments by a combination of forward speed and turning rate. This planter can be used with seed as small as pepper seed but it works best with coated seed.

Guidelines for Operation and Maintenance of Equipment

1. Check the planter for proper operation and replace worn parts during the off season.
2. Thoroughly understand the contents of the manufacturer's manual.
3. Make certain that the operator is trained to use the equipment and check its performance.
4. Double check your settings to be sure that you have the desired spacing and depth.
5. Make a trial run before moving to the field.
6. Operate the equipment at the recommended tractor speed. The number of seeds planted per minute at various speeds and spacings is shown in Table 5.13.
7. Check the seed drop of each unit periodically during the planting operation.

Adapted in part from D. C. Sanders, *Precision Seeding for Vegetable Crops* (North Carolina Cooperative Extension Service Publication HIL-36, 1997).

TABLE 5.13.   NUMBER OF SEEDS PLANTED PER MINUTE AT VARIOUS SPEEDS AND SPACINGS[1]

Planter Speed (mph)	In-row Spacing (in.)			
	2	3	4	6
2.5	1,320	880	660	440
3.0	1,584	1,056	792	528
4.0	2,112	1,408	1,056	704
5.0	2,640	1,760	1,320	880

[1] For most conditions, a planter speed of 2–3 mph will result in the greatest precision.

08

CONSERVATION TILLAGE AND NO-TILL SYSTEMS

Most vegetable production systems rely extensively on tillage for bed preparation, cultivation, and incorporation of crop residue. Conservation tillage (CT) or reduced tillage is an overarching term that includes several systems that reduce tillage intensity and soil disturbance, either by reducing the number of total tillage operations or by reducing the field area that is disturbed. Usually, CT systems rely on fall-planted cover crops that are terminated prior to planting the cash crop. Such systems offer several benefits: reduced fuel cost, reduced production of dust and particulates, and improvements in soil organic matter and soil structure. They present challenges relating to cover crop termination, managing residue and weeds, and soil temperature. Some basic definitions follow:

Standard tillage involves the full sequence of operations historically used to prepare a seedbed and produce a given crop.

Minimum tillage are systems in which the number of tillage passes are reduced relative to standard tillage.

No-till systems are ones in which the soil remains undisturbed from harvest to planting, aside from injecting fertilizers and adding seeds or transplants. The soil surface remains covered with residue from the previous crop throughout the growing season.

Strip tillage involves tilling a narrow strip around the seed line to remove surface plant debris and enhance soil warming. In some cases, this may include subsoiling—this is often called *Deep zone tillage*.

While CT systems are now the norm for some agronomic crops, their use in vegetable production has lagged behind, primarily due to lack of specialized transplanting and seeding equipment. In conventional reduced-tillage systems, the cover crops and any weeds that are present are terminated with herbicides prior to planting the cash crop. In organic reduced-tillage systems, cover crops are terminated mechanically, usually with a roller-crimper, or with a flail mower. Yield decreases compared with conventionally tilled organic systems have often been documented.

Key factors for success with reduced-tillage vegetable production are:

1. Establishing a dense and uniform cover crop that will provide an even and thick mulch.

2. Effectively terminating the cover crop, either chemically or mechanically.

- Chemical desiccation with contact herbicides should be done 2–5 weeks before planting. If glyphosate is being used, it should be applied at least 4 weeks prior to transplanting to avoid potential damage to the cash crop.
- Mechanical termination can be done just prior to planting, but the timing in relation to cover crop development is critical. It should take place once the cover crop has reached its' reproductive stage and is incapable of regrowth. For example, rye should be terminated at 50–75% anthesis, and hairy vetch should be terminated at late flowering to early pod set. Flail mowing effectively kills most crops, but a roller-crimper creates a mulch that is easier to plant through.

3. Establishing transplants or seeds into cover crops with a minimum of disturbance to surface residues and soil. Large-seeded crops (sweet corn, snap bean, lima bean, and cucurbits) may be direct-seeded provided the growing season is long enough and soil temperatures are warm enough to permit success. Other crops should be transplanted (*Brassica* spp., pepper, tomato). Specialized equipment that loosens the soil and removes a narrow strip of crop residue to facilitate planting is now available.

4. Lacking problematic perennial weed species, such as nutsedge, quackgrass, or morning glory.

5. Having a strategy for incorporation of amendments and fertility.

For more information, see:

J. P. Mitchell et al., Evolution of conservation tillage systems for processing tomato in California's Central Valley. *HortTechnology* 22(5):617–626 (2012).

R. D. Morse. No-till vegetable production—its time is now. *HortTechnology* 9(3):373–379 (1999).

R. L. Parish. Current developments in seeders and transplanters for vegetable crops. *HortTechnology* 15(2):346–351 (2005).

C. Halde et al., Organic no-till systems in Eastern Canada: a review. *Agriculture* 7:36, doi:10.3390/agriculture7040036 (2017).

TABLE 5.14. STORAGE OF PLANT PARTS USED FOR VEGETATIVE PROPAGATION

Plant Part	Temperature (°F)	Relative Humidity (%)	Comments
Asparagus crowns	30–32	85–90	Roots may be trimmed to 8 in. Prevent heating and excessive drying
Garlic bulbs	50	50–65	Fumigate for mites, if present. Hot-water-treat (120°F for 20 min) for control of stem and bulb nematode immediately before planting
Horseradish roots	32	85–90	Pit storage is used in cold climates
Onion sets	32	70–75	Sets may be cured naturally in the field, in trays, or artificially with warm, dry air
Potato tubers	36–40 (extended storage) 45–50 (short storage)	90	Cure at 60–65°F and 90–95% relative humidity for 10–14 days. Move to 60–65°F 10–14 days before planting
Rhubarb crowns	32–35	80–85	Field storage is satisfactory in cold climates
Strawberry plants	30–32	85–90	Store in crates lined with 1.5-mil polyethylene
Sweet potato roots	55–60	85–90	Cure roots at 85°F and 85–90% relative humidity for 6–8 days before storage
Witloof chicory roots	32	90–95	Prevent excessive drying

TABLE 5.15. FIELD REQUIREMENTS FOR VEGETATIVELY PROPAGATED CROPS

Vegetable	Plant Parts	Quantity/Acre[1]
Artichoke	Root sections	807–1,261
Asparagus	Crowns	5,808–10,890
Dasheen	Corms (2–5 oz)	9–18 cwt
Garlic	Cloves	8–20 cwt
Jerusalem artichoke	Tubers (2 oz)	10–12 cwt
Horseradish	Root cuttings	9,000–11,000
Onion	Sets	5–10 cwt
Potato	Tubers or tuber sections	13–26 cwt
Rhubarb	Crown divisions	4,000–5,000
Strawberry	Plants	6,000–50,000
Sweet potato	Roots for bedding	5–6 cwt

[1]Varies with field spacing, size of individual units, and vigor of stock.

TABLE 5.16. SEED POTATOES REQUIRED PER ACRE, WITH VARIOUS PLANTING DISTANCES AND SIZES OF SEED PIECE

Seed Piece Weights

Spacing of Rows and Seed Pieces	1 oz	1¼ oz	1½ oz	1¾ oz	2 oz
			(Pounds of Seed/Acre)		
Rows 30 in. Apart					
8-in. spacing	1,632	2,040	2,448	2,856	3,270
10-in. spacing	1,308	1,638	1,956	2,286	2,614
12-in. spacing	1,089	1,361	1,632	1,908	2,178
14-in. spacing	936	1,164	1,398	1,632	1,868
16-in. spacing	816	1,020	1,224	1,428	1,632
Rows 32 in. Apart					
8-in. spacing	1,530	1,914	2,298	2,682	3,066
10-in. spacing	1,224	1,530	1,836	2,142	2,448

TABLE 5.16. SEED POTATOES REQUIRED PER ACRE, WITH VARIOUS PLANTING DISTANCES AND SIZES OF SEED PIECE (Continued)

Seed Piece Weights

Spacing of Rows and Seed Pieces	1 oz	1¼ oz	1½ oz	1¾ oz	2 oz
12-in. spacing	1,020	1,278	1,536	1,788	2,040
14-in. spacing	876	1,092	1,314	1,530	1,752
16-in. spacing	768	960	1,152	1,344	1,536
Rows 34 in. Apart					
8-in. spacing	1,440	1,800	2,160	2,520	2,880
10-in. spacing	1,152	1,440	1,728	2,016	2,304
12-in. spacing	960	1,200	1,440	1,680	1,920
14-in. spacing	822	1,026	1,236	1,440	1,644
16-in. spacing	720	900	1,080	1,260	1,440
Rows 36 in. Apart					
8-in. spacing	1,362	1,704	2,040	2,382	2,724
10-in. spacing	1,086	1,362	1,632	1,902	2,178
12-in. spacing	906	1,134	1,362	1,590	1,812
14-in. spacing	780	972	1,164	1,362	1,554
16-in. spacing	678	852	1,020	1,188	1,362
18-in. spacing	606	756	906	1,056	1,212
Rows 42 in. Apart					
18-in. spacing	516	648	780	906	1,038
24-in. spacing	390	486	582	678	780
30-in. spacing	312	390	468	546	624
36-in. spacing	258	324	390	456	516
Rows 48 in. Apart					
18-in. spacing	456	570	678	792	906
24-in. spacing	342	426	510	594	678
30-in. spacing	270	342	408	474	546
36-in. spacing	228	282	342	396	456

10

POLYETHYLENE AND DEGRADABLE MULCHES

Polyethylene mulch has been used commercially on vegetables since the early 1960s. Three types of plastic agricultural mulch films are commercially available: (1) conventional plastic films made mostly of polyethylene, which must be removed from the field at the end of the cropping cycle; (2) biodegradable plastic mulches made of starch and synthetic inputs, which are designed to be tilled into the soil at the end of the cropping cycle, and (3) oxo-degradable mulches, which are made of any of a number of conventional plastics in combination with additives that cause the mulch to become brittle and break into fragments when exposed to UV light, heat, and/or oxygen. Lastly, fully biodegradable paper mulches are now commercially available. Very limited research suggests that such paper mulches offer some weed control advantages, but that they may not be comparable to plastic mulches in enhancing plant growth, and that their application and management require special considerations.

CONVENTIONAL POLYETHYLENE MULCHES

Three colors of mulch are commonly used commercially: black, clear, and white (or white-on-black). Specialized mulches, such as red, silver reflective, and infrared-transmitting (IRT) mulches, are used for specific circumstances, such as weed control, plant growth regulation, enhanced temperature accumulation, and insect repelling. Black mulch is used most widely because it suppresses weed growth, holds soil moisture, and warms the soil by contact. Clear polyethylene is used widely in the northern United States because it promotes warmer soil temperatures (by the greenhouse effect) than black mulch. Clear mulch requires use of labeled fumigants or herbicides underneath to prevent weed growth. White or white-on-black mulch is used for cool-season crops established under hot summer conditions. Soils under white mulch or white-on-black mulch remain cooler because less radiant energy is absorbed by the mulch. Some growers create their own "white" mulch by painting the surface of black-mulched beds with white latex paint.

BENEFITS OF MULCH

- *Increases early yields*. The largest benefit from polyethylene mulch is the increase in soil temperature in the bed, which promotes faster crop development and earlier yields.

- *Aids moisture retention.* Mulch reduces evaporation from the bed soil surface. As a result, a more uniform soil moisture regime is maintained and the frequency of irrigation is reduced slightly. Irrigation is still mandatory for mulched crops so that the soil under the mulch doesn't dry out excessively. Tensiometers (or other soil moisture sensors—see Part 7) placed in the bed between plants can help indicate when irrigation is needed.

- *Inhibits weed growth.* Black and white-on-black mulches greatly inhibit light penetration to the soil. Therefore, weed seedlings cannot survive under the mulch. Nutgrass can still be a problem, however. The nuts provide enough energy for the young nutgrass to puncture the mulch and emerge. Other pests, such as soilborne pathogens, insects, and nematodes, are not reduced by most mulches. Some benefit has been shown from high temperatures under clear mulch (solarization). Currently, the best measure for nutgrass and pest control under the mulch is labeled fumigation.

- *Reduces fertilizer leaching.* Fertilizer placed in the bed under the mulch is less subject to leaching by rainfall. As a result, the fertilizer program is more efficient and the potential exists for reducing traditional amounts of fertilizer. Heavy rainfall that floods the bed can still result in fertilizer leaching. This fertilizer can be replaced if the grower is using drip irrigation, or it can be replaced with a liquid fertilizer injection wheel.

- *Decreases soil compaction.* Mulch acts as a barrier to the action of rainfall, which can cause soil crusting, compaction, and erosion. Less-compacted soil provides a better environment for seedling emergence and root growth.

- *Protects fruits.* Mulch reduces rain-splashed soil deposits on fruits. In addition, mulch reduces fruit rot caused by soil-inhabiting organisms, because there is a protective barrier between the fruit and the organisms.

- *Aids fumigation.* Mulches increase the effectiveness of soil fumigant chemicals. Acting as a barrier to gas escape, mulches help keep gaseous fumigants in the soil. Recently, virtually or totally impermeable films (VIF or TIF) have been developed to help trap fumigants better and reduce amounts of fumigants needed.

NEGATIVE ASPECTS OF MULCH

- *Mulch removal and disposal.* The biggest problems associated with mulch use are removal and disposal. Since most mulches are not

187

biodegradable, they must be removed from the field after use. This usually may involve hand labor, although mulch lifting and removal machines are available. Disposal also presents a problem because of the quantity of waste generated.

- *Specialized equipment.* The mulch cultural system requires a small investment in some specialized equipment, including a bed press, mulch layer, and mulch transplanter or plug-mix seeder. Vacuum seeders are also available for seeding through mulch. This equipment is not very expensive, is easily obtained, and some can even be manufactured on the farm.

MULCH APPLICATION

Mulch is applied by machine for commercial operations. Machines that prepare beds, fertilize, fumigate, and mulch in separate operations or in combination are available. The best option is to complete all of these operations in one pass across the field. In general, all chemicals and fertilizers are applied to the soil before mulching. Nitrogen and potassium fertilizers can be injected through a drip irrigation system under the mulch.

When laying mulch, be sure the bed is pressed firmly and that the mulch is in tight contact with the bed. This helps transfer heat from mulch to bed and reduces flapping in the wind, which results in tears and blowing of mulch from the bed. The mulch layer should be adjusted so that the edges are buried sufficiently to prevent uplifting by wind.

DEGRADABLE MULCHES

Degradable plastic mulches have many of the properties and provide the same benefits of standard polyethylene mulches. One important difference is that degradable mulches begin to break down in certain environmental conditions. Because they are tilled into the soil at the end of the cropping cycle, the labor, time, and costs associated with removal and disposal are avoided. This partially offsets their higher purchase costs.

Oxo-degradable mulches are broken down by exposure to UV light. The edges covered by the soil retain their strength and break down only after being disced to the surface where they are exposed to UV light. Standard biodegradation tests have shown that oxo-degradable plastics only partially degrade in the environment, and that they do not meet the standards to be considered biodegradable or compostable, nor are they suitable for anaerobic digestion. As a result, these mulches add plastic fragments to the soil that will persist for a very long time.

Biodegradable mulches are broken down into fragments by the action of microorganisms in warm and humid conditions, and ultimately, decompose completely into carbon dioxide and water. Note that biodegradable is not the same as "bio-based," which means that feedstocks are renewable resources. Biodegradable films that are currently commercially available are made of starches and other biodegradable polymers, some of which are fossil-based. The term "biodegradable" should be accompanied by ASTM standards that specify how degradation was confirmed in lab-based tests (e.g. composting vs. soil environment, percent conversion into carbon dioxide, and the duration of time).

As they break down, degradable mulches become brittle and develops cracks, tears, and holes. Small sections of film may tear off and be blown around by the wind. Finally, the film breaks down into smaller and smaller pieces, and either biodegrades or persists in the environment (for oxo-degradable mulches). Under certain conditions, even biodegradable mulches will break down slowly, such that plastic residue fragments remain in the soil for the next crop. Seeding early crops in a field that had a long-term, degradable mulch the previous season should be avoided.

The following factors influence the time and rate of breakdown of degradable mulches:

- The formulation, thickness, and other specifications of the film, i.e., short-, intermediate-, or long-lasting film.
- Factors that influence soil microbial activity, such as moisture and temperature (for biodegradable mulches), and light exposure (for oxo-degradable mulches). Thus, the growth habit of the crop (vine or upright), the time of year the film is applied, the time between application and planting, crop vigor, weed growth, and length of time the mulch is left in the field after harvest can all influence rate and extent of breakdown.
- Physical factors, including wind, depressions in the bed, footprints, animal and tire tracks, trickle irrigation tubes under the film, and stress on the plastic resulting from making holes for plants and planting, all weaken the film and increase the rate of breakdown.

Suggestions for using biodegradable mulches:

- Store the mulch in a cool, dry environment, and store the rolls vertically, so that the weight of the roll pressing against itself is not creating pressure that could create holes before laying the mulch. Purchase only as much mulch as you will use in a single season.

- Make uniform beds, free from depressions and footprints.
- Minimize damage to the film and avoid unnecessary footprints, especially during planting and early in the growing season.
- Mulch-layers may require adjustment for biodegradable mulches to reduce tension when laying the mulch.
- Plant immediately after laying the mulch, and avoid transplanting mid-day in high temperatures.

Suggestions for using oxo-degradable mulches:

- Plant a cover crop to trap larger fragments and prevent them from blowing around. Plant a border strip of a tall-growing grass around the field to prevent fragments from blowing into neighboring areas.
- Lift the soil-covered edges before final harvest or as soon as possible after harvest, and mow the crop immediately after the last harvest. This will expose as much of the plastic as possible to UV light and start the breakdown process.

Planting into mulches:

Mechanical planters are now available that permit direct-seeding or to transplanting into plastic mulches. Special "mulch transplanters" feature heavy wheel-mounted spades that poke holes into the plastic mulch at preset intervals. The operator(s) set the transplants into the holes, and water is provided to each transplant. Mulch seeders are available that combine a vacuum seeder with wheel-mounted spades that open holes into the plastic and deposit seeds directly into the soil. These are available both as tractor-mounted and as hand-operated versions for smaller-scale producers.

Adapted from G. J. Hochmuth, R.C. Hochmuth, and S. M. Olson, *Polyethylene Mulching for Early Vegetable Production in North Florida* (Florida Cooperative Extension Service Circular 805, 2018), https://edis.ifas.ufl.edu/publication/CV213; W. L. Schrader, Plasticulture in California. Publ. 8016 (2000), https://anrcatalog.ucanr.edu/pdf/8016.pdf; Biodegradable Mulch Resource Page, https://ag.tennessee.edu/biodegradablemulch/ (accessed June 2021); T. Dupont, S. Guiser, J. Esslinger, A. Frankenfield, and M. Orzolek, Biodegradable Mulch Demonstrations (2016), https://extension.psu.edu/biodegradable-mulch-demonstrations; C. Miles, Oxo-degradable plastics risk environmental pollution. Report No. FA-2017-1 (2017), https://ag.tennessee.edu/biodegradablemulch/Documents/oxo%20plastics.pdf.

11
ROW COVERS

Row covers are defined by Wells and Loy (1993) as "flexible, transparent material (polyethylene, polyester, or polypropylene) that is hoop-supported or floated over a row or rows of crops at planting (seeds or transplants)." While the most common use of row covers is to promote early-season crop growth, row covers are also used widely for protection from insects, frosts, and freezes. Row covers are often used in conjunction with other season extension techniques such as raised beds and plastic mulches.

The first row covers developed were clear plastic (polyethylene), which provided good warming but also presented a risk over crop overheating and required daily management for ventilation. Slitted clear plastic coverings that are 1–1.5 miles thick provide automatic ventilation and offer the greatest warming potential. These usually feature ventilation slits 5 in. long and ¾ in. apart in two rows, on either side of the center. Nonwoven fabrics are now the most common row covers as they offer both insect protection and warming, and there are also woven nettings marketed specifically for insect exclusion. Row covers vary in weight from less than 0.45 oz/yd² (15 g/m²) to heavyweight blankets of 2 oz/yd², with light transmission ranging from 30 to 90%. They are manufactured in rolls of varying lengths and widths; very long and wide custom pieces can be purchased to cover relatively large areas easily. In general, the lighter weight coverings are used primarily for insect control, and heavier weight coverings offer much better frost or freeze protection. Row covers can usually be reused several times if stored carefully where rodents cannot access them; heavier fabrics tend to be more durable.

Row covers may be laid directly on the crop (floating) or supported (usually by hoops). Hoops are often made from no. 8 or no. 9 wire; 63 in. lengths are used for 5-ft wide covers. Hoops are installed over the crop so that the center of the hoop is 14–16 in. above the crop.

Floating row covers are applied directly over the row immediately after planting (seeds or transplants), and the edges are secured with soil, boards, bricks, or wire pins. Because the material is of such light weight, the plants will push up the fabric as the plants grow. Accordingly, enough slack should be provided to allow for the plants to reach maximum size during the time the material is left over the plants. For bean or tomato, about 12 in. of slack should be left. For a crop such as cucumber, 8 in. is sufficient. Supports should be considered in windy growing areas so plants are not damaged by cover abrasion.

Floating covers can be left over vegetables for 3–8 weeks, depending on the crop and the weather. For tomato and pepper, it can be left on for about

1 month but should be removed (at least partially) when the temperature under the covers reaches 86°F and is likely to remain that high for several hours. Blossoms of cucurbit crops can withstand very high temperatures, but the cover must be removed when the first female flowers appear so that bees can begin pollination.

Various systems exist to lay and to pick up floating row cover. Some (e.g., the Hiwer system) are equipped to lay and pick up long and wide pieces of row cover that is wound on large spools when not in use; others are similar to modified mulch layers and are designed to lay and bury edges of floating row cover over a single row. Supported row covers can also be mechanically applied over hoops with a high-clearance tractor and a modified mulch applicator. Recent efforts have developed mechanized systems to install row covers and hoops, and secure row cover edges at the same time (see videos listed under additional resources, below).

FROST PROTECTION

Frost protection with slitted and floating covers is not as good as with solid plastic covers, or with covers designed specifically for cold protection. A maximum of 3–4°F is all that can be expected, whereas with solid covers, frost protection of 5–7°F has been attained. Polypropylene floating row covers can be used for frost protection of vegetables and strawberries. Heavier covers 1.0–1.5 oz/yd^2 can protect strawberries to 23–25°F. Row covers should not be viewed merely as a frost-protection system but as a growth-intensifying system during cool spring weather. Therefore, do not attempt to plant very early and hope to be protected against heavy frosts. An earlier planting date of 10 days to 2 weeks would be more reasonable. The purpose of row covers is to increase productivity through an economical increase of early and perhaps total production per unit area.

Adapted from O. S. Wells and J. B. Loy, Rowcovers and high tunnels enhance crop production in the Northeastern United States. *HortTechnology* 3(1):92–95 (1993); G. J. Hochmuth, R. C. Hochmuth, S. Kostewicz, and W. Stall, Row covers for commercial vegetable culture in Florida. IFAS Publication #CIR728 (2018), https://edis.ifas.ufl.edu/publication/cv201; See also: Searching for Ways to Mechanize Deployment of Row Covers: Tunnel Layer, and Mechanizing Row Cover Deployment and Retrieval: Hiwer Machine. YouTube videos, Department of Plant Pathology and Microbiology, Iowa State University (2014), https://www.youtube.com/watch?v=MaiAhxEuSxM and https://www.youtube.com/watch?v=MaiAhxEuSxM.

Windbreaks are important considerations in an intensive vegetable production system. Use of windbreaks can result in increased yield and earlier crop production. The type and height of the windbreak determine its effect. Windbreaks can be living or nonliving.

Young plants are most susceptible to wind damage and "sand blasting." Rye or other tall-growing grass strips between rows can provide protection from wind and wind-borne sand (Figure 5.1). Windbreaks can improve early plant growth and earlier crop production, particularly with melons, cucumbers, squash, peppers, eggplant, tomatoes, and okra.

A major benefit of a windbreak is improved use of moisture. Reducing the wind speed reaching the crop reduces both the direct evaporation from the soil and the moisture transpired by the crop. This moisture advantage also improves conditions for seed germination. Seeds germinate more rapidly and young plants establish root systems more quickly. Improved moisture conditions continue to enhance crop growth and development throughout the growing season.

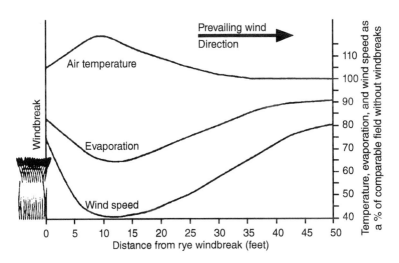

Figure 5.1. Air temperature, evaporation rate, and wind speed changes with distance from windbreak. Variables are expressed as percent of their level if the windbreak was not present.

In general, windbreaks should be as close as economically viable, for example, every three or four beds of melons. The windbreak should be oriented perpendicular to the prevailing wind direction. Rye strips should be planted prior to the crop to be protected, to obtain good plant establishment and to provide adequate time for plant growth prior to beginning the next production season. Fertilization and pest management of rye windbreaks may be necessary to encourage growth to the desired height.

Adapted from J. R. Schultheis, D. C. Sanders, and K. B. Perry. "Windbreaks and Drive Rows," in D. C. Sanders (ed.), *A Guide to Intensive Vegetable Systems* (North Carolina Cooperative Extension AG-502, p. 9, 1993); From R. Rouse and L. Hodges, "Windbreaks," in W. Lamont (ed.), *Production of Vegetables, Strawberries, and Cut-Flowers Using Plasticulture* (Ithaca, NY: NRAES, Cooperative Extension Service, pp. 57–66, 2004); See also J. Brandle and L. Hodges, Field Windbreaks (University of Nebraska Cooperative Extension Service EC1778), https://nfs.unl.edu/documents/fieldwindbreaks.pdf (accessed July 2021).

13
ENVIRONMENTAL MONITORING: SOIL, WATER, AND AIR

Environmental variables impact all aspects of crop growth, governing rate of growth, crop yields, and quality. Measuring (and forecasting) temperature, precipitation, and soil physical characteristics have been important for a long time, but dramatic changes in recent decades have greatly increased the amount of information available to growers, and upon which they can base decisions.

More data are available in real time than ever before. Sensor cost in general has decreased. Further, nearly everyone carries portable computers that contain a variety of sensors, in the form of smartphones. The limitation to using the copious amounts of data that are now easily generated has been and remains the capacity to analyze and process the data, to then turn it into something that can be readily used in decision-support systems. It is likely that the availability and use of "big data" will transform our ability to rapidly collect real-time information, combine it with prediction models (e.g., pest and disease development, weather forecasting, nutrient use, and soil health models), and use the information to make practical decisions that affect crop performance.

Two examples where technological advances are currently enhancing better use of environmental data for vegetable farmers follow. Weather station networks that incorporate real-time meteorological data collection with weather forecasting and models relevant to crop, disease, or insect development can help producers determine when to take particular action such as turn on frost protection or implement preventative pest management. Remotely monitoring temperature can alert growers to problems as they occur so they can take action to mitigate damage, such as when greenhouse heaters or storage coolers malfunction.

Weather station networks. There are several weather station networks throughout the United States that provide regional environmental and weather data relevant to horticultural crop producers. While the data are publicly available, the utility of the data depends on the proximity of the nearest weather station. In some cases, those who purchase a supported weather station can join the network and get even more localized information. An incomplete list of these networks follows:

- Agrimet (the pacific northwest)
- AgWeatherNet (Washington state)
- Agroclimate (the southeastern United States)
- Florida Automated Weather Network (FAWN; Florida)

- California Irrigation Management Information Systems (CIMIS; California))
- CoAgMet (Colorado)
- Delta Agricultural Weather Center (Mississippi)
- ND Ag Weather Network (NDAWN; North Dakota, Minnesota)
- Network for Environmental and Weather Applications (NEWA; the northeastern and north central United States)

Remote monitoring of temperature. Remote monitoring of temperature in different on-farm locations (fields, rooms, coolers, greenhouses) using the internet involves the sensing of conditions, transmitting that data to a gateway or logging device, transmitting that data again to a server or cloud service, and then retrieving that data on a web-enabled device. Notifications can also be part of this, i.e., "pushing" an alert to the user when predefined thresholds are exceeded. Monitoring systems typically include sensors, a data aggregator (network gateway), internet connectivity, data server (usually a subscription), and an application or website to access data.

In practice, there are many different ways to build these systems. Some sensors send data to the gateway with wireless communication. Others use wired connections. Some monitoring systems are simple switches that turn an alarm on. Others can be used to control system based on the sensed conditions. Wireless communication can be challenging through metal materials such as those commonly used in cooler box construction. Some wireless sensors have remote antennae that allow the sending probe to remain in the cooler, while the antenna is routed outside for improved communication reliability with the gateway.

The benefits of remote monitoring systems include:

1. Real-time monitoring with multiple sensors throughout the room can capture temperature variations that a typical thermostatic control or manual measurements will not,
2. Real-time alerts based on prescribed thresholds can help alert the operator to problems needing attention (e.g., doors left open, compressor cycling frequently or not enough or not at all, freezing conditions occurring),
3. Careful monitoring of environmental conditions can lead to improved crop storage, crop growth, crop quality, etc.

The challenges involved with remote monitoring systems include:

1. Wireless devices (transmitters) don't work well in metal boxes if the other parts (receivers) are outside the metal boxes. Moving the gateway into the location where the wireless sensors are located can improve local network reliability. But this requires separate attention to how the gateway has internet connectivity. Remote antennae for wireless sensors can also be helpful.

2. Internet connectivity is still very limited at many farms. Even if the office has internet and Wi-Fi, fields, cold rooms, and greenhouses may not. Wi-Fi extenders are useful for getting the signal further out into the field and outbuildings. Wi-Fi extenders are also helpful for converting a wireless signal to a wired ethernet connection using the ethernet ports on the back of them.

3. Cellular coverage is also very limited and otherwise challenging. In the absence of broadly distributed ethernet and Wi-Fi at some sites, cellular internet connections may be needed. This adds yet another part to an already complicated system. Cellular connections are not very reliable, even when using a stationary transmitter/receiver. The data plans are also still quite expensive, but some lower cost options specifically for these sorts of systems are becoming available.

Adapted from C. Callahan, Remote Monitoring University of Vermont Agricultural Engineering (2021), https://blog.uvm.edu/cwcallah/remote-monitoring/ (accessed Oct. 2021).

PART **6**

SOILS AND FERTILIZERS

Knott's Handbook for Vegetable Growers, Sixth Edition. George J. Hochmuth and Rebecca G. Sideman.
© 2023 John Wiley & Sons Ltd. Published 2023 by John Wiley & Sons Ltd.
Companion website: https://www.wiley.com/go/hochmuth/vegetablegrowers6e

201

NUTRIENT BEST MANAGEMENT PRACTICES (BMPs)

With the passage of the Federal Clean Water Act in 1972, states are required to assess the impact of nonpoint sources of pollution on surface and groundwaters and to establish programs to minimize these sources of pollution. This act also requires states to identify impaired water bodies and establish total maximum daily loads (TMDLs) for pollutants entering those water bodies. TMDLs are the maximum amounts of pollutants that can enter a water body and still allow it to meet its designated uses, such as swimming, potable water, and fishing. States have implemented various programs to address TMDLs. For example, Florida has adopted a best management practice (BMP) approach to addressing TMDLs whereby nutrient BMPs are adopted by state rule. The following definition of a BMP is taken from the Florida Department of Agriculture and Consumer Services handbook, *Water Quality/Quantity Best Management Practices for Florida Vegetable and Agronomic Crops*. BMPs are a practice or combination of practices determined by state agencies, based on research, field testing, and expert review, to be the most effective and practical on-location means, including economic and technical considerations, for improving water quality in agricultural and urban discharges. BMPs must be technically feasible, economically viable, socially acceptable, and based on sound science.

Some states' programs involve incentive measures for adopting BMPs, such as cost-share for certain management practices on the farm, for specific equipment, and other technical assistance. Agricultural producers who adopt approved BMPs, depending on the state and the program, may be "presumed to be in compliance" with state water quality standards and are eligible for cost-share funds to implement certain BMPS on the farm. States designate agencies for implementing the BMP programs and for verifying that the BMPs are effective at reducing pollutant loads.

The 4 Rs. This acronym has come into widespread use in the last few years to help focus on several important aspects of fertilizer management under the umbrella of Best Management Practices. Used by the Fertilizer Institute and widely by Cooperative Extension and various fertilizer trade organizations, it addresses four important management aspects of fertilizer use in cropping systems. These are the *right source* of fertilizer, the *right rate*, the *right timing*, and the *right placement*. These principles are addressed below in this chapter. One additional comment is important because it bears heavily on fertilizer management, and that is irrigation. We should think of irrigation at the 5th R: *right irrigation* program.

Nutrient budgets. Accounting for nutrient inputs and exports from a farm or crop can help fine-tune nutrient management practices. A nutrient budget measures the inputs of a particular nutrient, for example, nitrogen, from all sources including fertilizer, organic materials, air sources, and N in irrigation water. Then the exports are measured, including crop removal from the field or off the farm, leaching losses, runoff, and volatilization. Knowing where the nitrogen is going can help the farmer devise management strategies to minimize losses because losses represent money lost and could indicate pollution of the environment. Most Extension advisors can help growers develop budgets but some of the needed laboratory analyses can be costly.

Some information on BMP programs can be found at:

- *USDA Natural Resources Conservation Service Field Office Technical Guide*, http://www.nrcs.usda.gov (accessed 25 June 2022).
- Florida Department of Agriculture and Consumer Services Water Quality/Quantity Best Management Practices for Florida Vegetable and Agronomic Crops, https://www.fdacs.gov/Divisions-Offices/ Agricultural-Water-Policy (accessed 25 June 2022).
- *Farming for Clean Water in South Carolina: A Handbook of Conservation Practices* (S.C.: NRCS), http://www.sc.nrcs.usda.gov/ pubs.html (accessed 25 June 2022).
- Efficient Nitrogen Fertility and Irrigation Management of Cool-Season Vegetables in Coastal California, http://ciwr.ucanr.edu/ files/283981.pdf (accessed 25 June 2022).
- Soil and Fertilizer Management for Vegetable Production in Florida, https://edis.ifas.ufl.edu/publication/CV101 (accessed 25 June 2022).
- *Maryland Nutrient Management Manual,* https://mda.maryland.gov/ resource_conservation/Pages/nm_manual.aspx (accessed 25 June 2022).
- Natural Resources Conservation Service, https://www.nrcs.usda.gov/ wps/portal/nrcs/site/national/home/ (accessed 25 June 2022).

SOIL CONSERVATION PRACTICES

Soil is often referred to as the most important asset a farmer has. Keeping the soil in place on the farm is therefore important because the lost soil also takes nutrients with it and could result in pollution. Loss of soil to erosion from water or wind can be minimized by using various appropriate conservation practices. Installing soil conservation practices may be eligible for cost-share from various federal or state agencies. Soil conservation practices often go hand-in-hand with other soil quality improving practices discussed elsewhere. Soil conservation practices include:

- Regular cover cropping where the soil has a crop growing nearly all year-round. For example, growing a grass cover crop in between commercial crops to add organic matter to the soil or protect the soil from wind erosion.
- Grassed waterways. Farmers understand where in their fields the soil may be eroded by surface water movement from rain or melting snow. This usually happens on highpoints in fields or in the slopes or swales. Keeping these areas covered with a grass helps slow down water or wind movement thus erosion. Some farmers argue they need to give up crop land for these grassed areas, but the loss of fertile soil also represents an economic loss.
- Riparian strips. Planting trees, native shrubs, perennial grasses, etc. along water bodies help reduce the flow of soil and nutrients from an adjacent field into those water bodies.
- Wind breaks. Permanent or temporary wind breaks used in a crop field or on the perimeters can help slow down wind speed and reduce wind erosion.

02

ORGANIC MATTER

FUNCTION OF ORGANIC MATTER IN SOIL

Rapid decomposition of fresh organic matter contributes most effectively to the physical condition of a soil. Plenty of moisture, nitrogen, and a warm temperature speed the rate of decomposition.

Organic matter serves as a source of energy and nutrients for soil microorganisms and as a source of nutrients for plants. Organic matter holds the minerals in the soil against loss by leaching until they are released for plant uptake by the action of microorganisms. Bacteria thriving on the organic matter produce complex carbohydrates that cement soil particles into aggregates helping improve soil structure and quality. Acids produced in the decomposition of organic matter may make available mineral nutrients of the soil to crop plants. The entrance and percolation of water into and through the soil are facilitated. This reduces losses of soil by erosion. Penetration of roots through the soil is improved by good structure brought about by the decomposition of organic matter. The water-holding capacity of sands and sandy soils may be increased by the incorporation of organic matter. Aggregation in heavy soils may improve drainage. It is seldom possible to make a large permanent increase in the organic matter content of a soil because the soil organic matter is rapidly decomposing in agricultural fields.

Nutrient cycling and soil health. Nutrients important for crop growth will cycle among forms in the soil. For example, nitrogen can be added as fertilizer, then taken up by the plant, then the N returns to the soil as the leftover plant parts (after harvest) are returned to the soil. The plant materials then decompose via microbial activity, releasing the N back to the soil for uptake by the next crop. All essential mineral nutrients for plants have a cycling system. It would be helpful to know estimates of the amount of nutrients that can be returned to the soil from various crops. Most Extension sources have estimates of these numbers. *Soil health* is a quality term being used more recently to describe soils that have the physical, chemical, and biological properties that contribute to good crop production while maintaining the good soil quality. These soils have active microbial activity to aid in returning nutrients to the soil from organic matter. Good soil health results from deliberate action by the farmer with attention to soil and microbial properties.

ORGANIC SOIL AMENDMENTS

Animal manures, sludges, and plant materials have been used commercially for decades for vegetable production. Today, society demands efficient use of natural materials, so recycling of wastes to agriculture is viewed as important. Many municipalities are producing solid waste materials that can be used on the farm as soil amendments and sources of nutrients for plants. The technology of compost production and utilization is constantly developing. One challenge for the grower is to locate compost sources that yield consistent chemical and physical soil qualities. Incompletely composted waste, sometimes called *green compost,* can reduce crop growth because nitrogen is *robbed,* or used by the microorganisms to decompose the organic matter in the compost. Growers contemplating use of soil amendments should thoroughly investigate the quality of the product, including testing for nutrient content.

ENVIRONMENTAL ASPECTS OF ORGANIC SOIL AMENDMENTS

Although the addition of organic matter, such as manures, to the soil can have beneficial effects on crop performance, there are some potential challenges. As the nitrogen is released from the organic matter, it can be subject to leaching. Heavy applications of manure can contribute to groundwater pollution unless a crop is planted soon to utilize the nitrogen. This potential can be especially great in southern climates, where nitrogen release can be rapid and most nitrogen is released in the first season after application. Plastic mulch placed over the manured soil reduces the potential for nitrate leaching. In today's environmentally aware world, manures must be used carefully to manage the released nutrients. Growers contemplating using manures as soil amendments or crop nutrient sources should have the manure tested for nutrient content. The results from such tests can help determine the best rate of application so that excess nutrients such as N or P are not available for leaching or losses to erosion. A further potential issue with animal manures is the food safety question. Research is ongoing to determine implications of runoff from nearby animal feeding operations on fresh vegetable contamination. Land-Grant University Extension Services have recommendation guides for managing manures for crop production.

03

SOIL HEALTH AND SOIL-IMPROVING CROPS

Farmers can manage soil health by their farming practices, including avoiding compaction, maintaining soil organic matter, encouraging populations of beneficial organisms (earth worms, fungi, insects, etc.), and preventing loss of soil, among other practices. Healthy soils provide the best chance for good crops while minimizing costly chemical inputs. One of the most important practices for maintaining good soil health is growing soil-improving crops. These plants add organic matter and nutrients to the soil that benefit soil organisms and crops (Table 6.1).

TABLE 6.1. SEED REQUIREMENTS OF SOIL-IMPROVING CROPS AND AREAS OF ADAPTATION

Soil-Improving Crops	Seeding Rate (lb/acre)	Area in the U.S. Where Crop is Adapted
Winter Cover Crops		
Legumes		
Berseem (*Trifolium alexandrinum*)	15	West and southeast
Black medic (*Medicago lupulina*)	15	All
Black lupine (*Lupinus hirsutus*)	70	All
Clover		
Crimson (*Trifolium incarnatum*)	15	South and southeast
Bur, California (*Medicago hispida*)	25	South
Southern (*M. arabica*) unhulled	100	Southeast
Tifton (*M. rigidula*) unhulled	100	Southeast
Sour (*Melilotus indica*)	20	South
Sweet, hubam (*Melilotus alba*)	20	All
Fenugreek (*Trigonella foenumgraecum*)	30	Southwest

TABLE 6.1.   SEED REQUIREMENTS OF SOIL-IMPROVING CROPS AND AREAS OF ADAPTATION (*Continued*)

Soil-Improving Crops	Seeding Rate (lb/acre)	Area in the U.S. Where Crop is Adapted
Field pea (*Pisum sativum*)		
Canada	80	All
Austrian winter	70	All
Horse bean (*Vicia faba*)	100	Southwest and southeast
Rough pea (*Lathyrus hirsutus*)	60	Southwest and southeast
Vetch		
Bitter (*Vicia ervilia*)	30	West and southeast
Common (*V. sativa*)	50	West and southeast
Hairy (*V. villosa*)	30	All
Hungarian (*V. pannonica*)	50	West and southeast
Monantha (*V. articulata*)	40	West and southeast
Purple (*V. bengalensis*)	40	West and southeast
Smooth (*V. villosa* var. glabrescens)	30	All
Woollypod (*V. dasycarpa*)	30	Southeast
Nonlegumes		
Barley (*Hordeum vulgare*)	75	All
Mustard (*Brassica nigra*)	20	All
Oat (*Avena sativa*)	75	All
Rape (*Brassica napus*)	20	All
Rye (*Secale cereale*)	75	All
Wheat (*Triticum sativum*)	75	All
Summer Cover Crops		
Legumes		
Alfalfa (*Medicago sativa*)	20	All
Beggarweed (*Desmodium purpureum*)	10	Southeast

TABLE 6.1. SEED REQUIREMENTS OF SOIL-IMPROVING CROPS AND AREAS OF ADAPTATION (*Continued*)

Soil-Improving Crops	Seeding Rate (lb/acre)	Area in the U.S. Where Crop is Adapted
Clover		
Alyce (*Alysicarpus vaginalis*)	20	Southeast
Crimson (*Trifolium incartum*)	15	Southeast
Red (*T. pratense*)	10	All
Cowpea (*Vigna sinensis*)	90	South and southwest
Hairy indigo (*Indigofera hirsuta*)	10	Southern tier
Lezpedeza		
Common (*Lezpedeza striata*)	25	Southeast
Korean (*L. stipulacea*)	20	Southeast
Sesbania (*Sesbania exaltata*)	30	Southwest
Soybean (*Glycine max*)	75	All
Sweet clover, white (*Melilotus alba*)	20	All
Sweet clover (*M. officinalis*)	20	All
Velvet bean (*Stizolobium deeringianum*)	100	Southeast
Nonlegumes		
Buckwheat (*Fagopyrum esculentum*)	75	All
Pearl millet (*Pennisetum glaucum*)	25	Southern and southeast
Sorghum, Hegari (*Sorghum vulgare*)	40	Western half
Sudan grass (*Sorghum vulgare* var. *sudanese*)	25	All

Adapted from *Growing Summer Cover Crops* (USDA Farmer's Bulletin 2182, 1967);P. R. Henson and E. A. Hollowell, *Winter Annual Legumes for the South* (USDA Farmers Bulletin 2146, 1960); *NRCS Cover Crop Plant specification Guide*, https://www.nrcs.usda.gov/Internet/FSE_DOCUMENTS/stelprdb1081555.pdf; *Cornell cover crop Guide for NY Growers* (2019). http://covercrop.org/cover-crop-decision-tool.

DECOMPOSITION OF SOIL-IMPROVING CROPS

The normal carbon-nitrogen (C : N) ratio in soils is about 10 : 1. Turning under organic matter alters this ratio because most organic matter is richer in carbon than in nitrogen (Table 6.2). Unless the residue contains at least 1.5% nitrogen (dry matter basis), the decomposing organisms will utilize soil nitrogen as a nutrient source for the decomposition process. Soil organisms can tie up as much as 25 lb nitrogen per acre from the soil in the process of decomposition of carbon-rich fresh organic matter. Soil microbial activity is best with a C : N ratio of about 25:1. The C : N ratio in the organic matter to be added to the soil should be considered in the decision about a soil-improving crop. If possible, choose materials with less than a 50 : 1 C : N ratio.

A soil-improving crop may require N fertilization to increase biomass and raise the nitrogen content somewhat to improve later decomposition. Nitrogen may have to be added as the soil-improving crop is incorporated into the soil. This speeds the decomposition and prevents a temporary shortage of nitrogen for the succeeding vegetable crop. As a general rule, about 20 lb nitrogen may be needed for each ton of dry matter for a nonlegume green-manure crop.

TABLE 6.2 APPROXIMATE CARBON-TO-NITROGEN RATIOS OF COMMON ORGANIC MATERIALS

Material	C : N Ratio
Alfalfa	12 : 1
Sweet clover, young	12 : 1
Sweet clover, mature	24 : 1
Rotted manure	20 : 1
Chicken, pig manure	5 : 1
Oat, rye, wheat straw	75 : 1
Corn stalks	80 : 1
Timothy straw	80 : 1
Sawdust	300 : 1
Rye cover crop (before heading)	25 : 1
Red, crimson clover	20 : 1

211

MANURES AND OTHER ORGANIC SOIL AMMENDENTS

TYPICAL COMPOSITION OF MANURES

Manures vary greatly in their nutrient content (Table 6.3). The kind of feed used, the percentage and type of litter or bedding, the moisture content, and the age and degree of decomposition or drying all affect the composition. Some nitrogen is lost in the process of producing commercially dried, pulverized manures. Nitrogen can be lost during the application process (Table 6.4).

The following data are representative analyses from widely scattered reports.

TABLE 6.3. COMPOSITION OF MANURES

Source	Dry Matter (%)	Approximate Composition (% Dry Weight)		
		N	P_2O_5	K_2O
Dairy	15–25	0.6–2.1	0.7–1.1	2.4–3.6
Feedlot	20–40	1.0–2.5	0.9–1.6	2.4–3.6
Horse	15–25	1.7–3.0	0.7–1.2	1.2–2.2
Poultry	20–30	2.0–4.5	4.5–6.0	1.2–2.4
Sheep	25–35	3.0–4.0	1.2–1.6	3.0–4.0
Swine	20–30	3.0–4.0	0.4–0.6	0.5–1.0

TABLE 6.4. NITROGEN LOSSES FROM ANIMAL MANURE TO THE AIR BY METHOD OF APPLICATION

Application Method	Type of Manure	Nitrogen Loss (%)[1]
Broadcast without incorporation	Solid	15–30
	Liquid	10–25
Broadcast with incorporation	Solid	1–5
	Liquid	1–5
Injection (knifing)	Liquid	0–2
Irrigation	Liquid	30–40

University of California, *Soil Fertility Management for Organic Crops* (Vegetable Research and Information Center), http://agwaterstewards.org/wp-content/uploads/2016/08/Soil_Fertility_Management_for_Organic_Crops.pdf. Also University of Minnesota, https://extension.umn.edu/manure-management/manure-application-methods-and-nitrogen-losses.

[1] Loss within 3 days of application.

TYPICAL COMPOSITION OF SOME ORGANIC FERTILIZER MATERIALS

Under most environments, the nutrients in organic materials become available to plants slowly. However, mineralization of nutrients in organic matter can be hastened under warm, humid conditions. For example, in Florida, most usable nitrogen can be made available from poultry manure during one cropping season. There is considerable variation in nutrient content among samples of organic soil amendments. Commercial manure products should have a summary of the chemical analyses on the container. Growers should have any organic soil amendment tested for nutrient content so fertilization programs can be planned. The data below (Tables 6.5 and 6.6) are representative of many analyses noted in the literature and in reports of state analytical laboratories.

TABLE 6.5. COMPOSITION OF ORGANIC MATERIALS

Organic Materials	Percentage on a Dry Weight Basis		
	N	P_2O_5	K_2O
Bat guano	10.0	4.0	2.0
Blood	13.0	2.0	1.0
Bone meal, raw	3.0	22.0	—
Bone meal, steamed	1.0	15.0	—
Castor bean meal	5.5	2.0	1.0
Cottonseed meal	6.6	3.0	1.5
Fish meal	10.0	6.0	—
Garbage tankage	2.5	2.0	1.0
Peanut meal	7.0	1.5	1.2
Sewage sludge	1.5	1.3	0.4
Sewage sludge, activated	6.0	3.0	0.2
Soybean meal	7.0	1.2	1.5
Tankage	7.0	10.0	1.5

Another excellent source of information on nutrient content of organic materials can be found in the following source: R. Penhallegon, *Nitrogen–Phosphorus–Potassium Values of Organic Fertilizers* (Oregon State University Extension Service), https://extension.oregonstate.edu/crop-production/organic/nitrogen-phosphorus-potassium-values-organic-fertilizers.

TABLE 6.6. COMPOSITION OF CROP FORAGE MATERIALS

Materials	Moisture (%)	Approximate Pounds Per Ton of Dry Material		
		N	P_2O_5	K_2O
Alfalfa hay	10	50	11	50
Alfalfa straw	7	28	7	36
Barley hay	9	23	11	33
Barley straw	10	12	5	32
Bean straw	11	20	6	25
Beggarweed hay	9	50	12	56
Buckwheat straw	11	14	2	48
Clover hay				
Alyce	11	35	—	—
Bur	8	60	21	70
Crimson	11	45	11	67
Ladino	12	60	13	67
Sweet	8	60	12	38
Cowpea hay	10	60	13	36
Cowpea straw	9	20	5	38
Field pea hay	11	28	11	30
Field pea straw	10	20	5	26
Horse bean hay	9	43	—	—
Lespedeza hay	11	41	8	22
Lespedeza straw	10	21	—	—
Oat hay	12	26	9	20
Oat straw	10	13	5	33
Ryegrass hay	11	26	11	25
Rye hay	9	21	8	25
Rye straw	7	11	4	22
Sorghum stover, Hegari	13	18	4	—
Soybean hay	12	46	11	20
Soybean straw	11	13	6	15
Sudan grass hay	11	28	12	31
Sweet corn fodder	12	30	8	24
Velvet bean hay	7	50	11	53

TABLE 6.6.   COMPOSITION OF CROP FORAGE MATERIALS
(Continued)

Materials	Moisture (%)	Approximate Pounds Per Ton of Dry Material		
		N	P_2O_5	K_2O
Vetch hay				
Common	11	43	15	53
Hairy	12	62	15	47
Wheat hay	10	20	8	35
Wheat straw	8	12	3	19

Adapted from *Morrison Feeds and Feeding* (Ithaca, N.Y.: Morrison, 1948).

05

ORGANIC PRODUCTION SYSTEMS

Treadwell and Perez (reference below) describe organic production of crops as the following. "Organic farming can generally be described as a method of production that utilizes non-synthetic inputs and emphasizes biological and ecological processes, to improve soil quality, manage soil fertility, and optimize pest management." Uniform organic farming standards were established in 1990 to help farmers understand the requirements for production practices and standardized international trade policies. Individuals desiring to become an organic vegetable producer will need to go through a certification process for the farm and the production practices. Organic farms use approved inputs for fertilizers, pesticides, and other inputs on the farm. There is a national list of such inputs (The National List of Allowed and Prohibited Substances). In general, most synthetic substances are not allowed in organic production systems. The Organic Materials Review Institute (OMRI) also publishes lists of brand-name and generic products that are permitted for use in organic systems.

Organic production systems rely on natural biological processes to provide nutrients for crops. This includes the decomposition of soil amendments and manures being used as fertilizers. Organic growers pay attention to the same soil health and soil quality parameters as were discussed earlier. Planning and record-keeping are important facets of organic crop production. Cover crops play an important role in maintaining soil quality and providing plant nutrients. Organically produced seeds and transplants are recommended for organic producers. Pest management in organic systems takes several approaches. Growers should first use production practices that are known to reduce pest populations. Cultural and mechanical methods are used to reduce a pest population buildup. Finally, growers have access to biological products as well as approved synthetic products for use against pests. The idea is that growers should start with recommended cultural practices, then move to approved pesticides if the situation warrants it.

Organic growers have many Extension documents from their state Extension system, including basic guides for beginning farmers and advanced production guides. Some examples are provided below.

D. Treadwell and J. Perez, *Introduction to Organic Crop Production* (University of Florida IFAS Extension, 2017), https://edis.ifas.ufl.edu/publication/cv118 (accessed 25 June 2022).

University of California, *Soil Fertility Management for Organic Crops* (Vegetable Research and Information Center), http://agwaterstewards.org/wpcontent/uploads/2016/08/Soil_Fertility_Management_for_Organic_Crops.pdf (accessed 25 June 2022).

P. Coleman, *Guide for Organic Crop Producers* (2012), https://www.ams.usda.gov/sites/default/files/media/GuideForOrganicCropProducers.pdf (accessed 25 June 2022).

Sustainable Agriculture Research and Extension, *Transitioning to Organic Production*, https://www.sare.org/publications/transitioning-to-organic-production/what-is-organic-farming/ (accessed 25 June 2022).

Texas AgriLife Extension, *Vegetable Resources. Organic Crop Production Requirements*, https://aggie-horticulture.tamu.edu/vegetable/guides/organic-vegetable-production-guide/organic-crop-production-requirements/ (accessed 25 June 2022).

COMPOSTING

Composting is a practice where organic materials are decomposed to be later used as a mulch or soil amendment, or a source of plant nutrients. The composting process relies on microbes that use the organic matter as sources of water, carbon, and nutrients for their growth. In the process, the organic matter is decomposed to a condition useful for farmers as a soil amendment.

The rate of the decomposition process depends on several factors, including temperature, the carbon content of the organic matter, moisture, among others. Organic materials with a carbon to nitrogen (C : N) ratio of less than 30 will degrade faster. The best C : N ratio is 25:1. Organic materials with less than this ratio tend to decompose fast, creating a wet, slimy material that can have a bad odor. The other problem with high-carbon materials, like wood chips or sawdust, is that they are low in nitrogen content. Therefore, the composting microbes take the N they need for growth from other materials in the composting bin leaving little nitrogen for the resulting soil amendment (see Table 6.2).

Composting produces a rich, fertilizer material to use as a source of organic matter and nutrients. Composting provides a useful means for recycling organic material on the farm, thus reducing the cost of landfilling. More information on composting can be found in the following sources.

S. Hu, *Composting 101* (Natural Resources Defense Council, 2020), https://www.nrdc.org/stories/composting-101 (accessed 25 June 2022).

Compost (Sarasota County: University of Florida Extension), https://sfyl. ifas.ufl.edu/sarasota/natural-resources/waste-reduction/composting/ (accessed 25 June 2022).

SOIL TEXTURE

The particles of a soil are classified by size into sand, silt, and clay, and together they make up the soil texture (Table 6.7; Figure 6.1).

TABLE 6.7. CLASSIFICATION OF SOIL-PARTICLE SIZES

Soil Particle Size Classes (Diameter, mm)

2.0	0.02	0.002	0
Gravel	Sand	Silt	Clay
Particles visible with the naked eye	Particles visible under microscope	Particles visible under electron microscope	

SOIL TEXTURAL TRIANGLE

The percentage of sand, silt, and clay in a soil may be plotted on a diagram to determine the textural class of that soil. Soil texture is important in the following farming practices:

1. Rates of herbicides needed, less for sandy soils
2. Potential for nutrient leaching, more for sandy soils
3. Soil warming in the spring, faster for sandy soils
4. Ease of tillage, easier for sandy soils
5. Water-holding capacity, less for sandy soils
6. Organic matter loss, faster for sandy soils

Example: A soil containing 13% clay, 41% silt, and 46% sand would have a loam texture.

Information on soil type and chemical and physical characteristics for any given location in the United States can be obtained from the USDA Web Soil Survey at: https://websoilsurvey.sc.egov.usda.gov/App/HomePage.htm (accessed 25 June 2022).

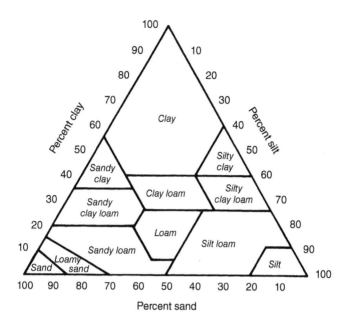

Figure 6.1. Soil textural triangle. From *Soil Conservation Service, Soil Survey Manual* (USDA Agricultural Handbook 18, 1951); *Natural Resources Conservation Service—Soils*, https://www.nrcs.usda.gov/wps/portal/nrcs/detail/soils/survey/?cid=nrcs142p2_054167 (accessed 25 June 2022).

08

SOIL REACTION

RELATIVE TOLERANCE OF VEGETABLE CROPS TO SOIL REACTION

Vegetables vary in their tolerance to soil acidity (Table 6.8). Vegetables in the slightly tolerant group can be grown successfully on soils that are on the alkaline or basic side of neutrality. They do well up to pH 7.6 if there is no deficiency of essential nutrients. Vegetables in the very tolerant group grow satisfactorily at a soil pH as low as 5.0. For the most part, even the most tolerant crops grow better at pH 6.0–6.8 than in more acidic soils.

TABLE 6.8.   TOLERANCE OF VEGETABLES TO SOIL ACIDITY

Slightly Tolerant (pH 6.8–6.0)	Moderately Tolerant (pH 6.8–5.5)	Very Tolerant (pH 6.8–5.0)
Asparagus	Bean	Chicory
Beet	Bean, lima	Dandelion
Broccoli	Brussels sprouts	Endive
Cabbage	Carrot	Fennel
Cantaloupe	Collards	Potato
Cauliflower	Cucumber	Rhubarb
Celery	Eggplant	Shallot
Chard, Swiss	Garlic	Sorrel
Chinese cabbage	Gherkin	Sweet potato
Cress	Horseradish	Watermelon
Leek	Kale	
Lettuce	Kohlrabi	
New Zealand spinach	Mustard	
Okra	Parsley	
Onion	Pea	
Orach	Pepper	
Parsnip	Pumpkin	
Salsify	Radish	
Soybean	Rutabaga	
Spinach	Squash	
Watercress	Strawberry	
	Sweet corn	
	Tomato	
	Turnip	

EFFECT OF SOIL REACTION ON AVAILABILITY OF NUTRIENTS

Soil reaction affects plants by influencing the availability of nutrients. Changes in soil reaction caused by liming or by the use of sulfur and acid-forming fertilizers may increase or decrease the supply of the nutrients available to the plants (Figure 6.2). Certain elements that may be toxic to plants such as aluminum or manganese may be more soluble and injurious to plants under acidic soil conditions. Also soil reaction affects certain soil processes such as microbial activity.

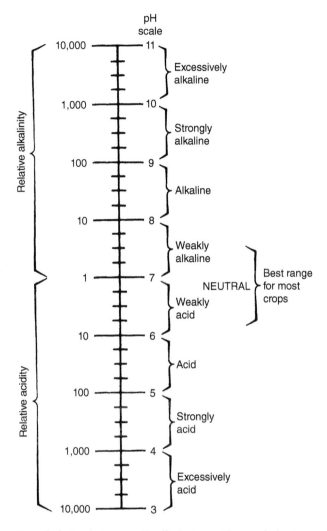

Figure 6.2. Relation between pH, alkalinity, acidity, and plant growth.

The general relationship between soil reaction and availability of plant nutrients in organic soils differs from that in mineral soils. The diagrams depict nutrient availability for both mineral and organic soils (Figures 6.3 and 6.4). The width of the band indicates the general availability of the nutrient. It does not indicate the actual amount present.

223

CORRECTION OF SOIL ACIDITY

Liming materials are used to change an unfavorable acidic soil reaction to a pH more favorable for crop production (Table 6.9). Likewise, acidifying materials can help reduce high pH due to overliming (Tables 6.10 and 6.11). Certain fertilizers can change the soil pH and should be factored into the management of soil reaction (Table 6.12). However, soil types differ in their response to liming, a property referred to as the soil's *pH buffering capacity*. Acidic soil reaction is caused by hydrogen ions present in the soil solution (*active acidity*) and those attached to soil particles or organic matter (*potential acidity*). Active acidity can be neutralized rapidly, whereas

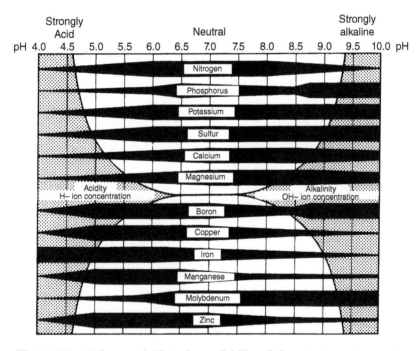

Figure 6.3. Influence of pH on the availability of plant nutrients in organic soils; widest parts of the shaded areas indicate maximum availability of each element. Adapted from R. E. Lucas and J. F. Davis, "Relationships Between pH Values of Organic Soils and Availability of 12 Plant Nutrients," *Soil Science* 92:177–182 (1961).

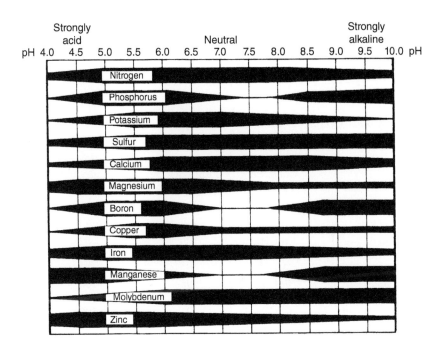

Figure 6.4. Influence of pH on the availability of plant nutrients in mineral soils; widest parts of the shaded areas indicate maximum availability of each element. Adapted from L. B. Nelson (ed.), *Changing Patterns in Fertilizer Use* (Madison, Wis.: Soil Science Society of America, 1968); *Effects of pH, Sodicity, and Salinity on Soil Fertility* (University of California, 2021), https://ucanr. edu/sites/Salinity/Salinity_Management/Effect_of_salinity_on_soil_ properties/Effect_of_pH_sodicity_and_salinity_on_soil_fertility_/ (accessed 25 June 2022).

potential acidity is neutralized over time as it is released. Soils vary in their relative content of these sources of acidity. Due to this complexity in soil pH, it is difficult to provide a rule of thumb for rates of liming materials. Most soil testing laboratories now use a *lime requirement test* to estimate the potential acidity and therefore provide a more accurate liming recommendation than could be done before. The lime requirement test treats the soil sample with a buffer solution to estimate the potential acidity, and thus provides a more accurate lime recommendation than can usually be obtained by treating the soil sample with water only. Soils with similar amounts of active acidity might have different amounts of potential acidity and thus require different lime recommendations even though the rule-of-thumb approach might have given similar lime recommendations. Soils with large potential acidity (clays and mucks) require more lime than sandy soils with a similar water pH.

TABLE 6.9. COMMON LIMING MATERIALS

Material	Chemical Formula	Pure $CaCO_3$ Equivalent (%)	Liming Material (lb) Necessary to Equal 100 lb Limestone
Burned lime	CaO	150	64
Hydrated lime	$Ca(OH)_2$	120	82
Dolomitic limestone	$CaCO_3$, $MgCO_3$	104	86
Limestone	$CaCO_3$	95	100
Marl	$CaCO_3$	95	100
Shell, oyster, etc.	$CaCO_3$	95	100

TABLE 6.10. COMMON ACIDIFYING MATERIALS[1]

Material	Chemical Formula	Sulfur (%)	Acidifying Material (lb) Necessary to Equal 100 lb Soil Sulfur
Soil sulfur	S	99.0	100
Sulfuric acid (98%)	H_2SO_4	32.0	306
Sulfur dioxide	SO_2	50.0	198
Lime-sulfur solution (32° Baumé)	CaSx + water	24.0	417
Iron sulfate	$FeSO_4 \cdot 7H_2O$	11.5	896
Aluminum sulfate	$Al_2(SO_4)_3$	14.4	694

[1] Certain fertilizer materials also markedly increase soil acidity when used in large quantities (see Table 6.12).

226

TABLE 6.11. APPROXIMATE QUANTITY OF SOIL SULFUR NEEDED TO INCREASE SOIL ACIDITY TO ABOUT pH 6.5

	Sulfur (lb/acre)		
Change in pH Desired	Sands	Loams	Clays
8.5–6.5	2,000	2,500	3,000
8.0–6.5	1,200	1,500	2,000
7.5–6.5	500	800	1,000
7.0–6.5	100	150	300

TABLE 6.12. **EFFECT OF SOME FERTILIZER MATERIALS ON THE SOIL REACTION**

| Materials | N (%) | Pounds Limestone (CaCO₃) | |
		Per lb N	Per 100 lb Fertilizer Material
			Needed to Counteract the Acidity Produced
Acidity-Forming			
Ammonium nitrate	33.5	1.80	60
Monoammonium phosphate	11	5.35	59
Ammonium phosphate sulfate	16	5.35	88
Ammonium sulfate	21	5.35	110
Anhydrous ammonia	82	1.80	148
Aqua ammonia	24	1.80	44
Aqua ammonia	30	1.80	54
Diammonium phosphate	16–18	1.80	70
Liquid phosphoric acid	52 (P_2O_5)	—	110
Urea	46	1.80	84
			Equivalents Produced
Alkalinity-Forming			
Calcium cyanamide	22	2.85	63
Calcium nitrate	15.5	1.35	20
Potassium nitrate	13	1.80	23
Sodium nitrate	16	1.80	29

Neutral

Ammonium nitrate-lime	Potassium sulfate
Calcium sulfate (gypsum)	Superphosphate
Potassium chloride	

Based on the method of W. H. Pierre, "Determination of Equivalent Acidity and Basicity of Fertilizers," *Industrial Engineering Chemical Analytical Edition,* 5:229–234 (1933).
Fertilizers and Soil Acidity Technical Bulletin (Fertiliser Technology Research Centre. The University of Adelaide), https://sciences.adelaide.edu.au/fertiliser/system/files/media/documents/2020-01/factsheet-fertilizers-and-soil-acidity.pdf (accessed 25 June 2022).

SOIL SALINITY

With an increase in soil salinity, plant roots extract water less easily from the soil solution (Table 6.13). This situation is more critical under hot and dry than under humid conditions. High soil salinity may result also in toxic concentrations of certain ions in plants and a reduction in yield. Soil salinity is determined by finding the electrical conductivity of the soil saturation extract (ECe). The electrical conductivity is measured in millimhos per centimeter (mmho/cm). One mmho/cm is equivalent to 1 decisiemen per meter (dS/m) and, on the average, to 640 ppm salt.

TABLE 6.13. CROP RESPONSE TO SALINITY

Salinity (Expressed as ECe, mmho/cm, or dS/m) Saturation Extract	Crop Responses
0–2	Salinity effects mostly negligible.
2–4	Yields of very sensitive crops may be restricted.
4–8	Yields of many crops restricted.
8–16	Only tolerant crops yield satisfactorily.
Above 16	Only a few very tolerant crops yield satisfactorily.

Adapted from L. Bernstein, *Salt Tolerance of Plants* (USDA Agricultural Information Bulletin 283, 1970).
University of California Salinity Management (2021), https://ucanr.edu/sites/Salinity/Salinity_Management/Effect_of_soil_salinity_on_crop_growth/Crop_salinity_tolerance_and_yield_function/.
E. V. Maas, *Testing Crops for Salinity Tolerance* (1993), https://www.ars.usda.gov/arsuserfiles/20360500/pdf_pubs/P1287.pdf (accessed 25 June 2022).

RELATIVE SALT EFFECTS OF FERTILIZER MATERIALS ON THE SOIL SOLUTION

When fertilizer materials are placed close to seeds or plants, they may increase the osmotic pressure of the soil solution and cause injury to the crop (Tables 6.14 and 6.15). The term *salt index* refers to the effect of a material in relation to that produced by sodium nitrate, which is given a rating of 100. The *partial index* shows the relationships per unit (20 lb) of the actual fertilizer supplied. Any material with a high salt index (>50) must be used with great care.

TABLE 6.14. SALT INDEX OF SEVERAL FERTILIZER MATERIALS

Material	Salt Index	Partial Salt Index Per 20 lb of Fertilizer
Anhydrous ammonia	47.1	0.572
Ammonium nitrate	104.7	2.990
Ammonium nitrate-lime (Cal-Nitro)	61.1	2.982
Ammonium sulfate	69.0	3.253
Calcium carbonate (limestone)	4.7	0.083
Calcium nitrate	52.5	4.409
Calcium sulfate (gypsum)	8.1	0.247
Diammonium phosphate	29.9	1.614^1
		0.637^2
Dolomite (calcium and magnesium carbonates)	0.8	0.042
Monoammonium phosphate	34.2	2.453^1
		0.485^2
Monocalcium phosphate	15.4	0.274
Nitrogen solution, 37%	77.8	2.104
Potassium chloride, 50%	109.4	2.189
Potassium chloride, 60%	116.3	1.936
Potassium nitrate	73.6	5.336^1
		1.580^3
Potassium sulfate	46.1	0.853
Sodium chloride	153.8	2.899
Sodium nitrate	100.0	6.060
Sulfate of potash-magnesia	43.2	1.971
Superphosphate, 20%	7.8	0.390
Superphosphate, 45%	10.1	0.224
Urea	75.4	1.618

Adapted from L. F. Rader, L. M. White, and C. W. Whittaker, "The Salt Index: A Measure of the Effect of Fertilizers on the Concentration of the Soil Solution," *Soil Science* 55:201–218 (1943).
J. J. Mortvedt, *Calculating Salt Index* (2001), http://www.fluidfertilizer.com/pastart/pdf/33P8-11.pdf.
[1] N.
[2] P_2O_5.
[3] K_2O.

TABLE 6.15. RELATIVE SALT TOLERANCE OF VEGETABLES

With most crops, there is little difference in salt tolerance among varieties. Boron tolerances may vary depending on climate, soil condition, and crop variety.

Vegetable	Maximum Soil Salinity Without Yield Loss (Threshold) (dS/m)	Decrease in Yield at Soil Salinities Above the Threshold (% Per dS/m)
Sensitive crops		
Bean	1.0	19
Carrot	1.0	14
Strawberry	1.0	33
Onion	1.2	16
Moderately sensitive		
Turnip	0.9	9
Radish	1.2	13
Lettuce	1.3	13
Pepper	1.5	14
Sweet potato	1.5	11
Broad bean	1.6	10
Corn	1.7	12
Potato	1.7	12
Cabbage	1.8	10
Celery	1.8	6
Spinach	2.0	8
Cucumber	2.5	13
Tomato	2.5	10
Broccoli	2.8	9
Squash, scallop	3.2	16
Moderately tolerant		
Beet	4.0	9
Squash, zucchini	4.7	9

Adapted from E. V. Maas, "Crop Tolerance," *California Agriculture* (Oct. 1984).
Note: 1 decisiemens per meter (dS/m) = 1 mmho/cm = approximately 640 mg/L salt.

10

FERTILIZERS

FERTILIZER DEFINITIONS

Grade or *analysis* means the minimum guarantee of the percentage of total nitrogen (N), available phosphoric acid (P_2O_5), and water-soluble potash (K_2O) in the fertilizer.

Example: 20-0-20 or 5-15-5

Ratio is the grade reduced to its simplest terms.

Example: A 20-0-20 has a ratio of 1-0-1, as does a 10-0-10.

Formula shows the actual pound and percentage composition of the ingredients or compounds that are mixed to make up a ton (2,000 lb) of fertilizer.

An *open-formula mix* carries the formula as well as the grade on the tag attached to each bag.

Carrier, simple, or *source* is the material or compound in which a given plant nutrient is found or supplied.

Example: Ammonium nitrate and urea are sources or carriers that supply nitrogen.

Unit means 1% of 1 ton or 20 lb. On the basis of a ton, the units per ton are equal to the percentage composition or the pounds per 100 lb.

Example: Ammonium sulfate contains 21% nitrogen, or 21 lb nitrogen/100 lb, or 21 units of nitrogen in a ton.

Primary nutrient refers to nitrogen, phosphorus, and potassium, which are used in considerable quantities by crops.

Secondary nutrient refers to calcium, magnesium, and sulfur, which are used in moderate quantities by crops.

Micronutrient, trace, or *minor element* refers to iron, boron, manganese, zinc, copper, molybdenum, and nickel, the essential plant nutrients used in relatively small quantities. Scientists now consider nickel (Ni) to be an essential (micro) nutrient.

GENERAL CATEGORIES OF FERTILIZERS

Fertilizer materials or nutrient sources can be thought of in several ways depending on the type of production system or philosophical approach to crop production. Some terms used are synthetic, conventional, chemical, dry, liquid, and organic.

SYNTHETIC LIQUID OR DRY FERTILIZERS

Synthetic fertilizers are most often supplied in dry form (Tables 6.16–6.19). Certain water-soluble materials, sometimes called salts and be dissolved in water and supplied in liquid form (Table 6.20). Dry fertilizers can be spread by broadcasting over the surface of the field or by banding along the side of the crop row. Liquid fertilizers can be banded along the row or injected into an irrigation system. Not all dry fertilizers can be dissolved in water for use as a liquid fertilizer. Table 6.17 provides information on the solubility of some fertilizer salts used to make liquid fertilizers and Table 6.20 provides guidance on calculating amounts of liquid fertilizers to achieve a needed rate of application. Table 6.21 contains some useful conversion factors for converting elemental forms and chemical compounds of fertilizers.

CONTROLLED-RELEASE FERTILIZERS

Controlled-release fertilizers are sometimes also referred to as slow-release, extended-release, coated fertilizers, etc. The idea behind these fertilizer materials is to provide nutrients, most often nitrogen, over a longer period in the growing cycle. An early slow-release fertilizer was sulfur-coated urea for supplying nitrogen over an extended period in the growing season. Providing nitrogen in slowly released amounts reduced the likelihood of large losses to leaching. Another advantage was the reduction in trips across the field to apply fertilizer. Early controlled-release fertilizers were expensive and accurately predicting nutrient release rates was difficult for field use. These fertilizers gained popularity in the greenhouse and nursery industries. With time, research and development led to polymer-coated controlled-release fertilizers where the nutrient release rate is governed by the type of coating and soil temperature. Various fertilizers with different release rates are now available and research is demonstrating economic and environmental benefits for vegetable crops. Less leaching of nitrogen results from the use of controlled-release fertilizers. The most benefit is with crops that do not use plastic mulch and fertigation. These crops would include potato, sweet corn, carrots, among others. Controlled-released fertilizers are also being tested for mulched crops where growers find cost competitiveness with liquid fertilizers/fertigation. Early polymer-coated materials were mostly nitrogen fertilizers but now there is more variety in the nutrient package in the fertilizer.

234

TABLE 6.16. APPROXIMATE COMPOSITION OF SOME CHEMICAL FERTILIZER MATERIALS[1]

Fertilizer Material	Total Nitrogen (% N)	Available Phosphorus (% P_2O_5)	Water-soluble Potassium (% K_2O)
Nitrogen			
Ammonium nitrate	33.5	—	—
Ammonium nitrate-lime (A-N-L, Cal-Nitro)	20.5	—	—
Monoammonium phosphate	11.0	48.0	—
Ammonium phosphate-sulfate	16.0	20.0	—
Ammonium sulfate	21.0	—	—
Anhydrous ammonia	82.0	—	—
Aqua ammonia	20.0	—	—
Calcium cyanamide	21.0	—	—
Calcium nitrate	15.5	—	—
Calcium ammonium nitrate	17.0	—	—
Diammonium phosphate	16–18	46.0–48.0	—
Potassium nitrate	13.0	—	44.0
Sodium nitrate	16.0	—	—
Urea	46.0	—	—
Urea formaldehyde	38.0	—	—
Phosphorus			
Phosphoric acid solution	—	52.0–54.0	—
Normal (single) superphosphate	—	18.0–20.0	—
Concentrated (triple or treble) superphosphate	—	45.0–46.0	—
Monopotassium phosphate	—	53.0	—
Potassium			
Potassium chloride	—	—	60.0–62.0
Potassium nitrate	13.0	—	44.0
Potassium sulfate	—	—	50.0–53.0
Sulfate of potash-magnesia	—	—	26.0
Monopotassium phosphate	—	—	34.0

[1] See Table 6.12 for effect of these materials on soil reaction.

TABLE 6.17.   SOLUBILITY OF FERTILIZER MATERIALS

Solubility of fertilizer materials is an important factor in preparing liquid starter solutions, foliar sprays, and solutions to be knifed into the soil or injected into an irrigation system. Hot water may be needed to dissolve certain chemicals.

Material	Solubility in Cold Water (lb/100 gal)
Primary Nutrients	
Ammonium nitrate	984
Ammonium sulfate	592
Calcium cyanamide	Decomposes
Calcium nitrate	851
Diammonium phosphate	358
Monoammonium phosphate	192
Potassium nitrate	108
Sodium nitrate	608
Superphosphate, single	17
Superphosphate, treble	33
Urea	651
Secondary Nutrients and Micronutrients	
Ammonium molybdate	Decomposes
Borax	8
Calcium chloride	500
Copper oxide	Insoluble
Copper sulfate	183
Ferrous sulfate	242
Magnesium sulfate	592
Manganese sulfate	876
Sodium chloride	300
Sodium molybdate	467
Zinc sulfate	625

TABLE 6.18. AMOUNTS OF CARRIERS NEEDED TO SUPPLY CERTAIN AMOUNTS OF NUTRIENTS PER ACRE[1]

Nutrients (lb/acre)							
20	40	60	80	100	120	160	200
Carriers Needed (lb)							

Nutrient in Carrier (%)	20	40	60	80	100	120	160	200
3	667	1,333	2,000					
4	500	1,000	1,500	2,000				
5	400	800	1,200	1,600	2,000			
6	333	667	1,000	1,333	1,667	2,000		
7	286	571	857	1,142	1,429	1,714		
8	250	500	750	1,000	1,250	1,500	2,000	
9	222	444	667	889	1,111	1,333	1,778	
10	200	400	600	800	1,000	1,200	1,600	2,000
11	182	364	545	727	909	1,091	1,455	1,818
12	166	333	500	666	833	1,000	1,333	1,666
13	154	308	462	615	769	923	1,231	1,538
15	133	267	400	533	667	800	1,067	1,333
16	125	250	375	500	625	750	1,000	1,250
18	111	222	333	444	555	666	888	1,111
20	100	200	300	400	500	600	800	1,000
21	95	190	286	381	476	571	762	952
25	80	160	240	320	400	480	640	800
30	67	133	200	267	333	400	533	667
34	59	118	177	235	294	353	471	588
42	48	95	143	190	238	286	381	476
45	44	89	133	178	222	267	356	444
48	42	83	125	167	208	250	333	417
50	40	80	120	160	200	240	320	400
60	33	67	100	133	167	200	267	333

[1]This table can be used in determining the acre rate for applying a material to supply a certain number of pounds of a nutrient.

Example: A carrier provides 34% of a nutrient. To get 200 lb of the nutrient, 588 lb of the material is needed, and for 60 lb of the nutrient, 177 lb of carrier is required.

237

TABLE 6.19. APPROXIMATE RATES OF MATERIALS TO PROVIDE CERTAIN QUANTITIES OF NITROGEN PER ACRE

Fertilizer Material	% N	15	30	45	60	75	100
N (lb/acre):		15	30	45	60	75	100
	% N	Material to Apply (lb/acre)					
Solids							
Ammonium nitrate	33	45	90	135	180	225	300
Ammonium phosphate (48% P_2O_5)	11	135	270	410	545	680	910
Ammonium phosphate-sulfate (20% P_2O_5)	16	95	190	280	375	470	625
Ammonium sulfate	21	70	140	215	285	355	475
Calcium nitrate	15.5	95	195	290	390	485	645
Potassium nitrate	13	115	230	345	460	575	770
Sodium nitrate	16	95	190	280	375	470	625
Urea	46	35	65	100	130	165	215
Liquids							
Anhydrous ammonia (approx. 5 lb/gal)[1]	82	20	35	55	75	90	120
Aqua ammonium phosphate (24% P_2O_5; approx. 10 lb/gal)	8	190	375	560	750	940	1250
Aqua ammonia (approx. 7½ lb/gal)[1]	20	75	150	225	300	375	500
Nitrogen solution (approx. 11 lb/gal)	32	50	100	150	200	250	330

[1] To avoid burning, especially on alkaline soils, these materials must be placed deeper and farther away from the plant row than dry fertilizers are placed.

TABLE 6.20. RATES OF APPLICATION FOR SOME NITROGEN SOLUTIONS

Nitrogen (lb/acre)	Nitrogen Solution Needed (gal/acre)		
	21% Solution	32% Solution	41% Solution
20	8.9	5.6	5.1
25	11.1	7.1	6.4
30	13.3	8.5	7.7
35	15.6	9.9	9.0
40	17.8	11.3	10.3
45	20.0	12.7	11.5
50	22.2	14.1	12.8
55	24.4	15.5	14.1
60	26.7	16.5	15.4
65	28.9	18.4	16.7
70	31.1	19.8	17.9
75	33.3	21.2	19.2
80	35.6	22.6	20.5
85	37.8	24.0	21.8
90	40.0	25.4	23.1
95	42.2	26.8	24.4
100	44.4	28.2	25.6
110	48.9	31.1	28.2
120	53.3	33.9	30.8
130	57.8	36.7	33.3
140	62.2	39.6	35.9
150	66.7	42.4	38.5
200	88.9	56.5	51.3

Adapted from C. W. Gandt, W. C. Hulburt, and H. D. Brown, *Hose Pump for Applying Nitrogen Solutions* (USDA Farmer's Bulletin 2096, 1956). Texas A and M University. Soil, Water and Forage Testing Laboratory.
Agricultural Liquid Fertilizer Calculator (Texas A and M University AgriLife Extension), http:// soiltesting.tamu.edu/Agliquidfertcalc/agliquidfertcalc.htm (accessed 25 June 2022).

11
FERTIGATION

The term fertigation refers to the process of applying fertilizers through the irrigation system simultaneously with the irrigation of the crop. Fertilizer can be injected through most any irrigation systems but most often it is done with center pivot systems and drip irrigation systems. Certain equipment is required for injecting fertilizers, including a nonleaking irrigation system, an injection pump made for handling fertilizer solutions, a tank to hold the fertilizer solution, certain backflow equipment to keep the fertilizers from draining backward into the irrigation well or other water source, and an irrigation controller to manage and monitor the process.

Using fertigation can save time by pairing irrigation and fertilization activities. Fertigation can increase the efficiency of fertilizer application because amounts of fertilizer can be applied just when the crops need the nutrients. This minimizes the risk of losing larger amounts of nutrients applied by side-dressing to rainfall.

TABLE 6.21. CONVERSION FACTORS FOR FERTILIZER
MATERIALS

Multiply	By	To Obtain Equivalent Nutrient
Ammonia—NH_3	4.700	Ammonium nitrate—NH_4NO_3
Ammonia—NH_3	3.879	Ammonium sulfate—$(NH_4)_2SO_4$
Ammonia—NH_3	0.823	Nitrogen—N
Ammonium nitrate—NH_4NO_3	0.350	Nitrogen—N
Ammonium sulfate— $(NH_4)_2SO_4$	0.212	Nitrogen—N
Borax—$Na_2B_4O_7 \cdot 10H_2O$	0.114	Boron—B
Boric acid—H_3BO_3	0.177	Boron—B
Boron—B	8.813	Borax—$Na_2B_4O_7 \cdot 10H_2O$
Boron—B	5.716	Boric acid—H_3BO_3
Calcium—Ca	1.399	Calcium oxide—CaO
Calcium—Ca	2.498	Calcium carbonate—$CaCO_3$
Calcium—Ca	1.849	Calcium hydroxide—$Ca(OH)_2$
Calcium—Ca	4.296	Calcium sulfate—$CaSO_4 \cdot 2H_2O$ (gypsum)
Calcium carbonate—$CaCO_3$	0.400	Calcium—Ca
Calcium carbonate—$CaCO_3$	0.741	Calcium hydroxide—$Ca(OH)_2$
Calcium carbonate—$CaCO_3$	0.560	Calcium oxide—CaO
Calcium carbonate—$CaCO_3$	0.403	Magnesia—MgO
Calcium carbonate—$CaCO_3$	0.842	Magnesium carbonate—$MgCO_3$
Calcium hydroxide—$Ca(OH)_2$	0.541	Calcium—Ca
Calcium hydroxide—$Ca(OH)_2$	1.351	Calcium carbonate—$CaCO_3$
Calcium hydroxide—$Ca(OH)_2$	0.756	Calcium oxide—CaO
Calcium oxide—CaO	0.715	Calcium—Ca
Calcium oxide—CaO	1.785	Calcium carbonate—$CaCO_3$
Calcium oxide—CaO	1.323	Calcium hydroxide—$Ca(OH)_2$
Calcium oxide—CaO	3.071	Calcium sulfate—$CaSO_4 \cdot 2H_2O$ (gypsum)
Gypsum—$CaSO_4 \cdot 2H_2O$	0.326	Calcium oxide—CaO
Gypsum—$CaSO_4 \cdot 2H_2O$	0.186	Sulfur—S
Magnesia—MgO	2.480	Calcium carbonate—$CaCO_3$

TABLE 6.21. **CONVERSION FACTORS FOR FERTILIZER**
MATERIALS (*Continued*)

Multiply	By	To Obtain Equivalent Nutrient
Magnesia—MgO	0.603	Magnesium—Mg
Magnesia—MgO	2.092	Magnesium carbonate—MgCO3
Magnesia—MgO	2.986	Magnesium sulfate—$MgSO_4$
Magnesia—MgO	6.114	Magnesium sulfate—$MgSO_4·7H_2O$ (Epsom salts)
Magnesium—Mg	4.116	Calcium carbonate—$CaCO_3$
Magnesium—Mg	1.658	Magnesia—MgO
Magnesium—Mg	3.466	Magnesium carbonate—$MgCO_3$
Magnesium—Mg	4.951	Magnesium sulfate—$MgSO_4$
Magnesium—Mg	10.136	Magnesium sulfate—$MgSO_4·7H_2O$ (Epsom salts)
Magnesium carbonate—$MgCO_3$	1.187	Calcium carbonate—$CaCO_3$
Magnesium carbonate—$MgCO_3$	0.478	Magnesia—MgO
Magnesium carbonate—$MgCO_3$	0.289	Magnesium—Mg
Magnesium sulfate—$MgSO_4$	0.335	Magnesia—MgO
Magnesium sulfate—$MgSO_4$	0.202	Magnesium—Mg
Magnesium sulfate—$MgSO_4·7H_2O$ (Epsom salts)	0.164	Magnesia—MgO
Magnesium sulfate—$MgSO_4·7H_2O$ (Epsom salts)	0.099	Magnesium—Mg
Manganese—Mn	2.749	Manganese(ous) sulfate—$MnSO_4$
Manganese—Mn	4.060	Manganese(ous) sulfate—$MnSO_4·4H_2O$
Manganese(ous) sulfate—$MnSO_4$	0.364	Manganese—Mn
Manganese(ous) sulfate—$MnSO_4·4H_2O$	0.246	Manganese—Mn
Nitrate—NO_3	0.226	Nitrogen—N
Nitrogen—N	1.216	Ammonia—NH_3

TABLE 6.21. CONVERSION FACTORS FOR FERTILIZER MATERIALS (*Continued*)

Multiply	By	To Obtain Equivalent Nutrient
Nitrogen—N	2.856	Ammonium nitrate—NH_4NO_3
Nitrogen—N	4.716	Ammonium sulfate—$(NH_4)2SO_4$
Nitrogen—N	4.426	Nitrate—NO_3
Nitrogen—N	6.068	Sodium nitrate—$NaNO_3$
Nitrogen—N	6.250	Protein
Phosphoric acid—P_2O_5	0.437	Phosphorus—P
Phosphorus—P	2.291	Phosphoric acid—P_2O_5
Potash—K_2O	1.583	Potassium chloride—KCl
Potash—K_2O	2.146	Potassium nitrate—KNO_3
Potash—K_2O	0.830	Potassium—K
Potash—K_2O	1.850	Potassium sulfate—K_2SO_4
Potassium—K	1.907	Potassium chloride—KCl
Potassium—K	1.205	Potash—K_2O
Potassium—K	2.229	Potassium sulfate—K_2SO_4
Potassium chloride—KCl	0.632	Potash—K_2O
Potassium chloride—KCl	0.524	Potassium—K
Potassium nitrate—KNO_3	0.466	Potash—K_2O
Potassium nitrate—KNO_3	0.387	Potassium—K
Potassium sulfate—K_2SO_4	0.540	Potash—K_2O
Potassium sulfate—K_2SO_4	0.449	Potassium—K
Sodium nitrate—$NaNO_3$	0.165	Nitrogen—N
Sulfur—S	5.368	Calcium sulfate—$CaSO_4 \cdot 2H_2O$ (gypsum)
Sulfur—S	2.497	Sulfur trioxide—SO_3
Sulfur—S	3.059	Sulfuric acid—H_2SO_4
Sulfur trioxide—SO_3	0.401	Sulfur—S
Sulfuric acid—H_2SO_4	0.327	Sulfur—S

Examples: 80 lb ammonia (NH_3) contains the same amount of N as 310 lb ammonium sulfate [$(NH_4)_2SO_4$], $80 \times 3.88 = 310$. Likewise, 1000 lb calcium carbonate multiplied by 0.400 equals 400 lb calcium. A material contains 20% phosphoric acid. This percentage (20) multiplied by 0.437 equals 8.74% phosphorus.

APPROXIMATE CROP CONTENT OF NUTRIENT ELEMENTS

Sometimes crop removal values (Table 6.22) are used to estimate fertilizer needs by crops. Removal values are obtained by analyzing plants and fruits for nutrient content and then expressing the results on an acre basis. It is risky to relate fertilizer requirements on specific soils to generalized listings of crop removal values. A major problem is that crop removal values are usually derived from analyzing plants grown on fertile soils where much of the nutrient content of the crop is supplied from soil reserves rather than from fertilizer application. Because plants can absorb larger amounts of specific nutrients than they require, crop removal values can overestimate the true crop nutrient requirement of a crop. Crop removal values can estimate the nutrient supply capacity on an unfertilized soil or may be a good starting point for estimating a nutrient mass budget, discussed earlier. The crop content (removal) values presented in the table are presented for information purposes and are not suggested for use in formulating fertilizer recommendations. The values were derived from various sources and publications. For example, a similar table was published in M. McVicker and W. Walker, *Using Commercial Fertilizer* (Danville, Ill.: Interstate Printers and Publishers, 1978).

TABLE 6.22. APPROXIMATE ACCUMULATION OF NUTRIENTS BY SOME VEGETABLE CROPS

Vegetable	Yield (cwt/acre)	N	P_2O_5	K_2O
		Nutrient Absorption (lb/acre)		
Broccoli	100 heads	20	2	45
	Other	145	8	165
		165	10	210
Brussels sprouts	160 sprouts	150	20	125
	Other	85	9	110
		235	29	235
Cantaloupe	225 fruits	95	17	120
	Vines	60	8	35
		155	25	155
Carrot	500 roots	80	20	200
	Tops	65	5	145
		145	25	345
Celery	1000 tops	170	35	380
	Roots	25	15	55
		195	50	435
Honeydew melon	290 fruits	70	8	65
	Vines	135	15	95
		205	23	160
Lettuce	350 plants	95	12	170
Onion	400 bulbs	110	20	110
	Tops	35	5	45
		145	25	155

TABLE 6.22. APPROXIMATE ACCUMULATION OF NUTRIENTS BY SOME VEGETABLE CROPS (Continued)

Vegetable	Yield (cwt/acre)	Nutrient Absorption (lb/acre)		
		N	P_2O_5	K_2O
Pea, shelled	40 peas	100	10	30
	Vines	70	12	50
		170	22	80
Pepper	225 fruits	45	6	50
	Plants	95	6	90
		140	12	140
Potato	400 tubers	150	19	200
	Vines	60	11	75
		210	30	275
Snap bean	100 beans	120	10	55
	Plants	50	6	45
		170	16	100
Spinach	200 plants	100	12	100
Sweet corn	130 ears	55	8	30
	Plants	100	12	75
		155	20	105
Sweet potato	300 roots	80	16	160
	Vines	60	4	40
		140	20	200
Tomato	600 fruits	100	10	180
	Vines	80	11	100
		180	21	280

14
PLANT ANALYSIS

Analyses of plant parts, most commonly leaves or petioles, can help diagnose potential nutrient deficiencies or toxicities. Most soil-testing labs provide plant tissue analyses. We have provided below several interpretation tables for various analytical methods (Tables 6.23–6.26). Growers using plant tissue testing to diagnose problems should be sure their sampling procedures and sample handling match the protocols of the lab they are using.

TABLE 6.23. PLANT ANALYSIS GUIDE FOR SAMPLING TIME, PLANT PART, AND NUTRIENT CONCENTRATION OF VEGETABLE CROPS (DRY WEIGHT BASIS)[1]

Crop	Time of Sampling	Plant Part	Source	Nutrient Concentration[2]	Nutrient Level Deficient	Sufficient
Asparagus	Midgrowth of fern	4-in. tip section of new fern branch	NO$_3$	N, ppm	100	500
			PO$_4$	P, ppm	800	1,600
				K, %	1	3
Bean, bush snap	Midgrowth	Petiole of 4th leaf from tip	NO$_3$	N, ppm	2,000	3,000
			PO$_4$	P, ppm	1,000	2,000
				K, %	3	5
	Early bloom	Petiole of 4th leaf from tip	NO$_3$	N, ppm	1,000	1,500
			PO$_4$	P, ppm	800	1,500
				K, %	2	4
Broccoli	Midgrowth	Midrib of young, mature leaf	NO$_3$	N, ppm	7,000	9,000
			PO$_4$	P, ppm	2,500	4,000
				K, %	3	5
	First buds	Midrib of young, mature leaf	NO$_3$	N, ppm	5,000	7,000
			PO$_4$	P, ppm	2,500	4,000
				K, %	2	4
Brussels sprouts	Midgrowth	Midrib of young, mature leaf	NO$_3$	N, ppm	5,000	7,000
			PO$_4$	P, ppm	2,000	3,500
				K, %	3	5

Crop	Growth stage	Plant part	Element	Form		
	Late growth	Midrib of young, mature leaf	N, ppm P, ppm K, %	NO$_3$ PO$_4$	2,000 1,000 2	3,000 3,000 4
Cabbage	At heading	Midrib of wrapper leaf	N, ppm P, ppm K, %	NO$_3$ PO$_4$	5,000 2,500 2	7,000 3,500 4
Cantaloupe	Early growth (short runners)	Petiole of 6th leaf from growing tip	N, ppm P, ppm K, %	NO$_3$ PO$_4$	8,000 2,000 4	12,000 3,000 6
	Early fruit	Petiole of 6th leaf from growing tip	N, ppm P, ppm K, %	NO$_3$ PO$_4$	5,000 1,500 3	8,000 2,500 5
	First mature fruit	Petiole of 6th leaf from growing tip	N, ppm P, ppm K, %	NO$_3$ PO$_4$	2,000 1,000 2	3,000 2,000 4
	Early growth	Blade of 6th leaf from growing tip	N, ppm P, ppm K, %	NO$_3$ PO$_4$	2,000 1,500 1	3,000 2,300 2.5
	Early fruit	Blade of 6th leaf from growing tip	N, ppm P, ppm K, %	NO$_3$ PO$_4$	1,000 1,300 1	1,500 1,700 2.0
	First mature fruit	Blade of 6th leaf from growing tip	N, ppm P, ppm K, %	NO$_3$ PO$_4$	500 1,000 1	800 1,500 1.8

TABLE 6.23. PLANT ANALYSIS GUIDE FOR SAMPLING TIME, PLANT PART, AND NUTRIENT CONCENTRATION OF VEGETABLE CROPS (DRY WEIGHT BASIS) (Continued)

Crop	Time of Sampling	Plant Part	Source	Nutrient Concentration[2]	Nutrient Level Deficient	Nutrient Level Sufficient
Chinese cabbage	At heading	Midrib of wrapper leaf	NO_3 PO_4	N, ppm P, ppm K, %	8,000 2,000 4	10,000 3,000 7
Carrot	Midgrowth	Petiole of young, mature leaf	NO_3 PO_4	N, ppm P, ppm K, %	5,000 2,000 4	7,500 3,000 6
Cauliflower	Buttoning	Midrib of young, mature leaf	NO_3 PO_4	N, ppm P, ppm K, %	5,000 2,500 2	7,000 3,500 4
Celery	Midgrowth	Petiole of newest fully elongated leaf	NO_3 PO_4	N, ppm P, ppm K, %	5,000 2,500 4	7,000 3,000 7
	Near maturity	Petiole of newest fully elongated leaf	NO_3 PO_4	N, ppm P, ppm K, %	4,000 2,000 3	6,000 3,000 5
Cucumber, pickling	Early fruit set	Petiole of 6th leaf from tip	NO_3 PO_4	N, ppm P, ppm K, %	5,000 1,500 3	7,500 2,500 5

250

Crop	Growth stage	Plant part	Nutrient	Units		
Cucumber, slicing	Early harvest period	Petiole of 6th leaf from growing tip	NO$_3$ PO$_4$	N, ppm P, ppm K, %	5,000 1,500 4	7,500 2,500 7
Eggplant	At first harvest	Petiole of young, mature leaf	NO$_3$ PO$_4$	N, ppm P, ppm K, %	5,000 2,000 4	7,500 3,000 7
Garlic	Early growth (prebulbing)	Newest fully elongated leaf	PO$_4$	P, ppm K, %	2,000 3	3,000 4
	Midseason (bulbing)	Newest fully elongated leaf	PO$_4$	P, ppm K, %	2,000 2	3,000 3
	Late season (postbulbing)	Newest fully elongated leaf	PO$_4$	P, ppm K, %	2,000 1	3,000 2
Lettuce	At heading	Midrib of wrapper leaf	NO$_3$ PO$_4$	N, ppm P, ppm K, %	4,000 2,000 2	6,000 3,000 4
	At harvest	Midrib of wrapper leaf	NO$_3$ PO$_4$	N, ppm P, ppm K, %	3,000 1,500 1.5	5,000 2,500 2.5
Onion	Early season	Tallest leaf	PO$_4$	P, ppm K, %	1,000 3	2,000 4.5
	Midseason	Tallest leaf	PO$_4$	P, ppm K, %	1,000 2	2,000 4
	Late season	Tallest leaf	PO$_4$	P, ppm K, %	1,000 2	2,000 3

TABLE 6.23. PLANT ANALYSIS GUIDE FOR SAMPLING TIME, PLANT PART, AND NUTRIENT CONCENTRATION OF VEGETABLE CROPS (DRY WEIGHT BASIS) (Continued)

Crop	Time of Sampling	Plant Part	Source	Nutrient Concentration[2]	Nutrient Level Deficient	Nutrient Level Sufficient
Pepper, chili	Early growth first bloom	Petiole of young, mature leaf	NO_3 PO_4	N, ppm P, ppm K, %	5,000 2,000 3	7,000 2,500 5
	Early fruit set	Petiole of young, mature leaf	NO_3 PO_4	N, ppm P, ppm K, %	1,000 1,500 2	1,500 2,000 4
	Fruits, full size	Petiole of young, mature leaf	NO_3 PO_4	N, ppm P, ppm K, %	750 1,500 1.5	1,000 2,000 3
	Early growth first bloom	Blade of young, mature leaf	NO_3 PO_4	N, ppm P, ppm K, %	1,500 1,500 3	2,000 2,000 5
	Early fruit set	Blade of young, mature leaf	NO_3 PO_4	N, ppm P, ppm K, %	500 1,500 2	800 2,000 4

Crop	Growth stage	Plant part		Element		
Pepper, sweet	Early growth, first flower	Petiole of young, mature leaf	NO₃	N, ppm	8,000	10,000
			PO₄	P, ppm	2,000	3,000
				K, %	4	6
	Early fruit set, 1 in. diameter	Petiole of young, mature leaf	NO₃	N, ppm	5,000	7,000
			PO₄	P, ppm	1,500	2,500
				K, %	3	5
	Fruit ¾ size	Petiole of young, mature leaf	NO₃	N, ppm	3,000	5,000
			PO₄	P, ppm	1,200	2,000
				K, %	2	4
	Early growth, first flower	Blade of young, mature leaf	NO₃	N, ppm	2,000	3,000
			PO₄	P, ppm	1,800	2,500
				K, %	3	5
	Early fruit set, 1 in. diameter	Blade of young, mature leaf	NO₃	N, ppm	1,500	2,000
			PO₄	P, ppm	1,500	2,000
				K, %	2	4
Potato	Early season	Petiole of 4th leaf from growing tip	NO₃	N, ppm	8,000	12,000
			PO₄	P, ppm	1,200	2,000
				K, %	9	11
	Midseason	Petiole of 4th leaf from growing tip	NO₃	N, ppm	6,000	9,000
			PO₄	P, ppm	800	1,600
				K, %	7	9
	Late season	Petiole of 4th leaf from growing tip	NO₃	N, ppm	3,000	5,000
			PO₄	P, ppm	500	1,000
				K, %	4	6
Spinach	Midgrowth	Petiole of young, mature leaf	NO₃	N, ppm	4,000	6,000
			PO₄	P, ppm	2,000	3,000
				K, %	2	4

TABLE 6.23. PLANT ANALYSIS GUIDE FOR SAMPLING TIME, PLANT PART, AND NUTRIENT CONCENTRATION OF VEGETABLE CROPS (DRY WEIGHT BASIS) (Continued)

Crop	Time of Sampling	Plant Part	Source	Nutrient Concentration[2]	Nutrient Level Deficient	Sufficient
Summer squash (zucchini)	Early bloom	Petiole of young, mature leaf	NO_3 PO_4	N, ppm P, ppm K, %	12,000 4,000 6	15,000 6,000 10
Sweet corn	Tasseling	Midrib of 1st leaf above primary ear	NO_3 PO_4	N, ppm P, ppm K, %	500 500 2	1,000 1,000 4
Sweet potato	Midgrowth	Petiole of 6th leaf from the growing tip	NO_3 PO_4	N, ppm P, ppm K, %	1,500 1,000 3	2,500 2,000 5
Tomato, cherry	Early fruit set	Petiole of 4th leaf from the growing tip	NO_3 PO_4	N, ppm P, ppm K, %	8,000 2,000 4	10,000 3,000 7
	Fruit ½ in. diameter	Petiole of 4th leaf from growing tip	NO_3 PO_4	N, ppm P, ppm K, %	5,000 2,000 3	7,000 3,000 5
	At first harvest	Petiole of 4th leaf from growing tip	NO_3 PO_4	N, ppm P, ppm K, %	1,000 2,000 2	2,000 3,000 4

Crop	Growth stage	Plant part	Form	Nutrient	Deficient	Adequate
Tomato, processing and determinate, fresh market	Early bloom	Petiole of 4th leaf from growing tip	NO$_3$	N, ppm	8,000	12,000
			PO$_4$	P, ppm	2,000	3,000
				K, %	3	6
	Fruit 1 in. diameter	Petiole of 4th leaf from growing tip	NO$_3$	N, ppm	4,000	6,000
			PO$_4$	P, ppm	1,500	2,500
				K, %	2	4
	First color	Petiole of 4th leaf from growing tip	NO$_3$	N, ppm	2,000	3,000
			PO$_4$	P, ppm	1,000	2,000
				K, %	1	3
Tomato, fresh market indeterminate	Early bloom	Petiole of 4th leaf from growing tip	NO$_3$	N, ppm	10,000	14,000
			PO$_4$	P, ppm	2,500	3,000
				K, %	4	7
	Fruit 1 in. diameter	Petiole of 4th leaf from growing tip	NO$_3$	N, ppm	8,000	12,000
			PO$_4$	P, ppm	2,500	3,000
				K, %	3	5
	Full ripe fruit	Petiole of 4th leaf from growing tip	NO$_3$	N, ppm	4,000	6,000
			PO$_4$	P, ppm	2,000	2,500
				K, %	2	4
Watermelon	Early fruit set	Petiole of 6th leaf from growing tip	NO$_3$	N, ppm	5,000	7,500
			PO$_4$	P, ppm	1,500	2,500
				K, %	3	5

[1] Adapted from H. M. Reisenauer (ed.), Soil and Plant Tissue Testing in California (University of California Division of Agricultural Science Bulletin 1879, 1983); and Plant Tissue Sampling (2017), https://apps1.cdfa.ca.gov/FertilizerResearch/docs/Plant_Tissue_Sampling.pdf.

[2] Two percent acetic acid-soluble NO$_3$–N and PO$_4$–P and total K (dry weight basis). Values represent conventionally fertilized crops. Organically managed crops may show lower petiole-nitrate (NO$_3$–N) concentrations. Total macronutrient concentration of whole leaves is the preferred method of evaluating nutrient sufficiency under organic fertility management.

TABLE 6.24. TOTAL NUTRIENT CONCENTRATION FOR DIAGNOSIS OF THE NUTRIENT LEVEL OF VEGETABLE CROPS

Crop	Time of Sampling	Plant Part	Nutrient	Nutrient Level (% Dry Weight)	
				Deficient	Sufficient
Asparagus	Early fern growth	4-in. tip section of new fern branch	N	4.00	5.00
			P	0.20	0.40
			K	2.00	4.00
	Mature fern	4-in. tip section of new fern branch	N	3.00	4.00
			P	0.20	0.40
			K	1.00	3.00
Bean, bush snap	Full bloom	Petiole: recent fully exposed trifoliate leaf	N	1.50	2.25
			P	0.15	0.30
			K	1.00	2.50
	Full bloom	Blade: recent fully exposed trifoliate leaf	N	1.25	2.25
			P	0.25	0.40
			K	0.75	1.50
Bean, lima	Full bloom	Oldest trifoliate leaf	N	2.50	3.50
			P	0.20	0.30
			K	1.50	2.25
Celery	Midgrowth	Petiole	N	1.00	1.50
			P	0.25	0.55
			K	4.00	5.00

Crop	Growth stage	Plant part	Nutrient		
Cantaloupe	Early growth	Petiole of 6th leaf from growing tip	N	2.50	3.50
			P	0.30	0.60
			K	4.00	6.00
	Early fruit	Petiole of 6th leaf from growing tip	N	2.00	3.00
			P	0.20	0.35
			K	3.00	5.00
	First mature fruit	Petiole of 6th leaf from growing tip	N	1.50	2.00
			P	0.15	0.30
			K	2.00	4.00
Garlic	Early season (prebulbing)	Newest fully elongated leaf	N	4.00	5.00
			P	0.20	0.30
			K	3.00	4.00
	Midseason (bulbing)	Newest fully elongated leaf	N	3.00	4.00
			P	0.20	0.30
			K	2.00	3.00
	Late season (postbulbing)	Newest fully elongated leaf	N	2.00	3.00
			P	0.20	0.30
			K	1.00	2.00
Lettuce	At heading	Leaves	N	1.50	3.00
			P	0.20	0.35
			K	2.50	5.00
	Nearly mature	Leaves	N	1.25	2.50
			P	0.15	0.30
			K	2.50	5.00

TABLE 6.24. TOTAL NUTRIENT CONCENTRATION FOR DIAGNOSIS OF THE NUTRIENT LEVEL OF VEGETABLE CROPS (Continued)

Crop	Time of Sampling	Plant Part	Nutrient	Nutrient Level (% Dry Weight)	
				Deficient	Sufficient
Onion	Early season	Tallest leaf	N	3.00	4.00
			P	0.10	0.20
			K	3.00	4.00
	Midseason	Tallest leaf	N	2.50	3.00
			P	0.10	0.20
			K	2.50	4.00
	Late season	Tallest leaf	N	2.00	2.50
			P	0.10	0.20
			K	2.00	3.00
Pepper, sweet	Full bloom	Blade and petiole	N	3.00	4.00
			P	0.15	0.25
			K	1.50	2.50
	Full bloom, fruit ¾ size	Blade and petiole	N	2.50	3.50
			P	0.12	0.20
			K	1.00	2.00

258

Crop	Growth stage	Plant part		Value 1	Value 2
Potato	Early, plants 12 in. tall	Petiole of 4th leaf from tip	N	2.50	3.50
			P	0.20	0.30
			K	9.00	11.00
	Midseason	Petiole of 4th leaf from tip	N	2.25	2.75
			P	0.10	0.20
			K	7.00	9.00
	Late, nearly mature	Petiole of 4th leaf from tip	N	1.50	2.25
			P	0.08	0.15
			K	4.00	6.00
	Early, plants 12 in. tall	Blade of 4th leaf from tip	N	4.00	6.00
			P	0.30	0.60
			K	3.50	5.00
	Midseason	Blade of 4th leaf from tip	N	3.00	5.00
			P	0.20	0.40
			K	2.50	3.50
	Late, nearly mature	Blade of 4th leaf from tip	N	2.00	4.00
			P	0.10	0.20
			K	1.50	2.50
Southern pea (cowpea)	Full bloom	Blade and petiole	N	2.00	3.50
			P	0.20	0.30
			K	1.00	2.00
Spinach	Midgrowth	Mature leaf blade and petiole	N	2.00	4.00
			P	0.20	0.40
			K	3.00	6.00
	At harvest	Mature leaf blade and petiole	N	1.50	3.00
			P	0.20	0.35
			K	2.00	5.00

TABLE 6.24. TOTAL NUTRIENT CONCENTRATION FOR DIAGNOSIS OF THE NUTRIENT LEVEL OF VEGETABLE CROPS (*Continued*)

Crop	Time of Sampling	Plant Part	Nutrient	Nutrient Level (% Dry Weight)	
				Deficient	Sufficient
Sweet corn	Tasseling	Sixth leaf from base of plant	N	2.75	3.50
			P	0.18	0.28
			K	1.75	2.25
	Silking	Leaf opposite first ear	N	1.50	2.00
			P	0.20	0.30
			K	1.00	2.00
Tomato (determinate)	Flowering	Leaf blade and petiole	N	2.50	3.50
			P	0.20	0.30
			K	1.50	2.50
	First ripe fruit	Leaf blade and petiole	N	1.50	2.50
			P	0.15	0.25
			K	1.00	2.00

Adapted from H. M. Reisenauer (ed.), *Soil and Plant Tissue Testing in California* (University of California Division of Agricultural Science Bulletin 1879, 1983).

TABLE 6.25. CRITICAL (DEFICIENCY) VALUES, ADEQUATE RANGES, HIGH VALUES, AND TOXICITY VALUES FOR PLANT NUTRIENT CONCENTRATION OF VEGETABLES

Crop	Plant Part[1]	Time of Sampling	Status	%						ppm					
				N	P	K	Ca	Mg	S	Fe	Mn	Zn	B	Cu	Mo
Bean, snap	MRM trifoliate leaf	Before bloom	Deficient	<3.0	0.25	2.0	0.8	0.20	0.20	25	20	20	15	5	—
			Adequate	3.0	0.25	2.0	0.8	0.20	0.40	25	20	20	15	5	0.4
			Range	4.0	0.45	3.0	1.5	0.45	0.40	200	100	40	40	10	1.0
			High	>4.1	0.46	3.1	1.6	0.45	0.40	200	100	40	40	10	—
			Toxic (>)	—	—	—	—	—	—	—	1,000	—	150	—	—
	MRM trifoliate leaf	First bloom	Deficient	<3.0	0.25	2.0	0.8	0.25	0.20	25	20	20	15	5	—
			Adequate	3.0	0.25	2.0	0.8	0.26	0.21	25	20	20	15	5	0.4
			Range	4.0	0.45	3.0	1.5	0.45	0.40	200	100	40	40	10	1.0
			High	>4.1	0.46	3.1	1.6	0.45	0.40	200	100	40	40	10	—
			Toxic (>)	—	—	—	—	—	—	—	1,000	—	150	—	—
	MRM trifoliate	Full bloom	Deficient	<2.5	0.20	1.5	0.8	0.25	0.20	25	20	20	15	5	—
			Adequate	2.5	0.20	1.6	0.8	0.26	0.21	25	20	20	15	5	0.4
			Range	4.0	0.40	2.5	1.5	0.45	0.40	200	100	40	40	10	1.0
			High	>4.1	0.41	2.5	1.6	0.45	0.40	200	100	40	40	10	—
			Toxic (>)	—	—	—	—	—	—	—	1,000	—	150	—	—
Beet, table	Leaf blades	5 weeks after seeding	Deficient	<3.0	0.22	2.0	1.5	0.25	—	40	30	15	30	5	0.05
			Adequate	3.0	0.25	2.0	1.5	0.25	0.60	40	30	15	30	5	0.20
			Range	5.0	0.40	6.0	2.0	1.00	0.80	200	200	30	80	10	0.60
			High	>5.0	0.40	6.0	2.0	1.00	—	—	—	—	80	10	—
			Toxic (>)	—	—	—	—	—	—	—	—	—	650	—	—
Beet, table	Leaf blades	9 weeks after seeding	Deficient	<2.5	0.20	1.7	1.5	0.30	—	—	—	15	30	5	0.05
			Adequate	2.6	0.20	1.7	1.5	0.30	0.60	—	70	15	60	5	0.60
			Range	4.0	0.30	4.0	3.0	1.00	0.80	—	200	30	80	10	—
			High	>4.0	0.30	4.0	3.0	1.00	—	—	—	—	80	10	—
			Toxic (>)	—	—	—	—	—	—	—	—	—	650	—	—

Crop	Plant Part[1]	Time of Sampling	Status	%						ppm					
				N	P	K	Ca	Mg	S	Fe	Mn	Zn	B	Cu	Mo
Broccoli	MRM leaf	Heading	Deficient	<3.0	0.30	1.1	0.8	0.23	0.20	40	20	25	20	3	0.04
			Adequate	3.0	0.30	1.5	1.2	0.23	—	40	25	45	30	5	0.04
			Range	4.5	0.50	4.0	2.5	0.40	—	300	150	95	50	10	0.16
			High	>4.5	0.50	4.0	2.5	0.40	—	300	150	100	100	10	—
Brussels sprouts	MRM leaf	At early sprouts	Deficient	<2.2	0.20	2.4	0.4	0.20	0.20	50	20	20	20	4	0.04
			Adequate	2.2	0.20	2.4	0.4	0.20	0.20	50	20	20	30	5	0.16
			Range	5.0	0.60	3.5	2.0	0.40	0.80	150	200	80	70	10	0.16
			High	>5.0	0.60	3.5	2.0	0.40	0.80	150	200	80	70	—	—
Cabbage	MRM leaf	5 weeks after transplanting	Deficient	<3.2	0.30	2.8	0.5	0.25	—	30	20	30	20	3	0.3
			Adequate	3.2	0.30	2.8	1.1	0.25	0.30	30	20	30	20	3	0.3
			Range	6.0	0.60	5.0	2.0	0.60	—	60	40	50	40	7	0.6
			High	>6.0	0.60	5.0	2.0	0.60	—	100	40	50	40	10	—
	MRM leaf	8 weeks after transplanting	Deficient	<3.0	0.30	2.0	0.5	0.20	—	30	20	30	20	3	0.3
			Adequate	3.0	0.30	2.0	1.5	0.25	0.30	30	20	30	20	3	0.3
			Range	6.0	0.60	4.0	2.0	0.60	—	60	40	50	40	7	0.6
			High	>6.0	0.60	4.0	2.0	0.60	—	100	40	50	40	10	—
	Wrapper leaf	Heads ½ grown	Deficient	<3.0	0.30	1.7	0.5	0.25	—	20	20	20	30	4	0.3
			Adequate	3.0	0.30	2.3	1.5	0.25	0.30	20	20	20	30	4	0.3
			Range	4.0	0.50	4.0	2.0	0.45	—	40	40	30	50	8	0.6
			High	>4.0	0.50	4.0	2.0	0.45	—	100	40	40	50	10	—
	Wrapper leaf	At harvest	Deficient	<1.8	0.26	1.2	0.5	0.25	—	20	20	20	30	4	0.3
			Adequate	1.8	0.26	1.5	1.5	0.25	0.30	20	20	20	30	4	0.3
			Range	3.0	0.40	3.0	2.0	0.45	—	40	40	30	50	8	0.6
			High	>3.0	0.40	3.0	2.0	0.45	—	100	40	40	50	10	—

Crop	Tissue	Stage	Status												
Cantaloupe	MRM leaf	12-in. vines	Deficient	<4.0	0.40	5.0	3.0	0.35	—	40	20	20	20	5	0.6
			Adequate	4.0	0.40	5.0	3.0	0.35	0.20	40	20	20	20	5	0.6
			Range	5.0	0.70	7.0	5.0	0.45	0.50	100	100	60	80	10	1.0
			High	>5.0	0.70	7.0	5.0	0.45	—	100	100	60	80	10	1.0
			Toxic (>)	—	—	—	—	—	—	—	900	—	150	—	—
	MRM leaf	Early fruit set	Deficient	<3.5	0.25	1.8	1.8	0.30	—	40	20	20	20	5	0.6
			Adequate	3.5	0.25	1.8	1.8	0.30	0.20	40	20	20	20	5	0.6
			Range	4.5	0.40	4.0	5.0	0.40	0.50	100	100	60	80	10	1.0
			High	>4.5	0.40	4.0	5.0	0.40	—	100	100	60	80	10	1.0
			Toxic (>)	—	—	—	—	—	—	—	900	—	150	—	—
Carrot	MRM leaf	60 days after seeding	Deficient	<1.8	0.20	2.0	1.0	0.15	—	30	30	20	20	4	—
			Adequate	1.8	0.20	2.0	2.0	0.20	—	30	30	20	20	4	—
			Range	2.5	0.40	4.0	3.5	0.50	—	60	60	60	40	10	—
			High	>2.5	0.40	4.0	3.5	0.50	—	60	60	60	40	10	—
	MRM leaf	Harvest	Deficient	<1.5	0.18	1.0	1.0	0.25	—	20	20	20	20	4	—
			Adequate	1.5	0.18	1.4	1.0	0.40	—	20	30	20	20	4	—
			Range	2.5	0.40	4.0	1.5	0.50	—	30	30	60	40	10	—
			High	>2.5	0.40	4.0	1.5	0.50	—	60	60	60	40	10	—
Cauliflower	MRM leaf	Buttoning	Deficient	<3.0	0.40	2.0	0.8	0.25	0.60	30	30	30	30	5	—
			Adequate	3.0	0.40	2.0	0.8	0.25	0.60	30	30	30	30	5	—
			Range	5.0	0.70	4.0	2.0	0.60	1.00	60	60	50	50	10	—
			High	>5.0	0.70	4.0	2.0	0.60	—	100	100	50	50	10	—
	MRM leaf	Heading	Deficient	<2.2	0.30	1.5	1.0	0.25	—	30	30	30	30	5	—
			Adequate	2.2	0.30	1.5	1.0	0.25	—	30	30	30	30	5	—
			Range	4.0	0.70	3.0	2.0	0.60	—	60	80	50	50	10	—
			High	>4.0	0.70	3.0	2.0	0.60	—	80	100	50	50	10	—
Celery	Outer petiole	6 weeks after transplanting	Deficient	<1.5	0.30	6.0	1.3	0.30	—	30	50	20	15	4	—
			Adequate	1.5	0.30	6.0	1.3	0.30	—	30	50	20	15	4	—
			Range	1.7	0.60	8.0	2.0	0.60	—	60	80	40	25	6	—
			High	>1.7	0.60	8.0	2.0	0.60	—	100	100	60	25		—
	Outer petiole	At maturity	Deficient	<1.5	0.30	5.0	1.3	0.30	—	20	5	20	20	1	—
			Adequate	1.5	0.30	5.0	1.3	0.30	—	20	5	20	20	1	—
			Range	1.7	0.60	7.0	2.0	0.60	—	30	10	40	40	3	—
			High	>1.7	0.60	7.0	2.0	0.60	—	100	20	60	40	3	—

TABLE 6.25. CRITICAL (DEFICIENCY) VALUES, ADEQUATE RANGES, HIGH VALUES, AND TOXICITY VALUES FOR PLANT NUTRIENT CONCENTRATION OF VEGETABLES (Continued)

Crop	Plant Part[1]	Time of Sampling	Status	%						ppm					
				N	P	K	Ca	Mg	S	Fe	Mn	Zn	B	Cu	Mo
Chinese cabbage (heading)	Oldest undamaged leaf	8-leaf stage	Deficient	<4.5	0.50	7.5	4.5	0.35	—	—	8	30	15	5	—
			Adequate	4.5	0.50	7.5	4.5	0.35	—	—	14	30	15	5	—
			Range	5.0	0.60	8.5	5.0	0.45	—	—	20	50	25	10	—
			High	>5.0	0.60	8.5	5.0	0.45	—	—	20	50	25	10	—
	Oldest undamaged leaf	At maturity	Deficient	<3.5	0.30	3.0	—	0.40	—	—	7	20	30	4	—
			Adequate	3.5	0.30	3.0	3.7	0.40	—	—	13	20	30	4	—
			Range	4.0	0.60	6.5	6.0	0.50	—	—	19	40	50	6	—
			High	>4.0	0.60	6.5	6.0	0.50	—	—	20	40	50	6	—
Collards	Tops	Young plants	Deficient	<4.0	0.30	3.0	1.0	0.40	—	40	40	25	25	5	—
			Adequate	4.0	0.30	3.0	1.0	0.40	—	40	40	25	25	5	—
			Range	5.0	0.60	5.0	2.0	1.00	—	100	100	50	50	10	—
			High	>5.0	0.60	5.0	2.0	1.00	—	100	100	50	50	10	—
	MRM leaf	Harvest	Deficient	<3.0	0.25	2.5	1.0	0.35	—	40	40	20	25	5	—
			Adequate	3.0	0.25	2.5	1.0	0.35	—	40	40	20	25	5	—
			Range	5.0	0.50	4.0	2.0	0.10	—	100	100	40	50	10	—
			High	>5.0	0.50	4.0	2.0	0.10	—	100	100	40	50	10	—
Cucumber	MRM leaf	Before bloom	Deficient	<3.5	0.30	1.6	2.0	0.58	0.30	40	30	20	20	5	0.2
			Adequate	3.5	0.30	1.6	2.0	0.58	0.30	40	30	20	20	5	0.3
			Range	6.0	0.60	3.0	4.0	0.70	0.80	100	100	50	60	20	1.0
			High	>6.0	0.60	3.0	4.0	0.70	0.80	100	100	50	60	20	2.0
	MRM leaf	Early bloom	Deficient	<2.5	0.25	1.6	1.3	0.30	0.30	40	30	20	20	5	0.2
			Adequate	2.5	0.25	1.6	1.3	0.30	0.30	40	30	20	20	5	0.3
			Range	5.0	0.60	3.0	3.5	0.60	0.80	100	100	50	60	20	1.0

Crop	Plant part	Growth stage	Status												
Eggplant	MRM leaf	Early fruit set	High	>5.0	0.60	3.0	3.5	0.60	0.80	100	100	50	60	20	2.0
			Toxic (>)	—	—	—	—	—	—	50	900	950	150	—	—
Endive	Oldest undamaged leaf	8-leaf stage	Deficient	<4.2	0.30	3.5	0.8	0.25	0.40	50	50	20	20	5	0.5
			Adequate	4.2	0.30	3.5	0.8	0.25	0.40	50	50	20	20	5	0.5
			Range	5.0	0.60	5.0	1.5	0.60	0.60	100	100	40	40	10	0.8
			High	>6.0	0.60	5.0	1.5	0.60	0.60	100	100	40	40	10	0.8
	Oldest undamaged leaf	Maturity	Deficient	<4.5	0.45	4.5	2.0	0.25	—	—	15	30	25	5	—
			Adequate	4.5	0.45	4.5	2.0	0.25	—	—	15	30	25	5	—
			Range	6.0	0.80	6.0	4.0	0.60	—	—	25	50	35	10	—
			High	>6.0	0.80	6.0	4.0	0.60	—	—	25	50	35	10	—
Escarole	Oldest undamaged leaf	8-leaf stage	Deficient	<3.5	0.40	4.0	1.8	0.30	—	—	15	20	30	5	—
			Adequate	3.5	0.40	4.0	1.8	0.30	—	—	15	20	30	5	—
			Range	3.5	0.60	6.0	3.0	0.40	—	—	20	40	40	10	—
			High	>4.2	0.60	6.0	3.0	0.40	—	—	20	40	40	10	—
	Oldest undamaged leaf	Maturity	Deficient	<4.2	0.45	5.7	1.7	0.25	—	—	15	30	20	4	—
			Adequate	4.2	0.45	5.7	1.7	0.25	—	—	15	30	20	4	—
			Range	5.0	0.60	6.5	2.2	0.35	—	—	25	50	30	6	—
			High	>5.0	0.60	6.5	2.2	0.35	—	—	25	50	30	6	—
Lettuce, Boston	Oldest undamaged leaf	8-leaf stage	Deficient	<3.0	0.35	5.5	2.0	0.25	—	—	15	20	30	4	—
			Adequate	3.0	0.35	5.5	2.0	0.25	—	—	15	20	30	4	—
			Range	4.5	0.45	6.5	3.0	0.35	—	—	25	50	45	6	—
			High	>4.5	0.45	6.5	3.0	0.35	—	—	25	50	45	6	—
	Oldest undamaged leaf	8-leaf stage	Deficient	<4.0	0.40	5.0	1.0	0.40	0.40	50	10	40	15	5	0.1
			Adequate	4.0	0.40	5.0	1.7	0.40	0.40	50	10	40	15	5	0.1
			Range	6.0	0.60	6.0	2.0	0.60	0.60	100	20	60	25	10	0.2
			High	>6.0	0.60	6.0	2.0	0.60	0.60	100	20	60	25	10	0.4
			Toxic (>)	—	—	—	—	—	—	—	250	—	100	—	—

TABLE 6.25. CRITICAL (DEFICIENCY) VALUES, ADEQUATE RANGES, HIGH VALUES, AND TOXICITY VALUES FOR PLANT NUTRIENT CONCENTRATION OF VEGETABLES (Continued)

Crop	Plant Part[1]	Time of Sampling	Status	% N	% P	% K	% Ca	% Mg	% S	ppm Fe	ppm Mn	ppm Zn	ppm B	ppm Cu	ppm Mo
	Oldest undamaged leaf	Maturity	Deficient	<3.0	0.35	5.0	1.0	0.30	—	50	10	20	15	5	0.1
			Adequate	3.0	0.35	5.0	1.7	0.30	—	50	10	20	15	5	0.1
			Range	4.0	0.45	6.0	2.0	0.60	—	100	20	40	25	10	0.2
			High	>4.0	0.45	6.0	2.0	0.60	—	100	20	40	25	10	0.4
			Toxic (>)	—	—	—	—	—	—	—	250	—	100	—	—
Lettuce, cos	Oldest undamaged leaf	8-leaf stage	Deficient	<4.0	0.50	4.0	1.7	0.30	—	40	10	40	20	5	—
			Adequate	4.0	0.50	4.0	1.7	0.30	—	40	10	40	20	5	—
			Range	5.0	0.60	6.0	2.0	1.70	—	100	20	60	40	10	—
			High	>5.0	0.60	6.0	2.0	1.70	—	100	20	60	40	10	—
	Oldest undamaged leaf	Maturity	Deficient	<3.0	0.40	4.0	1.7	0.30	—	20	10	20	20	5	—
			Adequate	3.0	0.40	4.0	1.7	0.30	—	20	10	20	20	5	—
			Range	4.0	0.60	6.0	2.0	0.70	—	50	20	40	40	10	—
			High	>4.0	0.60	6.0	2.0	0.70	—	50	20	40	40	10	—
Lettuce, crisphead	MRM	8-leaf stage	Deficient	<4.0	0.40	5.0	1.0	0.30	0.30	50	20	40	40	10	—
			Adequate	4.0	0.40	5.0	1.0	0.30	0.30	50	20	25	15	5	—
			Range	5.0	0.60	7.0	2.0	0.50	0.50	150	40	50	30	10	—
			High	>5.0	0.60	7.0	2.0	0.50	—	150	40	50	30	10	—
	Wrapper leaf	Heads ½ size	Deficient	<2.5	0.40	4.5	1.4	0.30	—	50	20	25	15	5	—
			Adequate	2.5	0.40	4.5	1.4	0.30	0.30	50	20	25	15	5	—
			Range	4.0	0.60	8.0	2.0	0.70	0.50	150	40	50	30	10	—
			High	>4.0	0.60	8.0	2.0	0.70	—	150	40	50	30	10	—
	Wrapper leaf	Maturity	Deficient	<2.0	0.25	2.5	1.4	0.30	—	50	20	25	15	5	—
			Adequate	2.0	0.25	2.5	1.4	0.30	0.30	50	20	25	15	5	—
			Range	3.0	0.50	5.0	2.0	0.70	0.50	150	40	50	30	10	—

Crop	Plant part	Growth stage	Status												
Lettuce, romaine	Oldest undamaged leaf	8-leaf stage	High	>3.0	0.50	5.0	2.0	0.70	—	150	40	50	30	10	—
	Oldest undamaged leaf	Maturity	Deficient	<5.0	0.35	5.0	2.0	0.25	—	—	15	20	30	5	—
			Adequate	5.0	0.35	5.0	2.0	0.25	—	—	15	20	30	5	—
			Range	6.0	0.80	6.0	3.0	0.35	—	—	25	50	45	10	0.1
			High	>6.0	0.80	6.0	3.0	0.35	—	—	25	50	45	10	0.1
Okra	MRM leaf	30 days after seeding	Deficient	<3.5	0.35	5.0	2.0	0.25	—	—	15	20	30	5	0.4
			Adequate	3.5	0.35	5.0	2.0	0.25	—	—	15	20	30	5	—
			Range	4.5	0.60	6.0	3.0	0.40	—	—	25	50	45	10	—
			High	>4.5	0.60	6.0	3.0	0.40	—	—	25	50	45	10	—
	MRM leaf	Prior to harvest	Deficient	<3.5	0.30	2.0	0.5	0.25	—	50	30	30	25	5	—
			Adequate	3.5	0.30	2.0	0.5	0.25	—	50	30	30	25	5	—
			Range	5.0	0.60	3.0	0.8	0.50	—	100	100	50	50	10	—
			High	>5.0	0.60	3.0	0.8	0.50	—	100	100	50	50	10	—
	MRM leaf	Prior to harvest	Deficient	<2.5	0.30	2.0	1.0	0.25	—	50	30	30	25	5	—
			Adequate	2.5	0.30	2.0	1.0	0.25	—	50	30	30	25	5	—
			Range	3.0	0.60	3.0	1.5	0.50	—	100	100	50	50	10	—
			High	>3.0	0.60	3.0	1.5	0.50	—	100	100	50	50	10	—
Onion, sweet	MRM leaf	Just prior to bulb initiation	Deficient	<2.0	0.20	1.5	0.6	0.15	0.20	—	10	15	10	5	—
			Adequate	2.0	0.20	1.5	0.6	0.15	0.20	—	10	15	10	5	—
			Range	3.0	0.50	3.0	0.8	0.30	0.60	—	20	20	25	10	—
			High	>3.0	0.50	3.0	0.8	0.30	0.60	—	20	20	25	10	—
			Toxic (>)	—	—	—	—	—	—	—	—	—	100	—	—
Pepper	MRM leaf	Prior to blossoming	Deficient	<4.0	0.30	5.0	0.9	0.35	0.30	30	30	25	20	5	—
			Adequate	4.0	0.30	5.0	0.9	0.35	0.30	30	30	25	20	5	—
			Range	5.0	0.50	6.0	1.5	0.60	0.60	150	100	80	50	10	—
			High	>5.0	0.50	6.0	1.5	0.60	0.60	150	100	80	50	10	—
			Toxic (>)	—	—	—	—	—	—	—	—	—	350	—	—
	MRM leaf	First blossoms open	Deficient	<3.0	0.30	2.5	0.9	0.30	0.30	30	30	25	20	5	—
			Adequate	3.0	0.30	2.5	0.9	0.30	0.30	30	30	25	20	5	—
			Range	5.0	0.50	5.0	1.5	0.50	0.60	150	100	80	50	10	—
			High	>5.0	0.50	5.0	1.5	0.50	0.60	150	100	80	50	10	—
			Toxic (>)	—	—	—	—	—	—	—	1,000	—	350	—	—

TABLE 6.25. CRITICAL (DEFICIENCY) VALUES, ADEQUATE RANGES, HIGH VALUES, AND TOXICITY VALUES FOR PLANT NUTRIENT CONCENTRATION OF VEGETABLES (Continued)

Crop	Plant Part[1]	Time of Sampling	Status	%						ppm					
				N	P	K	Ca	Mg	S	Fe	Mn	Zn	B	Cu	Mo
	MRM leaf	Early fruit set	Deficient	<2.9	0.25	2.5	1.0	0.30	0.30	30	30	25	20	5	—
			Adequate	2.9	0.25	2.5	1.0	0.30	0.30	30	30	25	20	5	—
			Range	4.0	0.40	4.0	1.5	0.40	0.40	150	100	80	50	10	—
			High	>4.0	0.40	4.0	1.5	0.40	0.40	150	100	80	50	10	—
			Toxic (>)	—	—	—	—	—	—	—	—	—	350	—	—
	MRM leaf	Early harvest	Deficient	<2.5	0.20	2.0	1.0	0.30	0.30	30	30	25	20	5	0.1
			Adequate	2.5	0.20	2.0	1.0	0.30	0.30	30	30	25	20	5	0.1
			Range	3.0	0.40	3.0	1.5	0.40	0.40	150	100	80	50	10	0.2
			High	>3.0	0.40	3.0	1.5	0.40	0.40	150	100	80	50	10	—
			Toxic (>)	—	—	—	—	—	—	—	—	—	350	—	—
Potato	MRM leaf	Plants 8–10 in. tall	Deficient	<3.0	0.20	3.5	0.6	0.30	0.25	40	30	30	20	5	0.1
			Adequate	3.0	0.20	3.5	0.6	0.30	0.25	40	30	30	20	5	0.1
			Range	6.0	0.80	6.0	2.0	0.60	0.50	150	60	60	60	10	0.2
			High	>6.0	0.80	6.0	2.0	0.60	0.50	150	60	60	60	10	—
	MRM leaf	First blossom	Deficient	<3.0	0.20	3.0	0.6	0.25	0.20	40	30	30	20	5	0.1
			Adequate	3.0	0.20	3.0	0.6	0.25	0.20	40	30	30	20	5	0.1
			Range	4.0	0.50	5.0	2.0	0.60	0.50	150	100	60	30	10	0.2
			High	>4.0	0.50	5.0	2.0	0.60	0.50	150	100	60	30	10	—
	MRM leaf	Tubers ½ grown	Deficient	<2.0	0.20	2.5	0.6	0.25	0.20	40	20	30	20	5	0.1
			Adequate	2.0	0.20	2.5	0.6	0.25	0.20	40	20	30	20	5	0.1
			Range	4.0	0.40	4.0	2.0	0.60	0.50	150	100	60	30	10	0.2
			High	>4.0	0.40	4.0	2.0	0.60	0.50	150	100	60	30	10	—
	MRM leaf	At tops-down	Deficient	<2.0	0.16	1.5	0.6	0.20	0.20	40	20	30	20	5	0.1
			Adequate	2.0	0.16	1.5	0.6	0.20	0.20	40	20	30	20	5	0.1

Crop	Tissue	Stage	Category												
Pumpkin	MRM leaf		Range	3.0	0.40	3.0	2.0	0.50	0.50	150	100	60	30	10	0.2
			High	>3.0	0.40	3.0	2.0	0.50	0.50	150	100	60	30	10	—
		5 weeks after seeding	Deficient	<3.0	0.30	2.3	0.9	0.35	0.20	40	40	20	25	5	0.3
			Adequate	3.0	0.30	2.3	0.9	0.35	0.20	40	40	20	25	5	0.3
			Range	6.0	0.50	4.0	1.5	0.60	0.40	100	100	50	40	10	0.5
			High	>6.0	0.50	4.0	1.5	0.60	0.40	100	100	50	40	10	—
	MRM leaf	8 weeks from seeding	Deficient	<3.0	0.25	2.0	0.9	0.30	0.20	40	40	20	20	5	0.3
			Adequate	3.0	0.25	2.0	0.9	0.30	0.20	40	40	20	20	5	0.3
			Range	4.0	0.40	3.0	1.5	0.50	0.40	100	100	50	40	10	0.5
			High	>4.0	0.40	3.0	1.5	0.50	0.40	100	100	50	40	10	—
Radish	MRM leaf	At harvest	Deficient	<3.0	0.25	1.5	1.0	0.30	—	30	20	30	15	3	0.1
			Adequate	3.0	0.25	1.5	1.0	0.30	—	30	20	30	15	3	0.1
			Range	4.5	0.40	3.0	2.0	0.50	—	50	40	50	30	10	2.0
			High	>4.5	0.40	3.0	2.0	0.50	—	50	40	50	30	10	2.0
			Toxic (>)	—	—	—	—	—	—	—	—	—	85	—	—
Southern pea	MRM leaf	Before bloom	Deficient	<3.5	0.30	2.0	1.0	0.30	—	30	30	20	15	5	—
			Adequate	3.5	0.30	2.0	1.0	0.30	—	30	30	20	15	5	—
			Range	5.0	0.80	4.0	1.5	0.50	—	100	100	40	25	10	—
			High	>5.0	0.80	4.0	1.5	0.50	—	100	100	40	25	10	—
	MRM leaf	First bloom	Deficient	<2.5	0.20	2.0	1.0	0.30	—	30	30	20	15	5	4.0
			Adequate	2.5	0.20	2.0	1.0	0.30	—	30	30	20	15	5	4.0
			Range	4.0	0.40	4.0	1.5	0.50	—	100	100	40	25	10	6.0
			High	>4.0	0.40	4.0	1.5	0.50	—	100	100	40	25	10	6.0
Spinach	MRM leaf	30 days after seeding	Deficient	<3.0	0.30	3.0	0.6	1.00	—	—	50	50	20	5	0.1
			Adequate	3.0	0.30	3.0	0.6	1.00	—	—	50	50	20	5	0.1
			Range	4.5	0.50	4.0	1.0	1.60	—	—	100	70	40	7	1.0
			High	>5.0	0.50	4.0	1.0	1.60	—	—	100	70	40	7	1.0
	MRM leaf	Harvest	Deficient	<3.0	0.25	2.5	0.6	1.00	—	—	30	50	20	5	0.1
			Adequate	3.0	0.25	2.5	0.6	1.00	—	—	30	50	20	5	0.1
			Range	4.0	0.50	3.5	1.0	1.60	—	—	50	70	40	7	1.0
			High	>4.0	0.50	4.0	1.0	1.60	—	—	80	70	40	7	1.0

TABLE 6.25. CRITICAL (DEFICIENCY) VALUES, ADEQUATE RANGES, HIGH VALUES, AND TOXICITY VALUES FOR PLANT NUTRIENT CONCENTRATION OF VEGETABLES (Continued)

Crop	Plant Part[1]	Time of Sampling	Status	%						ppm					
				N	P	K	Ca	Mg	S	Fe	Mn	Zn	B	Cu	Mo
Squash	MRM	Early fruit	Deficient	<3.0	0.25	2.0	1.0	0.30	0.20	40	40	20	25	5	0.3
			Adequate	3.0	0.25	2.0	1.0	0.30	0.20	40	40	20	25	5	0.3
			Range	5.0	0.50	3.0	2.0	0.50	0.50	100	100	50	40	20	0.5
			High	>5.0	0.50	3.0	2.0	0.50	0.50	100	100	50	40	20	0.5
Strawberry	MRM leaf	Transplants	Deficient	<2.8	0.25	1.5	0.3	0.30	—	50	30	25	25	5	—
			Adequate	2.8	0.25	1.5	0.3	0.30	—	50	30	25	25	5	—
			Range	3.5	0.40	3.0	1.5	0.60	—	100	100	40	40	10	—
			High	>3.5	0.40	3.0	1.5	0.60	—	100	100	40	40	10	—
	MRM leaf	Initial flower	Deficient	<3.0	0.20	1.5	0.4	0.25	—	50	30	20	20	5	—
			Adequate	3.0	0.20	1.5	0.4	0.25	—	50	30	20	20	5	—
			Range	4.0	0.40	3.0	1.5	0.50	—	100	100	40	40	10	—
			High	>4.0	0.40	3.0	1.5	0.50	—	100	100	40	40	10	—
	MRM leaf	Initial harvest	Deficient	<3.0	0.20	1.5	0.4	0.25	—	50	30	20	20	5	—
			Adequate	3.0	0.20	1.5	0.4	0.25	—	50	30	20	20	5	—
			Range	3.5	0.40	2.5	1.5	0.50	—	100	100	40	40	10	—
			High	>3.5	0.40	2.5	1.5	0.50	—	100	100	40	40	10	—
			Toxic (>)	—	—	—	—	—	—	—	800	—	—	—	—
	MRM leaf	Midseason	Deficient	<2.8	0.20	1.1	0.4	0.20	0.8	50	25	20	20	5	0.5
			Adequate	2.8	0.20	1.1	0.4	0.20	0.8	50	25	20	20	5	0.5
			Range	3.0	0.40	2.5	1.5	0.40	1.0	100	100	40	40	10	0.8
			High	>3.0	0.40	2.5	1.5	0.40	1.0	100	100	40	40	10	0.8
			Toxic (>)	—	—	—	—	—	—	—	800	—	—	—	—
	MRM leaf	End of season	Deficient	<2.5	0.20	1.1	0.4	0.20	—	50	25	20	20	5	—
			Adequate	2.5	0.20	1.1	0.4	0.20	—	50	25	20	20	5	—

Crop	Tissue	Stage	Category												
Sweet corn	Whole seedlings	3-leaf stage	Range	3.0	0.30	2.0	1.5	0.40	—	100	100	40	40	10	—
			High	>3.0	0.30	2.0	1.5	0.40	—	100	100	40	40	10	—
			Deficient	<3.0	0.35	2.5	0.6	0.25	0.4	50	40	30	10	5	0.1
			Adequate	3.0	0.35	2.5	0.6	0.25	0.4	50	40	30	10	5	0.1
			Range	4.0	0.50	4.0	0.8	0.50	0.6	100	100	40	30	10	0.2
			High	>4.0	0.50	4.0	0.8	0.50	0.6	100	100	40	30	10	0.2
			Toxic (>)	—	—	—	—	—	—	—	—	—	—	—	—
	Whole seedlings	6-leaf stage	Deficient	<3.0	0.25	2.5	0.5	0.25	0.4	50	40	30	10	5	0.1
			Adequate	3.0	0.25	2.5	0.5	0.25	0.4	50	40	30	10	5	0.1
			Range	4.0	0.50	4.0	0.8	0.50	0.6	100	100	40	30	10	0.2
			High	>4.0	0.50	4.0	0.8	0.50	0.6	100	100	40	30	10	0.2
			Toxic (>)	—	—	—	—	—	—	—	—	—	—	—	—
	MRM leaf	30 in. tall	Deficient	<2.5	0.20	2.5	0.5	0.20	0.2	40	40	25	10	4	0.1
			Adequate	2.5	0.20	2.5	0.5	0.20	0.2	40	40	25	10	4	0.1
			Range	4.0	0.40	4.0	0.8	0.40	0.4	100	100	40	30	10	0.2
			High	>4.0	0.40	4.0	0.8	0.40	0.4	100	100	40	30	10	0.2
			Toxic (>)	—	—	—	—	—	—	—	—	—	—	—	—
	MRM leaf	Just prior to tassel	Deficient	<2.5	0.20	2.0	0.3	0.15	0.2	30	30	20	10	4	0.1
			Adequate	2.5	0.20	2.0	0.3	0.15	0.2	30	30	20	10	4	0.1
			Range	4.0	0.40	3.5	0.6	0.40	0.4	100	100	40	20	10	0.2
			High	>4.0	0.40	3.5	0.6	0.40	0.4	100	100	40	20	10	0.2
			Toxic (>)	—	—	—	—	—	—	—	—	—	—	—	—
	Ear leaf	Tasseling	Deficient	<1.5	0.20	1.2	0.3	0.15	0.20	30	20	20	10	4	0.1
			Adequate	1.5	0.20	1.2	0.3	0.15	0.20	30	20	20	10	4	0.1
			Range	2.5	0.40	2.0	0.6	0.40	0.40	100	100	40	20	10	0.2
			High	>2.5	0.40	2.0	0.6	0.40	0.40	100	100	40	20	10	0.2
			Toxic (>)	—	—	—	—	—	—	—	—	—	—	—	—
Sweet potato	MRM leaf	Early vining	Deficient	<4.0	0.30	2.5	0.8	0.40	0.20	40	40	25	20	5	—
			Adequate	4.0	0.30	2.5	0.8	0.40	0.20	40	40	25	20	5	—
			Range	5.0	0.50	4.0	1.6	0.80	0.60	100	100	50	50	10	—
			High	>5.0	0.50	4.0	1.6	0.80	0.60	100	100	50	50	10	—

TABLE 6.25. CRITICAL (DEFICIENCY) VALUES, ADEQUATE RANGES, HIGH VALUES, AND TOXICITY VALUES FOR PLANT NUTRIENT CONCENTRATION OF VEGETABLES (Continued)

Crop	Plant Part[1]	Time of Sampling	Status	%						ppm					
				N	P	K	Ca	Mg	S	Fe	Mn	Zn	B	Cu	Mo
	MRM leaf	Midseason before root enlargement	Deficient	<3.0	0.20	2.0	0.8	0.25	0.20	40	40	25	25	5	—
			Adequate	3.0	0.20	2.0	0.8	0.25	0.20	40	40	25	25	5	—
			Range	4.0	0.30	4.0	1.8	0.50	0.40	100	100	40	40	10	—
			High	>4.0	0.30	4.0	1.8	0.50	0.40	100	100	40	40	10	—
	MRM leaf	Root enlargement	Deficient	<3.0	0.20	2.0	0.8	0.25	0.20	40	40	25	20	5	—
			Adequate	3.0	0.20	2.0	0.8	0.25	0.20	40	40	25	20	5	—
			Range	4.0	0.30	4.0	1.6	0.50	0.60	100	100	50	50	10	—
			High	>4.0	0.30	4.0	1.6	0.50	0.60	100	100	50	50	10	—
	MRM leaf	Just before harvest	Deficient	<2.8	0.20	2.0	0.8	0.25	0.20	40	40	25	20	5	—
			Adequate	2.8	0.20	2.0	0.8	0.25	0.20	40	40	25	20	5	—
			Range	3.5	0.30	4.0	1.6	0.50	0.60	100	100	50	50	10	—
			High	>3.5	0.30	4.0	1.6	0.50	0.60	100	100	50	50	10	—
Tomato	MRM leaf	5-leaf stage	Deficient	<3.0	0.30	3.0	1.0	0.30	0.30	40	30	25	20	5	0.2
			Adequate	3.0	0.30	3.0	1.0	0.30	0.30	40	30	25	20	5	0.2
			Range	5.0	0.60	5.0	2.0	0.50	0.80	100	100	40	40	15	0.6
			High	>5.0	0.60	5.0	2.0	0.50	0.80	100	100	40	40	15	0.6
	MRM leaf	First flower	Deficient	<2.8	0.20	2.5	1.00	0.30	0.30	40	30	25	20	5	0.2
			Adequate	2.8	0.20	2.5	1.00	0.30	0.30	40	30	25	20	5	0.2
			Range	4.0	0.40	4.0	2.00	0.50	0.80	100	100	40	40	15	0.6
			High	>4.0	0.40	4.0	2.00	0.50	0.80	100	100	40	40	15	0.6
			Toxic (>)	—	—	—	—	—	—	—	1,500	300	250	—	—
	MRM leaf	Early fruit set	Deficient	<2.5	0.20	1.0	1.0	0.25	0.30	40	30	20	20	5	0.2
			Adequate	2.5	0.20	1.0	1.0	0.25	0.30	40	30	20	20	5	0.2
			Range	4.0	0.40	2.0	2.0	0.50	0.60	100	100	40	40	10	0.6

Crop	Leaf	Growth stage	Status												
			High	>4.0	0.40	4.0	2.0	0.50	0.60	100	100	40	40	10	0.6
			Toxic (>)	—	—	—	—	—	—	—	—	—	250	—	—
	MRM leaf	First ripe fruit	Deficient	<2.0	0.20	2.0	1.0	0.25	0.30	40	30	20	20	5	0.2
			Adequate	2.0	0.20	2.0	1.0	0.25	0.30	40	30	20	20	5	0.2
			Range	3.5	0.40	4.0	2.0	0.50	0.60	100	100	40	40	10	0.6
			High	>3.5	0.40	4.0	2.0	0.50	0.60	100	100	40	40	10	0.6
	MRM leaf	During harvest period	Deficient	<2.0	0.20	1.5	1.0	0.25	0.30	40	30	20	20	5	0.2
			Adequate	2.0	0.20	1.5	1.0	0.25	0.30	40	30	20	20	5	0.2
			Range	3.0	0.40	2.5	2.0	0.50	0.60	100	100	40	40	10	0.6
			High	>3.0	0.40	2.5	2.0	0.50	0.60	100	100	40	40	10	0.6
Turnip greens	MRM leaf	Hypocotyl 1-in. diameter	Deficient	<3.0	0.25	2.5	0.8	0.25	0.20	30	30	20	20	5	—
			Adequate	3.0	0.25	2.5	0.8	0.25	0.20	30	30	20	20	5	—
			Range	5.0	0.80	4.0	1.5	0.60	0.60	100	100	40	40	10	—
			High	>5.0	0.80	4.0	1.5	0.60	0.60	100	100	40	40	10	—
Watermelon	MRM leaf	Layby (last cultivation)	Deficient	<3.0	0.25	3.0	1.0	0.25	0.20	30	20	20	20	5	—
			Adequate	3.0	0.25	3.0	1.0	0.25	0.20	30	20	20	20	5	—
			Range	4.0	0.50	4.0	2.0	0.50	0.40	100	100	40	40	10	—
			High	>4.0	0.50	4.0	2.0	0.50	0.40	100	100	40	40	10	—
			Toxic (>)	—	—	—	—	—	—	—	800	—	—	—	—
	MRM leaf	First flower	Deficient	<2.5	0.25	2.7	1.0	0.25	0.20	30	20	20	20	5	—
			Adequate	2.5	0.25	2.7	1.0	0.25	0.20	30	20	20	20	5	—
			Range	3.5	0.50	3.5	2.0	0.50	0.40	100	100	40	40	10	—
			High	>3.5	0.50	3.5	2.0	0.50	0.40	100	100	40	40	10	—
	MRM leaf	First fruit	Deficient	<2.0	0.25	2.3	1.0	0.25	0.20	30	20	20	20	5	—
			Adequate	2.0	0.25	2.3	1.0	0.25	0.20	30	20	20	20	5	—
			Range	3.0	0.50	3.5	2.0	0.50	0.40	100	100	40	40	10	—
			High	>3.0	0.50	3.5	2.0	0.50	0.40	100	100	40	40	10	—
	MRM leaf	Harvest period	Deficient	<2.0	0.25	2.0	1.0	0.25	0.20	30	20	20	20	3	—
			Adequate	2.0	0.25	2.0	1.0	0.25	0.20	30	20	20	20	3	—
			Range	3.0	0.50	3.0	2.0	0.50	0.40	100	100	40	40	10	—
			High	>3.0	0.50	3.0	2.0	0.50	0.40	100	100	40	40	10	—

Adapted from G. Hochmuth, D. Maynard, C. Vavrina, E. Hanlon, and E. Simonne, *Plant Tissue Analysis and Interpretation for Vegetable Crops in Florida* (Florida Cooperative Extension Service, 2018), https://edis.ifas.ufl.edu/publication/EP081.

[1] *MRM leaf* is the most recently matured whole leaf blade plus petiole.

TABLE 6.26. UNIVERSITY OF FLORIDA GUIDELINES FOR LEAF PETIOLE FRESH SAP NITRATE-NITROGEN AND POTASSIUM TESTING

Crop	Development Stage/Time	Fresh Petiole Sap Concentration (ppm)	
		NO_3–N	K
Eggplant	First fruit (2 in. long)	1,200–1,600	4,500–5,000
	First harvest	1,000–1,200	4,000–4,500
	Midharvest	800–1,000	3,500–4,000
Pepper	First flower buds	1,400–1,600	3,200–3,500
	First open flowers	1,400–1,600	3,000–3,200
	Fruits half-grown	1,200–1,400	3,000–3,200
	First harvest	800–1,000	2,400–3,000
	Second harvest	500–800	2,000–2,400
Potato	Plants 8 in. tall	1,200–1,400	4,500–5,000
	First open flowers	1,000–1,400	4,500–5,000
	50% flowers open	1,000–1,200	4,000–4,500
	100% flowers open	900–1,200	3,500–4,000
	Tops falling over	600–900	2,500–3,000
Strawberry[1]	November	800–900	3,000–3,500
	December	600–800	3,000–3,500
	January	600–800	2,500–3,000
	February	300–500	2,000–2,500
	March	200–500	1,800–2,500
	April	200–500	1,500–2,000

Crop	Growth stage		
Tomato	First buds	1,000–1,200	3,500–4,000
	First open flowers	600–800	3,500–4,000
	Fruits 1-in. diameter	400–600	3,000–3,500
	Fruits 2-in. diameter	400–600	3,000–3,500
	First harvest	300–400	2,500–3,000
	Second harvest	200–400	2,000–2,500
Watermelon—Seedless Cultivars	Vines six inches in length	1,200–1,500	4,000–5,000
	Fruits two inches in length	900–1000	4,000–5,000
	Fruits one-half mature	600–800	3,500–4,000
	At first harvest	400–600	3,000–3,500

Adapted from G. Hochmuth and Robert Hochmuth, "Plant Petiole Sap-testing Guide for Vegetable Crops," (Florida Cooperative Extension Service Circular 1144, 2022), https://edis.ifas.ufl.edu/publication/CV004.

[1] Annual hill production system.

Analyses for *total* amounts of nutrients in the soil are of limited value in predicting fertilizer needs. Consequently, various methods and extractants (soil tests) have been developed to estimate available soil nutrients and to serve as a basis for predicting fertilizer needs. Proper interpretation of the results of soil analyses is essential in recommending fertilizer needs. Soil testing procedure and extractants must be correlated with crop response to be of value for predicting crop need for fertilizer (Table 6.27). One soil test procedure developed for one soil and crop condition in one area of the country may not apply in another area. Growers should consult their crop and soils experts about soil testing procedures appropriate for their growing area.

Some references for soil testing:

T. K. Hartz, *Soil Testing for Nutrient Availability: Procedures and Interpretation for California Vegetable Crop Production*, https://vric.ucdavis. edu/pdf/fertilization/fertilization_Soiltestingfornutrientavailability2007.pdf (accessed 25 June 2022).

G. Hochmuth, R. Mylavarapu, and E. Hanlon, *Soil Testing for Plant-available Nutrients—What Is It and Why Do We Use It?* (University of Florida IFAS Extension, 2017), https://edis.ifas.ufl.edu/pdf/SS/SS62100.pdf (accessed 25 June 2022).

G. Hochmuth and E. Hanlon, *Principles of Sound Fertilizer Recommendations* (University of Florida IFAS Extension, 2019), https:// edis.ifas.ufl.edu/publication/ss527 (accessed 25 June 2022).

G. Hochmuth, R. Mylavarapu, and E. Hanlon, *Developing a Soil Test Extractant: The Correlation and Calibration Processes* (2017), https://edis. ifas.ufl.edu/publication/SS622 (accessed 25 June 2022).

DETERMINING THE KIND AND QUANTITY OF FERTILIZER TO USE

Many states issue suggested rates of application of fertilizers for specific vegetables. These recommendations are sometimes made according to the type of soil—that is, light or heavy soils, sands, loams, clays, peats, and mucks. Other factors often used in establishing these rates are whether manure or soil-improving crops are employed and whether an optimum moisture supply can be maintained. The nutrient requirements of each crop must be considered, as must the past fertilizer and cropping history. The season of the year affects nutrient availability. Broad recommendations are at best only a point from which to make adjustments to suit individual conditions. Each field may require a different fertilizer program for the same vegetable.

Calibrated soil testing can provide an estimate of the concentration of essential elements that will be available to the crop from the soil during the season and predict the amount of fertilizer needed, in addition to the soil nutrients, to produce a crop (Tables 6.27–6.31). Various extraction solutions are used by soil testing labs around the country to estimate the nutrient-supplying capacity of the soil, and not all solutions are calibrated for all soils. Therefore, growers must exercise care in selecting a lab to analyze soil samples, using only those labs that employ analytical procedures calibrated with yield response in specific soil types and growing regions. Even though several labs might differ in lab procedures, if all procedures are calibrated, then fertilizer recommendations should be similar among labs. If unclear about specific soil testing practices, growers should consult their Cooperative Extension Service and the specific analytical lab.

Phosphorus. This element is not very mobile in most agricultural soils. Phosphorus is fixed in soils with basic reactions (high pH) and large quantities of calcium, or in acidic soils containing aluminum or iron. Even though phosphorus can be fixed, if a calibrated soil test predicts no response to phosphorus fertilization, then growers need not add large amounts of phosphorus because enough phosphorus will be made available to the crop during the growing season, even though the soil has a high phosphorus-fixing capacity. Sometimes crops might respond to small amounts of starter phosphorus supplied to high phosphorus testing soils in cold planting seasons.

Potassium. Although not generally considered a mobile element in soils, potassium can leach in coarse, sandy soils. Clay soils and loamy soils

TABLE 6.27. PREDICTED RESPONSES OF CROPS TO RELATIVE AMOUNTS OF EXTRACTED PLANT NUTRIENTS BY SOIL TEST

Soil Test Interpretation	Predicted Crop Response
Very high	No crop response predicted to fertilization with a particular element.
High	No crop response predicted to fertilization with a particular element.
Medium	75–100% maximum expected yield predicted without fertilization.
Low	50–75% maximum expected yield predicted without fertilization.
Very low	25–50% maximum expected yield predicted without fertilization.

277

TABLE 6.28. INTERPRETATION OF SOIL TEST RESULTS FOR PHOSPHORUS BY THE OLSEN BICARBONATE EXTRACTION METHOD, FOR POTASSIUM AND MAGNESIUM BY THE AMMONIUM ACETATE EXTRACTION METHOD, AND FOR ZINC BY THE DPTA EXTRACTION METHOD

Nutrient Need	Extractable Amount in Soil (ppm)			
	Phosphorus[1] (PO$_4$–P)	Potassium[2] (K)	Magnesium[2] (Mg)	Zinc[3] (Zn)
Deficient levels for most vegetables	0–10	0–60	0–25	0–0.3
Deficient for susceptible vegetables	10–20	60–120	25–50	0.3–0.6
A few susceptible crops may respond	20–40	120–200	50–100	0.6–1.0
No crop response	Above 40	Above 200	Above 100	Above 1.0
Levels are excessive and could cause problems	Above 150	Above 2,000	Above 1,000	Above 3.0

Adapted from H. M. Reisenauer (ed.), *Soil and Plant Tissue Testing in California* (University of California Division of Agricultural Science Bulletin 1879, 1978).

[1] Olsen (0.5M, pH 8.5) sodium bicarbonate extractant.
[2] Exchangeable with 1N ammonium acetate extractant.
[3] DTPA (diethyienetriaminepentaacetic acid) extractable Zn.

TABLE 6.29. INTERPRETATION OF SOIL TEST RESULTS OBTAINED BY THE MEHLICH-1 DOUBLE ACID (0.05N HCL, 0.025N H$_2$SO$_4$) AND MEHLICH-3 SOIL EXTRACTANTS FOR THE MID-ATLANTIC REGION

Interpretations of soil tests by Mehlich-1 and Mehlich-3 in Mid-Atlantic states. Soil test indexes in table are in "lb/acre" and can be divided by 2 to get parts per million (ppm).

Nutrient Application Soil Test Category	Phosphorus (P)	Potassium (K)	Magnesium (Mg)	Calcium (Ca)[1]
Mehlich 3 soil test value (lb/acre)[2,3]				
Deficient (very low)	0–24	0–40	0–45	0–615
Deficient (low)	25–45	41–81	46–83	616–1007
Deficient (medium)	46–71	82–145	84–143	1008–1400
Optimum (high)	72–137	146–277	144–295	1401–1790
Exceeds crop needs (very high)	138+	278+	296+	1791+
Mehlich 1 soil test value (lb/acre)[2]				
Below optimum (very low)	0–3	0–15	0–24	0–240
Below optimum (low)	4–11	16–75	25–72	241–720
Below optimum (medium)	12–35	76–175	73–144	721–1440
Optimum (high)	36–110	176–310	145–216	1441–2160
Above optimum (very high)	111+	311+	217+	2161+

Adapted from the *2020/2021 Mid-Atlantic Commercial Vegetable Production Recommendations*.

[1] Calcium values are for sandy loam soils. Multiply the calcium values in the table above by 0.625 to use for loamy sand soils; by 1.25 for loam soils; by 1.5 for silt loam soils, and by 1.75 for clay loam soils.

[2] Values are reported in elemental forms. Divide lb/acre by 2 to get ppm.

[3] Soil tests that are based on Bray-1 extractable P and neutral, 1N ammonium acetate extractable, K, Ca, and Mg are very similar to the Mehlich-3 extractable concentrations of these nutrients.

TABLE 6.30. INTERPRETDATION OF THE MEHLICH-III EXTRACTANT USED BY THE UNIVERSITY OF FLORIDA FOR P, K, and Mg ON SANDY MINERAL SOILS

Element	Low	Medium	High
P	≤25	26–45	>45
K	≤35	36–60	>60
Mg	≤20	21–40	>40

Values are in ppm. Adapted from Mylavarapu et al., *Extraction of Soil Nutrients Using Mehlich – 3 Reagent for Acid-Mineral Soil of Florida* (University of Florida IFAS Extension, 2020), https://edis.ifas.ufl.edu/publication/SS620 (accessed 25 June 2022).

often contain adequate amounts of available potassium and may not need fertilization with potassium. Coarse, sandy soils usually test medium or low in extractable potassium, and crops growing on these soils respond to potassium fertilization.

Nitrogen. Most soil testing labs have no calibrated soil test for nitrogen because nitrogen is highly mobile in most soils and predicting a crop's response to nitrogen fertilization from a soil test is risky. However, some labs do predict the nitrogen-supplying capacity of a soil from a determination of soil organic matter. Estimates vary from 20 to 40 lb nitrogen made available during the season for each percent soil organic matter. Another soil nitrogen estimation procedure used by some labs is the pre-sidedress soil nitrate test. This test predicts the likelihood of need for sidedressed nitrogen during the season, especially for manured soils, but is relatively insensitive for predicting exact amounts of sidedress nitrogen.

PRE-SIDEDRESS NITROGEN TEST FOR SWEET CORN

The Pre-sidedress Nitrate Test was developed to aid farmers in the prediction of nitrogen needs by corn at the time when sidedress applications are normally made (Table 6.32). This test takes into consideration nitrogen released from organic nutrient sources (such as manure, compost, cover crops, and soil organic matter) in addition to nitrogen fertilizer.

Sampling Procedure for Nitrogen Soil Test

1. Sample soil when corn is 8–12 in. tall.
2. Collect 15–20 soil cores per field to a depth of 12 in., if possible. If not, sample as deeply as possible. Avoid areas where starter fertilizer bands were applied, areas where manure was stacked, and areas where starter fertilizer applications were unusually heavy or light.

280

Vegetable	Nutrient[1]	Vegetable Yield Response to Fertilizer Application	
		Likely (Soil ppm Less Than)	Not Likely (Soil ppm More Than)
Lettuce	P	15	25
	K	50	80
	Zn	0.5	1.0
Cantaloupe	P	8	12
	K	80	100
	Zn	0.4	0.6
Onion	P	8	12
	K	80	100
	Zn	0.5	1.0
Potato (mineral soils)	P	12	25
	K	100	150
	Zn	0.3	0.7
Tomato	P	8	12
	K	100	150
	Zn	0.3	0.7
Warm-season vegetables	P	8	12
	K	50	70
	Zn	0.2	0.5
Cool-season vegetables	P	20	30
	K	50	80
	Zn	0.5	1.0

Adapted from *Soil and Plant Tissue Testing in California* (University of California Division Agricultural Science Bulletin 1879, 1983). Updated 1996, personal communication, T. K. Hartz (University of California—Davis).
[1] Soil extracts: PO_4–P: 0.5M pH 8.5 sodium bicarbonate ($NaHCO_3$); K: 1.0M ammonium acetate (NH_4OAc); Zn: 0.005M diethyienetriaminepentaaacetic acid (DTPA).

TABLE 6.32. PSNT SWEET CORN NITROGEN TEST

Soil Test Results (ppm NO_3–N)	Recommended Nitrogen Sidedressing (lb Actual N/acre)
0–10	130
11–20	100
21–25	50
25+	0

Adapted from Using the PSNT test to manage N fertilization of vegetable crops by V. Grubinger (University of Vermont Cooperative Extension Service), http://www.uvm.edu/vtvegandberry/factsheets/PSNT.html (updated 2003).

3. Combine the cores for each field and mix completely. Take a subsample of approximately 1 cup and dry it immediately. Soil can be dried in an oven at about 200°F. Samples can also be air dried if spread out thinly on a nonabsorbent material in a dry, well-ventilated area. A fan reduces drying time. Do not put wet samples on absorbent material because it will absorb some nitrate with the moisture. The longer the delay in drying the sample, the less accurate the results will be.

Calculating Fertilizer Rates for Plastic-mulched Cropping Systems

Growers often ask if the amount of fertilizer per acre needs to change when the amount of crop planted per acre changes. Tables 6.33a and 6.33b were made to help growers calculate the amount of fertilizer needed if their production system differs from the traditional planting system. To determine the correct fertilization rate in lb nutrient per 100 linear bed feet (LBF), choose the crop and its typical bed spacing from Table 33a. Locate that typical bed spacing value in the Table 33b. Then locate the desired value for recommended fertilizer rate. Read down the column under recommended fertilizer rate until you reach the value in the row for your typical bed spacing.

FERTILIZER RATES FOR SELECTED REGIONS IN THE UNITED STATES

Each state's Extension Service publishes fertilizer recommendations for vegetables crop production in that state. The recommendations result from on-going research with vegetables and are updated regularly. Tables 6.34–6.36 present fertilizer recommendations for several major vegetable producing areas. Similar tables are available from corresponding recommendations in all states.

TABLE 6.33a. CONVERSION OF FERTILIZER RATES FROM APPLICATION ON A PER-ACRE BASIS TO RATES BASED ON LINEAR BED FEET FOR FULL-BED MULCHED CROPS

Typical Bed Spacing for Mulched Vegetables Grown in Florida:

Vegetable	Typical Spacing (ft)[1]	Rows of Plants Per Bed	Vegetable	Typical Spacing (ft)	Rows of Plants Per Bed
Broccoli	6	2	Lettuce	4	2
Cabbage	6	2	Pepper	6	2
Cantaloupe	5	1	Summer squash	6	2
Cauliflower	6	2	Strawberry	4	2
Cucumber	6	2	Tomato	6	1
Eggplant	6	1	Watermelon	8	1

[1] Spacing between the centers of two adjacent beds.

TABLE 6.33b. CONVERSION OF FERTILIZER RATES FROM APPLICATION ON A PER-ACRE BASIS TO RATES BASED ON LINEAR BED FEET FOR FULL-BED MULCHED CROPS

Typical Bed Spacing (ft)	Recommended Fertilizer (N, P_2O_5, or K_2O) (lb/acre)								
	20	40	60	80	100	120	140	160	180
	Resulting Fertilizer Rate (N, P_2O_5, or K_2O) (lb/100 LBF)								
3	0.14	0.28	0.41	0.55	0.69	0.83	0.96	1.10	1.24
4	0.18	0.37	0.55	0.73	0.92	1.10	1.29	1.47	1.65
5	0.23	0.46	0.69	0.92	1.15	1.38	1.61	1.84	2.07
6	0.28	0.55	0.83	1.10	1.38	1.65	1.93	2.20	2.48
8	0.37	0.73	1.10	1.47	1.84	2.20	2.57	2.94	3.31

G. Hochmuth and E. Hanlon, *Calculating Recommended Fertilizer Rates for Vegetables Grown in Raised-Bed Mulched Cultural Systems* (2018), https://edis.ifas.ufl.edu/publication/ss516. (accessed 6/25/22)

TABLE 6.34. RATES OF FERTILIZERS RECOMMENDED FOR VEGETABLE CROPS IN MID-ATLANTIC (DELAWARE, MARYLAND, NEW JERSEY, PENNSYLVANIA, AND VIRGINIA) STATES BASED ON SOIL ANALYSES[1]

Crop	N (lb/acre)	Soil Phosphorus Test Level				Soil Potassium Test Level			
		P_2O_5 (lb/acre)				K_2O (lb/acre)			
		Low	Med	High (Opt)	Very High	Low	Med	High (Opt)	Very High
Asparagus, cutting beds	75–100	200	150	100	0	300	225	150	0
Snap bean	40–80	80	60	40	0	80	60	40	0
Lima bean	60–90	100	60	20	0	140	100	60	0
Beet	75–100	150	100	50	0	150	100	50	0
Carrot	50–80	150	100	50	0	150	100	50	0
Celery	150–175	250	150	100	0	250	150	100	0
Cole crops, broccoli	150	200	100	50	0	200	100	50	0
Cole crops, Brussels sprouts, cabbage, cauliflower	100–150	200	100	50	0	200	100	50	0
Cole crops, kale, collards	100–200	200	100	50	0	200	100	50	0
Cole crops, kohlrabi	25–50	0	0	0	0	0	0	0	0
Cucumber	80–150	150	100	50	0	200	150	100	0
Eggplant	125–150	250	150	100	0	250	150	100	0
Garlic	125	150	150	150	0	150	150	150	0
Greens, Asian, mustard, turnip	50–170	150	100	50	0	150	100	50	0

TABLE 6.34. RATES OF FERTILIZERS RECOMMENDED FOR VEGETABLE CROPS IN MID-ATLANTIC (DELAWARE, MARYLAND, NEW JERSEY, PENNSYLVANIA, AND VIRGINIA) STATES BASED ON SOIL ANALYSES[1] (Continued)

Crop	N (lb/acre)	Soil Phosphorus Test Level				Soil Potassium Test Level			
		Low	Med	High (Opt)	Very High	Low	Med	High (Opt)	Very High
		P_2O_5 (lb/acre)				K_2O (lb/acre)			
Horseradish	150–200	200	150	100	0	200	150	100	0
Leeks	100–125	200	150	100	0	200	150	100	0
Lettuce—leaf lettuce, endive, escarole	100–125	200	150	100	0	200	150	100	50
Lettuce—iceberg	60–80	200	150	100	0	200	150	100	0
Muskmelon and mixed melons	75–150	150	100	50	0	200	150	100	0
Okra	100–150	250	150	100	0	250	150	100	0
Onion, bulb	75–100	200	100	50	0	200	100	50	0
Onion, green	150–200	200	100	50	0	200	100	50	0
Parsley	150–175	200	150	100	0	200	150	100	50
Parsnips	50–75	150	100	50	0	150	100	50	0
Peas, English	40–8	120	80	40	0	120	80	40	0
Pepper	100–180	200	150	100	0	200	150	100	0
Potato	150–180	200	150	100	0	300	200	100	0
Pumpkins, winter squash	50–100	150	100	50	0	200	150	100	0

Radish, rutabaga, turnip	50	150	100	50	0	150	100	50	0
Spinach	100–230	200	150	100	0	200	150	100	0
Strawberry, annual system	90–120	100	70	40	0–30	165	115	65	0
Summer squash	75–100	150	100	50	0	200	150	100	0
Sweet corn, fresh	125–175	160	120	80	0	160	120	80	0
Sweet corn, processing	150–200	160	120	80	0	160	120	80	0
Sweet potato	50–75	200	100	50	0	300	200	100	0
Tomato, fresh market, mulched	150–210	200	150	100	0	300	200	100	0
Tomato, processing	50–75	200	150	100	0	250	150	100	0
Watermelon, irrigated	125–150	150	100	50	0	200	150	100	0

Adapted from *2020–2021 Mid-Atlantic Commercial Vegetable Production Recommendations* (University of Maryland Extension Publ. EB137). Consult this guide for more details on management of fertilization for vegetables.

[1] A common recommendation is to broadcast and work deeply into the soil one-third to one-half of the fertilizer at planting and to apply the balance as a side dressing in one or two applications after the crop is fully established.

TABLE 6.35. RATES OF FERTILIZERS RECOMMENDED FOR VEGETABLE CROPS IN FLORIDA ON SANDY SOILS BASED ON MEHLICH-III SOIL TEST RESULTS

Target pH	Nitrogen (lb/acre)	Target pH	Nitrogen (lb/acre)
Tomato, pepper, potato, celery, sweet corn, crisphead lettuce, endive, escarole, romaine, lettuce, eggplant		Snapbean, lima, pole bean	
6.0 (potato) and 6.5	200	6.5	100
Broccoli, cauliflower, cabbage, Brussels sprouts, collards, Chinese cabbage, carrots		Radish and spinach	
6.5	175	6.5	90
Cucumber, squash, pumpkin, muskmelon, leaf lettuce, sweet bulb onion, watermelon, strawberry		Southern pea, snow pea, English pea, sweet potato	
6.0 (watermelon) and 6.5	150	6.5	60
Kale, turnip, mustard, parsley, okra, bunching onion, leek, beet			
6.5	120		

Fertilizer recommendations

P_2O_5 (lb/acre/crop season)				
Very low	Low	Medium	High	Very high
200	150	100	0	0

Celery

K_2O (lb/acre/crop season)				
Very low	Low	Medium	High	Very high
250	150	100	0	0

Crop										
Eggplant	160	130	100	130	160	0	100	130	0	0
Broccoli, cauliflower, Brussels sprouts, cabbage, collards, Chinese cabbage, carrots, kale, turnip, mustard, parsley, okra, muskmelon, leaf lettuce, sweet bulb onion, watermelon, pepper, sweet corn, crisphead lettuce, endive, escarole, strawberry, and romaine lettuce	120	100	100	120	120	0	100	0	0	0
Tomato	150	120	100	150	225	0	100	150	0	0
Cucumber, squash, pumpkin, snapbean, lima bean, pole bean, beet, radish, spinach, and sweet potato	120	100	80	100	120	0	80	100	0	0
Bunching onion and leek	120	100	100	100	120	0	100	150	0	0
Potato	120	120	60	120	150	0	60	150	0	0
Southern pea, snowpea, and English pea	80	80	60	80	80	0	60	80	0	0

Adapted from: G. D. Liu, E. H. Simonne, K. T. Morgan, and G. J. Hochmuth, "Soil and Fertilizer Management for Vegetable Production in Florida," in K. T. Morgan (ed.), *Nutrient Management of Vegetable and Row Crops Handbook* (University of Florida IFAS Extension Bull. No. SP500, 2018), https://edis.ifas.ufl.edu/publication/SS639.

TABLE 6.36. FERTILIZATION RECOMMENDATIONS FOR NEW ENGLAND VEGETABLES

Crop	Nitrogen (N) lb/acre	Phosphorus (P) lb P_2O_5/acre				Potassium (K) lb K_2O/acre			
Soil Test Results:		Very Low	Low	Optimum	Above Optimum	Very Low	Low	Optimum	Above Optimum
Asparagus established	75	50	90	30–60	0	300	150	75	0
Bean: snap, lima, dry	50	100	75	0–50	0	100	75	0–50	0
Beet and Swiss Chard	105–130	150	100	50	0	300	150	75–100	0
Cabbage, broccoli, cauliflower and other Brassica crops	160	150	100	50	0	175	125	50	0
Carrot and parsnip	110–150	150	100	50	0	300	200	75–100	0
Celery and celeriac	180	180	120	30–60	0	240	180	45–90	0
Sweet corn, ornamental, popcorn	100–130	140	80	40	0–40	180	120	0–30	0
Cucumber, muskmelon, watermelon	80–130	150	100	25–50	0	180	120	30–70	0
Eggplant	80–100	150	100	25–50	0	150	100	50	0
Garlic	120	150	100	25–50	0	150	100	50	0
Globe artichoke	120	75	50	0–25	0	150	100	50	0
Leek	130–150	150	100	25–50	0	150	100	50	0
Lettuce, endive, escarole	80–125	180	120	30–60	0–30	180	120	30–60	0
Okra	130	150	100	25–50	0	200	100	25–50	0
Onion, scallion, shallot	130–150	150	100	25–50	0	175	150	50	0

Pea	50–75	150	100	50	25–50	0	150	100	50
Pepper	140	150	100	25–50	0	200	150	50	0
Potato	120–180	200	120	30–60	0–30	300	200	100	0–50
Pumpkin, squash, gourd	110–140	150	100	0–40	0	200	150	40–80	0
Radish	50	150	100	25–50	0	125	100	50	0–25
Rutabaga, turnip	50	150	100	25–50	0	100	75	25	0
Salad mix, mesclun	50–80	180	120	30–60	0–30	180	120	30–60	0
Spinach	90–110	180	120	30–60	0–30	180	120	30–60	0
Sweet potato	50–75	200	120	30–60	0–30	300	200	50–100	0–50
Tomato (outdoor)	140–160	180	120	0–60	0	250	150	50–100	0–50

Adapted from K. Campbell-Nelson (general editor), *2020–2021 New England Vegetable Management Guide*, https://nevegetable.org/. Please see guide for more details regarding fertilizer management suggestions.

16

NUTRIENT DEFICIENCIES

Deficiencies of essential plant nutrients often are manifest in various patterns on the leaves of plants. Sometimes the patterns mimic pathogenic diseases, so care should be applied in diagnosing nutrient deficiencies. Table 6.37 provides a guide to diagnosing plant nutrient deficiencies.

TABLE 6.37. A KEY TO NUTRIENT DEFICIENCY SYMPTOMS

Nutrient	Plant Symptoms	Occurrence
Primary		
Nitrogen	Stems are thin, erect, and hard. Leaves are smaller than normal, pale green or yellow; lower leaves are affected first, but all leaves may be deficient in severe cases. Plants grow slowly	Excessive leaching on light soils
Phosphorus	Stems are thin and shortened. Leaves develop purple coloration, first on undersides and later throughout. Plants grow slowly, and maturity is delayed	On acid soils; temporary deficiencies on cold, wet soils
Potassium	Older leaves develop gray or tan areas near the margins. Eventually a scorch around the entire leaf margin may occur. Chlorotic areas may develop throughout leaf	Excessive leaching on light soils
Secondary nutrients and micronutrients		
Boron	Growing points die; stems are shortened and hard; leaves are distorted. Specific deficiencies include browning of cauliflower, cracked stem of celery, blackheart of beet, and internal browning of turnip	On soils with a pH above 6.8 or on crops with a high boron requirement

TABLE 6.37. A KEY TO NUTRIENT DEFICIENCY SYMPTOMS
(Continued)

Nutrient	Plant Symptoms	Occurrence
Calcium	Stem elongation restricted by death of the growing point. Root tips die, and root growth is restricted. Specific deficiencies include blossom-end rot of tomato, brownheart of escarole, celery blackheart, and carrot cavity spot.	On acid soils, following leaching rains, on soils with very high potassium levels, or on very dry soils
Copper	Yellowing of leaves. Leaves may become elongated. Onion bulbs are soft, with thin, pale-yellow scales	Most cases of copper deficiency occur on muck or peat soils.
Iron	Distinct yellow or white areas appear between the veins on the youngest leaves	On soils with a pH above 6.8
Magnesium	Initially, older leaves show yellowing between the veins; continued deficiency causes younger leaves to become affected. Older leaves may fall with prolonged deficiency	On acid soils, on soils with very high potassium levels, or on very light soils subject to leaching
Manganese	Yellow mottled areas, not as intense as with iron deficiency, appear on the youngest leaves. This finally results in an overall pale appearance. In beet, foliage becomes densely red. Onion and corn show narrow striping of yellow	On soils with a pH above 6.7
Molybdenum	Pale, distorted, very narrow leaves with some interveinal yellowing on older leaves. Whiptail of cauliflower; small, open, loose curds	On very acid soils
Zinc	Small reddish-brown spots on cotyledon leaves of bean. Green and yellow broad striping at base of leaves of corn. Interveinal yellowing with marginal burning on beet	On wet soils in early spring; often related to heavy phosphorus fertilization

17

MICRONUTRIENTS

Micronutrients are so named because these nutrients are required by plants in small, often miniscule amounts by plants. However small, these nutrients are essential for normal plant growth and functioning. Micronutrient-supplying capacity of soils can be determined by a calibrated soil test (Tables 6.38–6.39). Most state's Extension Services provide recommendations for micronutrients for vegetables, including crop sensitivities and toxicities (Tables 6.40–6.43).

TABLE 6.38. INTERPRETATION OF MICRONUTRIENT SOIL TESTS

Element	Method	Range in Critical Level (ppm)[1]
Boron (B)	Hot H_2O	0.1–0.7
Copper (Cu)	$NH_4C_2H_3O_2$ (pH 4.8)	0.2
	0.5M EDTA	0.1–0.7
	0.43N HNO_3	3–4
	Biological assay	2–3
Iron (Fe)	$NH_4C_2H_3O_2$ (pH 4.8)	2
	DTPA + $CaCl_2$ (pH 7.3)	2.5–4.5
Manganese (Mn)	0.05N HCl + 0.025N H_2SO_4	5–9
	0.1N H_3PO_4 and 3N $NH_4H_2PO_4$	15–20
	Hydroquinone + $NH_4C_2H_3O_2$	25–65
	H_2O	2
Molybdenum (Mo)	$(NH_4)_2C_2O_4$ (pH 3.3)	0.04–0.2
Zinc (Zn)	0.1N HCl	1.0–7.5
	Dithizone + $NH_4C_2H_3O_2$	0.3–2.3
	EDTA + $(NH_4)_2CO_3$	1.4–3.0
	DTPA + $CaCl_2$ (pH 7.3)	0.5–1.0

Reprinted with permission from S. S. Mortvedt, P. M. Giordano, and W. L. Lindsay (eds.), *Micronutrients in Agriculture* (Madison, Wis.: Soil Science Society of America, 1972).

[1] Deficiencies are likely to occur when concentrations are below the critical level. Consult the state's Extension Service for the latest interpretations.

TABLE 6.39. CRITICAL MEHLICH-3 SOIL TEST LEVELS FOR MICRONUTRIENTS IN NEW JERSEY

Micronutrients	Units	Critical Level	High
Zinc	ppm soil	1.0	50
Copper	ppm soil	0.5	20
Boron	ppm soil	0.5	20
Iron	ppm soil	50	100
Manganese	pH dependent: calculate an activity index $MnAI = 101.7 + 3.75Mn - 15.2pH$	25	100

Adapted from *New Jersey Agricultural Experiment Station*, https://njaes.rutgers.edu/soil-testing-lab/relative-levels-of-nutrients.php.

Micronutrient categories are less well defined than macronutrient categories. Values below the "critical level" should be considered deficient; values above "high" should be considered a warning. Certain micronutrients can be toxic to plants at excessive levels.

TABLE 6.40. BORON REQUIREMENTS OF VEGETABLES ARRANGED IN APPROXIMATE ORDER OF DECREASING REQUIREMENTS

High Requirement (More Than 0.5 ppm in Soil)	Medium Requirement (0.1–0.5 ppm in Soil)	Low Requirement (Less Than 0.1 ppm in Soil)
Beet	Tomato	Corn
Turnip	Lettuce	Cucumber
Cabbage	Sweet potato	Pea
Broccoli	Carrot	Bean
Cauliflower	Onion	Lima bean
Asparagus	Parsnip	Potato
Radish	Spinach	
Brussels sprouts		
Celery		
Rutabaga		

Adapted from K. C. Berger, "Boron in Soils and Crops," *Advances in Agronomy*, vol. 1, (New York: Academic Press, 1949), 321–351. And R. Goldy, *Boron in Vegetables* (Michigan University Extension, 2013), https://www.canr.msu.edu/news/boron_in_vegetables_not_too_little_not_too_much.

TABLE 6.41. RELATIVE TOLERANCE OF VEGETABLES TO BORON, ARRANGED IN ORDER OF INCREASING SENSITIVITY

Tolerant	Semitolerant	Sensitive
Asparagus	Celery	Jerusalem artichoke
Artichoke	Potato	Bean
Beet	Tomato	
Cantaloupe	Radish	
Broad bean	Corn	
Onion	Pumpkin	
Turnip	Bell pepper	
Cabbage	Sweet potato	
Lettuce	Lima bean	
Carrot		

Adapted from L. V. Wilcox, *Determining the Quality of Irrigation Water* (USDA Agricultural Information Bulletin 197, 1958).

TABLE 6.42. SOIL AND FOLIAR APPLICATION OF SECONDARY AND TRACE ELEMENTS

Vegetables differ in their requirements for these secondary nutrients. Availability in the soil is influenced by soil reaction and soil type. Use higher rates on muck and peat soils than on mineral soils and lower rates for band application than for broadcast. Foliar application is one means of correcting an evident deficiency that appears while the crop is growing.

Element	Application Rate (Per Acre Basis)	Source	Composition
Boron	0.5–3.5 lb (soil)	Borax ($Na_2B_4O_7 \cdot 10H_2O$)	11% B
		Boric acid (H_3BO_3)	17% B
		Sodium pentaborate ($Na_2B_{10}O_{16} \cdot 10H_2O$)	18% B
		Sodium tetraborate ($Na_2B_4O_7$)	21% B

TABLE 6.42. SOIL AND FOLIAR APPLICATION OF SECONDARY AND TRACE ELEMENTS (Continued)

Element	Application Rate (Per Acre Basis)	Source	Composition
Calcium	2–5 lb (foliar)	Calcium chloride ($CaCl_2$)	36% Ca
		Calcium nitrate ($CaNO_3 \cdot 2H_2O$)	20% Ca
		Liming materials and gypsum supply calcium when used as soil amendments	
Copper	2–6 lb (soil)	Cupric chloride ($CuCl_2$)	47% Cu
		Copper sulfate ($CuSO_4 \cdot H_2O$)	35% Cu
		Copper sulfate ($CuSO_4 \cdot 5H_2O$)	25% Cu
		Cupric oxide (CuO)	80% Cu
		Cuprous oxide (Cu_2O)	89% Cu
		Copper chelates	8–13% Cu
Iron	2–4 lb (soil)	Ferrous sulfate ($FeSO_4 \cdot 7H_2O$)	20% Fe
	0.5–1 lb (foliar)	Ferric sulfate [$Fe_2(SO_4)_3 \cdot 9H_2O$]	20% Fe
		Ferrous carbonate ($FeCO_3 \cdot H_2O$)	42% Fe
		Iron chelates	5–12% Fe
Magnesium	25–30 lb (soil)	Magnesium sulfate ($MgSO_4 \cdot 7H_2O$)	10% Mg
	2–4 lb (foliar)	Magnesium oxide (MgO)	55% Mg
		Dolomitic limestone	11% Mg
		Magnesium chelates	2–4% Mg
Manganese	20–100 lb (soil)	Manganese sulfate ($MnSO_4 \cdot 3H_2O$)	27% Mn
	2–5 lb (foliar)	Manganous oxide (MnO)	41–68% Mn
		Manganese chelates (Mn EDTA)	12% Mn
Molybdenum	25–400 g (soil)	Ammonium molybdate [$(NH_4)_6MO_7O_{24} \cdot 4H_2O$]	54% Mo
	25 g (foliar)	Sodium molybdate ($Na_2MoO_4 \cdot 2H_2O$)	39% Mo

TABLE 6.42. SOIL AND FOLIAR APPLICATION OF SECONDARY AND TRACE ELEMENTS (*Continued*)

Element	Application Rate (Per Acre Basis)	Source	Composition
Sulfur	20–50 lb (soil)	Sulfur (S)	100% S
		Ammonium sulfate [(NH$_4$)2SO$_4$]	24% S
			18% S
		Potassium sulfate (K$_2$SO$_4$)	16–18% S
		Calcium sulfate (CaSO$_4$)	18–19% S
		Ferric sulfate [Fe$_2$(SO$_4$)$_3$]	
Zinc	2–10 lb (soil)	Zinc oxide (ZnO)	80% Zn
	0.25 lb (foliar)	Zinc sulfate (ZnSO$_4$·7H$_2$O)	23% Zn
		Zinc chelates (Na$_2$Zn EDTA)	14% Zn

TABLE 6.43. BORON RECOMMENDATIONS BASED ON SOIL TESTS FOR VEGETABLE CROPS

Interpretation of Boron Soil Tests

ppm	lb/acre	Relative Level	Crops That Often Need Additional Boron	Boron Recommendation (lb/acre)
0.0–0.35	0.0–0.70	Low	Beet, broccoli, Brussels sprouts, cabbage, cauliflower, celery, rutabaga, turnip	3
			Asparagus, carrot, eggplant, horseradish, leek, onion, muskmelon, okra, parsnip, radish, squash, strawberry, sweet corn, tomato, white potato	2
			Pepper, sweet potato	1

TABLE 6.43. BORON RECOMMENDATIONS BASED ON SOIL TESTS FOR VEGETABLE CROPS (*Continued*)

Interpretation of Boron Soil Tests

ppm	lb/acre	Relative Level	Crops That Often Need Additional Boron	Boron Recommendation (lb/acre)
0.36–0.70	0.71–1.40	Medium	Beet, broccoli, Brussels sprouts, cabbage, cauliflower, celery, rutabaga, turnip	$1\frac{1}{2}$
			Asparagus, carrot, eggplant, horseradish, leek, onion, muskmelon, okra, parsnip, radish, squash, strawberry, sweet corn, tomato	1
>0.70	>1.40	High	All crops	0

Adapted from *2020/2021 Mid-Atlantic Commercial Vegetable Production Recommendations* (University of Maryland Extension Publ. EB137). Consult this guide for more details on management of fertilization for vegetables.

SECONDARY NUTRIENTS

Calcium, magnesium, and sulfur are often referred to as "secondary nutrients" even though a deficiency can cause major problems and crop limitations (Table 6.44). Some states include a test for calcium or magnesium along with their tests for P and K (see Section 15 on soil tests above).

TABLE 6.44. TOLERANCE OF VEGETABLES TO A DEFICIENCY OF SOIL MAGNESIUM

Tolerant	Not Tolerant
Bean	Cabbage
Beet	Cantaloupe
Chard	Corn
Lettuce	Cucumber
Pea	Eggplant
Radish	Pepper
Sweet potato	Potato
	Pumpkin
	Rutabaga
	Tomato
	Watermelon

Adapted from W. S. Ritchie and E. B. Holland, *Minerals in Nutrition* (Massachusetts Agricultural Experiment Station Bulletin 374, 1940); J. Kemble and J. Pickens, *Magnesium Important to Vegetable Growth* (Alabama A & M and Auburn Extension, 2020), https://www.aces.edu/blog/topics/crop-production/magnesium-important-to-vegetable-growth/.

FERTILIZER SPREADERS

ADJUSTMENT OF FERTILIZER SPREADERS

Each time a distributor is used, it is important to ensure that the proper quantity of fertilizer is being dispensed. Fertilizers vary greatly in the way they flow through the equipment. Movement is influenced by the humidity of the atmosphere as well as the degree of granulation of the material. Fertilizer spreaders should be calibrated frequently to make sure the correct amount of fertilizer is being applied (Tables 6.45–6.47).

TABLE 6.45. ADJUSTMENT OF ROW CROP SPREADER

1. Disconnect from one hopper the downspout or tube to the furrow opener for a row.
2. Attach a can just below the fertilizer hopper.
3. Fill the hopper under which the can is placed.
4. Engage the fertilizer attachment and drive the tractor the suggested distance according to the number of inches between rows.

Distance Between Rows (in.)	Distance to Pull the Spreader (ft)
20	261
24	218
30	174
36	145
38	138
40	131
42	124

5. Weigh the fertilizer in the can. Each pound in it equals 100 lb/acre. Each tenth of a pound equals 10 lb/acre.
6. Adjust the spreader for the rate of application desired, and then adjust the other spreader or distributors to the same setting.

TABLE 6.46. ADJUSTMENT OF GRAIN-DRILL-TYPE SPREADER

1. Remove four downspouts or tubes.
2. Attach a paper bag to each of the four outlets.
3. Fill the part of the drill over the bagged outlets.
4. Engage the spreader and drive the tractor the suggested distance according to the inches between the drill rows.

Distance Between Drill Rows (in.)	Distance to Pull the Drill (ft)
7	187
8	164
10	131
12	109
14	94

5. Weigh total fertilizer in the four bags. Each pound equals 100 lb/acre. Each tenth of a pound equals 10 lb/acre.

TABLE 6.47. CALIBRATION OF FERTILIZER DRILLS

Set drill at opening estimated to give the desired rate of application. Mark level of fertilizer in the hopper. Operate the drill for 100 ft Weigh a pail full of fertilizer. Refill hopper to marked level and again weigh pail. The difference is the pounds of fertilizer used in 100 ft. Consult the column under the row spacing being used. The left-hand column opposite the amount used shows the rate in pounds per acre at which the fertilizer has been applied. Adjust setting of the drill, if necessary, and recheck.

	Distance Between Rows (in.)				
	18	20	24	36	48
Rate (lb/acre)	Approximate Amount of Fertilizer (lb/100 ft of Row)				
250	0.9	1.1	1.4	1.7	2.3
500	1.7	2.3	2.9	3.5	4.6
750	2.6	3.4	4.3	5.2	6.9
1,000	3.5	4.6	5.8	6.9	9.2
1,500	5.2	6.8	8.6	10.4	13.8
2,000	6.8	9.2	11.6	13.0	18.4
3,000	10.5	14.0	17.5	21.0	28.0

Knott's Handbook for Vegetable Growers, Sixth Edition. George J. Hochmuth and Rebecca G. Sideman.
© 2023 John Wiley & Sons Ltd. Published 2023 by John Wiley & Sons Ltd.
Companion website: https://www.wiley.com/go/hochmuth/vegetablegrowers6e

01
SUGGESTIONS FOR SUPPLYING WATER TO VEGETABLES

About 98% of the water taken up by a plant is normally transpired from the plant leaves. The remainder is used in various growth processes in the plant. Plants in hot, dry areas lose more moisture to the air than those in cooler or more humid areas. Vegetables utilize more water in the later stages of growth when size and leaf area are greater. The root system becomes deeper and more widespread as the plant ages, exploring more soil for water and nutrients.

Some vegetables, especially lettuce and sweet corn, have sparse root systems that do not come into contact with all the soil moisture in their root zone. Cool-season vegetables normally root to a shallower depth than do warm-season vegetables and perennials.

When applying water, use enough to bring the soil moisture content of the effective rooting zone of the crop up to field capacity. This is the quantity of water that the soil holds against the pull of gravity.

TABLE 7.1. IRRIGATION MANAGEMENT IS RELATED CLOSELY TO THE SOIL, PLANT CHARACTERISTICS, CLIMATE/ WEATHER, AND WATER AVAILABILITY AND WATER QUALITY

Soil	Plant	Climate and Weather	Water Availability and Quality
Soil texture	Water requirements	Temperature	Quantity of water available
Soil compaction zones	Plant age and leaf area	Humidity	Distance from field
Soil structure	Rooting depth	Cloudiness	Chemical and biological contents of water source
Soil salinity		Rainfall amounts and pattern	
Water-holding capacity			

Adapted from T. F. Scherer, D. Franzen, and L. Cihacek, *Soil, Water, and Plant Characteristics Important to Irrigation* (North Dakota State University Publ. AE 1675, 2017), https://www.ag.ndsu. edu/publications/crops/soil-water-and-plant-characteristics-important-to-irrigation#section-14 (accessed 11 Nov. 2021).

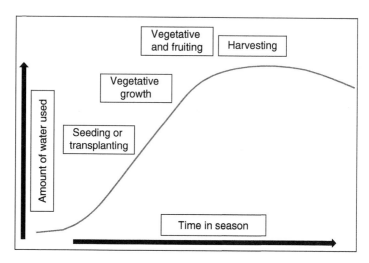

Figure 7.1. Generalized figure for crop growth and water use in the growing season. Drawing by George Hochmuth.

The frequency of irrigation depends on the total supply of available moisture reached by the roots and the rate of water use (Table 7.1 and Figure 7.1). The first is affected by soil type, depth of wetted soil, and the depth and dispersion of roots. The latter is influenced by weather conditions and the age of the crops. Add water when the available moisture in the root zone has been used to about the halfway point. Do not wait until vegetables show signs of wilting or develop color or texture changes that indicate they are not growing rapidly. A general rule is that vegetables need an average of 1 in. water/week from rain or supplemental irrigation to grow vigorously. In arid regions, about 2 in./week may be required (due to a higher evaporation rate from the soil). These amounts of water may vary from 0.5 in./week early in the season to more than 1 in. later in the season.

IRRIGATION MANAGEMENT AND NUTRIENT LEACHING

Irrigation management is critical to success in nutrient management for mobile nutrients such as nitrogen. It has been said that, "If farmers learn to first manage irrigation, then they can manage fertilizer." Irrigation management is particularly important in vegetable production in sandy soils where nitrogen is highly prone to leaching from the root zone with heavy rainfall or excessive irrigation. Leaching can occur with all irrigation systems

308

if more water is applied than the soil can hold at one time. If the water-holding capacity of the soil is exceeded with any irrigation event, nutrient leaching can occur. Information is provided here to assist growers in understanding the rooting zone for crops and water-holding capacity of soils, as well as application rates for various irrigation systems. Optimum irrigation management involves attention to these factors, knowing crop water needs, and keeping an eye on soil moisture levels during the season. These factors vary for the crop being grown, the soil used, the season, and the climate, among other factors. Please consult your local Extension Service for specific information for your production area.

02

ROOTING OF VEGETABLES

ROOTING DEPTH OF VEGETABLES

Vegetables vary in their rooting depth (Table 7.2). The depth of rooting of vegetables also is influenced by the soil profile. If there is a clay pan, hard pan, compacted layer, or other dense formation, the normal depth of rooting is not possible. Also, some transplanted vegetables may not develop root systems as deep as those of seeded crops. Although vegetables may root as deep as 18–24 in., most of the active root system for water uptake may be between 8 and 12 in. deep.

TABLE 7.2. CHARACTERISTIC MAXIMUM ROOTING DEPTHS OF VARIOUS VEGETABLES

Shallow (18–24 in.)	Moderately Deep (36–48 in.)	Deep (More Than 48 in.)
Broccoli	Bean, bush	Artichoke
Brussels sprouts	Bean, pole	Asparagus
Cabbage	Beet	Bean, lima
Cauliflower	Cantaloupe	Parsnip
Celery	Carrot	Pumpkin
Chinese cabbage	Chard	Squash, winter
Corn	Cucumber	Sweet potato
Endive	Eggplant	Tomato
Garlic	Mustard	Watermelon
Leek	Pea	
Lettuce	Pepper	
Onion	Rutabaga	
Parsley	Squash, summer	
Potato	Turnip	
Radish		
Spinach		
Strawberry		

03
SOIL MOISTURE AND IS MONITORING

Soil moisture can be monitored by a number of methods, varying in degree of technology employed. It is important to learn to estimate soil moisture, not only for the health of the crop from under-watering but also to minimize the chances of over-irrigation that would cost money and may lead to leaching of nutrients. Over-irrigation may damage the root system by saturation of the soil in the root zone. The simplest method for estimating soil moisture is by the appearance or feel method. More sophisticated methods include the use of soil moisture measuring devices or probes. Soils differ greatly in their water-holding capacity and depth of movement of water through the soil profile.

Farmers are rapidly moving away from managing irrigation by older general rules-of-thumb (e.g., "1 in./week") and manual control methods (hand turning of valves). Growers are adopting more sophisticated methods of monitoring irrigation (Tables 7.3 – 7.6 and Figure 7.2). More careful attention to irrigation is important for the following reasons:

1. Water is costly to pump and apply to a crop.
2. Under- or over-irrigation can cause crop losses.
3. Over-irrigation can increase risk for plant diseases.
4. Over-irrigation can lead to nutrient leaching which can cause reduced yields, pollution of local water bodies, and wasted fertilizer.

There are increasing numbers of irrigation management specialists in the university system because the Extension Service recognizes the importance of water management on the farm. In addition, there are many private consulting companies that provide the soil moisture monitoring equipment and the expertise. Some of these technologies are computer- and web-cloud-based allowing the farmer to monitor soil moisture in real-time in the office or from the truck by smart phone applications. Computer-assisted irrigation operating systems can control on and off cycles and collect data for irrigation records and management. The computer can control the frequency and length of fertilizer injection into the irrigation system. These systems make irrigation management more accurate and save considerable man-hours allowing the farmer to focus attention on other important tasks. This is not to say these systems do not need some checking and monitoring. They do.

DETERMINING MOISTURE IN SOIL BY APPEARANCE OR FEEL

A shovel serves to obtain a soil sample from a shallow soil or when a shallow-rooted crop is being grown. A soil auger or soil tube is necessary to draw samples from greater depths in the root zone.

Squeeze the soil sample in the hand and compare its appearance and behavior with those of the soils listed in the Practical Soil-Moisture Interpretation Chart (Table 7.3) to get a rough idea of its moisture content.

TABLE 7.3. PRACTICAL SOIL-MOISTURE INTERPRETATION CHART

	Soil Type			
Amount of Readily Available Moisture Remaining for the Plant	Sand (Gritty When Moist, Almost Like Beach Sand)	Sandy Loam (Gritty When Moist; Dirties Fingers; Contains Some Silt and Clay)	Clay Loam (Sticky and Plastic when Moist)	Clay (Very Sticky When Moist; Behaves Like Modeling Clay)
Close to 0%. Little or no moisture available	Dry, loose, single-grained; flows through fingers	Dry, loose, flows through fingers	Dry clods that break down into powdery condition	Hard, baked, cracked surface. Hard clods difficult to break, sometimes have loose crumbs on surface
50% or less. Approaching time to irrigate	Still appears to be dry; will not form a ball with pressure	Still appears to be dry; will not form a ball	Somewhat crumbly, but will hold together with pressure	Somewhat pliable; will ball under pressure
50–75%. Enough available moisture	Same as sand under 50%	Tends to ball under pressure but seldom holds together	Forms a ball, somewhat plastic; sometimes sticks slightly with pressure	Forms a ball, ribbons out between thumb and forefinger

TABLE 7.3. PRACTICAL SOIL-MOISTURE INTERPRETATION CHART (Continued)

Soil Type

Amount of Readily Available Moisture Remaining for the Plant	Sand (Gritty When Moist, Almost Like Beach Sand)	Sandy Loam (Gritty When Moist; Dirties Fingers; Contains Some Silt and Clay)	Clay Loam (Sticky and Plastic when Moist)	Clay (Very Sticky When Moist; Behaves Like Modeling Clay)
75% to field capacity. Plenty of available moisture	Tends to stick together slightly, sometimes forms a very weak ball under pressure	Forms weak ball, breaks easily, does not become slick	Forms a ball and is very pliable; becomes slick readily if high in clay	Easily ribbons out between fingers; feels slick
At field capacity. Soil will not hold any more water (after draining)	Upon squeezing, no free water appears; moisture is left on hand	Same as sand	Same as sand	Same as sand
Above field capacity. Unless water drains out, soil will be water-logged	Free water appears when soil is bounced in hand	Free water is released with kneading	Can squeeze out free water	Puddles and free water form on surface

Adapted from R. W. Harris and R. H. Coppock (eds.), "Saving Water in Landscape Irrigation," (University of California Division of Agricultural Science Leaflet 2976, 1978). Another very useful fact sheet with photos is *Estimating Soil Moisture by Feel and Appearance* (USDA/NRCS Program Aid Number 1619), https://www.wcc.nrcs.usda.gov/ftpref/wntsc/waterMgt/irrigation/EstimatingSoilMoisture.pdf (accessed 11 Nov. 2021).

TABLE 7.4. FIELD DEVICES FOR MONITORING SOIL MOISTURE

Method	Advantages	Disadvantages
Neutron moderation	Inexpensive per location, large sensing volume, not affected by salinity, stabile	Safety hazard, cumbersome, expensive, slow
Time domain Reflectometry (TDR)	Accurate, easily expanded, insensitive to normal salinity, soil-specific calibration not needed	Expensive, problems Under high salinity, small sensing volume
Frequency domain (FD) capacitance and FDR	Accurate after specific-soil calibration, better than TDR in saline soils, more flexible in probes than TDR, less expensive than TDR (some devices)	Small sensing sphere, needs careful installation, needs specific soil calibration
Tensiometer	Direct reading, minimal skill, inexpensive, not affected by salinity	Limited suction range, slow response time, frequent maintenance, requires intimate contact with soil
Resistance blocks	No maintenance, simple, inexpensive	Low resolution, slow reaction time, not suited for clays, block properties change, with time
Granular matrix sensors	No maintenance, simple, inexpensive, reduced problems compared to gypsum blocks	Low resolution, slow reaction time, not suited for clays, need to resaturate in dry soils

Adapted from R. Munoz-Carpena, *Field Devices for Monitoring Soil Water Content* (University of Florida Extension Service Bulletin 343, 2021), https://edis.ifas.ufl.edu/publication/ae266 (accessed 11 Nov. 2021).

TABLE 7.5. PERCENT OF AVAILABLE WATER DEPLETED FROM SOILS AT VARIOUS TENSIONS

Tension—Less Than—(Bars)[1]	Loamy Sand	Sandy Loam	Loam	Clay
0.3	55	35	15	7
0.5	70	55	30	13
0.8	77	63	45	20
1.0	82	68	55	27
2.0	90	78	72	45
5.0	95	88	80	75
15.0	100	100	100	100

Adapted from *Cooperative Extension* (University of California Soil and Water Newsletter No. 26, 1975). Also: *Soil Water Basics* (Irrometer, Riverside, CA), https://www.irrometer.com/basics.html (accessed 11 Nov. 2021).
[1] 1 bar = 100 kPa.

TO DETERMINE THE WATER NEEDED TO WET VARIOUS DEPTHS OF SOIL

Example: You wish to wet a loam soil to a 12-in. depth when half of the available water in that zone is gone. Using Figure 7.2, move across the chart from the left on the 12-in. line. Stop when you reach the diagonal line marked "loams." Move upward from that point to the scale at the top of the chart where you will read the depth of water required, in inches, based on depletion of about half the available water in the effective root zone. For our example, you will see that about ¾ in. of water is needed.

TABLE 7.6. APPROXIMATE SOIL WATER CHARACTERISTICS FOR TYPICAL SOIL CLASSES

Characteristic	Sandy Soil	Loamy Soil	Clay Soil
Dry weight of 1 cu ft	90 lb	80 lb	75 lb
Field capacity—% of dry weight	10%	20%	35%
Permanent wilting percentage	5%	10%	19%
Percent available water	5%	10%	16%
Water available to plants			
lb/cu ft	4 lb	8 lb	12 lb
in./ft depth	¾ in.	1½ in.	2¼ in.
gal/cu ft	½ gal	1 gal	1½ gal
Approximate depth of soil that will be wetted by each 1 in. water applied if half the available water has been used	24 in.	16 in.	11 in.
Suggested lengths of irrigation sprinkler runs	330 ft	660 ft	1,320 ft

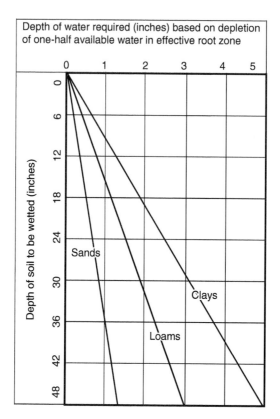

Figure 7.2. Chart for determining the amount of water needed to wet various depths of soil.

04

SURFACE IRRIGATION

RATES OF WATER APPLICATION FOR VARIOUS SURFACE IRRIGATION METHODS

Surface irrigation methods include furrow and basin irrigation. Furrow irrigation is mostly for crops in rows where the water is applied to one end of the furrow and flows toward the other end, seeping into the root zone as it flows. Basin irrigation is mostly used on crops that can be periodically flooded, such as some rice and some tree crops.

The infiltration rate has an important bearing on the intensity and frequency with which water should be applied by any method of irrigation. Normally, sandy soils have a high infiltration rate and clay soils have a low one. The irrigation rate is affected by soil texture, structure, dispersion, and the depth of the water table (Table 7.7). The longer the water is allowed to run, the more the infiltration rate decreases.

TABLE 7.7. APPROXIMATE TIME REQUIRED TO APPLY VARIOUS DEPTHS OF WATER PER ACRE WITH DIFFERENT FLOWS[1]

| Flow of Water | | | Approximate Time Required Per Acre for a Depth of: | | | | | | | |
| | | | 1 in. | | 2 in. | | 3 in. | | 4 in. | |
gpm	sec-ft	in/hr	hr	min	hr	min	hr	min	hr	min
50	0.11	1/8	9	03	18	06	27	09	36	12
100	0.22	1/4	4	32	9	03	13	35	18	06
150	0.33	5/16	3	01	6	02	9	03	12	04
200	0.45	7/16	2	16	4	32	6	47	9	03
250	0.56	9/16	1	49	3	37	5	26	7	14
300	0.67	11/16	1	31	3	01	4	32	6	02
350	0.78	3/4	1	18	2	35	3	53	5	10
400	0.89	7/8	1	08	2	16	3	24	4	32
450	1.00	1	1	00	2	01	3	01	4	01
500	1.11	1 1/8		54	1	49	2	43	3	37
550	1.23	1 3/16		49	1	39	2	28	3	18
600	1.34	1 5/16		45	1	31	2	16	3	01
650	1.45	1 7/16		42	1	24	2	05	2	48
700	1.56	1 9/16		39	1	18	1	56	2	35
750	1.67	1 21/32		36	1	12	1	49	2	24
800	1.78	1 3/4		34	1	08	1	42	2	16

850	1.89	1 7/8	32	1	04	1	36	2	08
900	2.01	2	30	1	00	1	31	2	01
950	2.12	2 3/32	29		57	1	26	1	54
1,000	2.23	2 3/16	27		54	1	21	1	49
1,050	2.34	2 5/16	26		52	1	18	1	44
1,100	2.45	2 7/16	25		49	1	14	1	38
1,150	2.56	2 1/2	24		47	1	11	1	34
1,200	2.67	2 5/8	23		45	1	08	1	31
1,300	2.90	2 7/8	21		42	1	03	1	24
1,400	3.12	3 1/16	20		39		58	1	18
1,500	3.34	3 5/16	18		36		54	1	12

[1] If a sprinkler system is used, the time required should be increased by 2–10% to compensate for the water that will evaporate before reaching the soil.

FURROW IRRIGATION

With furrows, use a flow of water initially 2–3 times that indicated to fill the run as quickly as possible (Tables 7.8 – 7.11 and Figures 7.3 – 7.5). Then cut back the flow to the indicated amount. This prevents excessive penetration at the head and equalizes the application of water throughout the whole furrow.

Beds intended for two rows are usually on 36-, 40-, or 42-in. centers, with the bed surface 4–6 in. above the bottom of the furrow. The depth of penetration of an equal quantity of water varies with the class of soil as indicated.

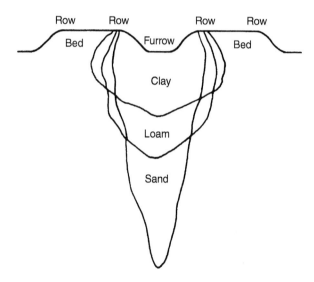

Figure 7.3. Arrangement of beds for furrow irrigation.

TABLE 7.8. APPROXIMATE FLOW OF WATER PER FURROW AFTER WATER REACHES THE END OF THE FURROW

		Slope of Land (%)		
		0–0.2	0.2–0.5	0.5–1
Infiltration Rate of Soil (in./hr)	Length of Furrow (ft)	Flow of Water Per Furrow (gal/min)		
High (1.5 or more)	330	9	4	3
	660	20	9	7
	1,320	45	20	15
Medium (0.5–1.5)	330	4	3	1.5
	660	10	7	3.5
	1,320	25	15	7.5
Low (0.1–0.5)	330	2	1.5	1
	660	4	3.5	2
	1,320	9	7.5	4

TABLE 7.9. APPROXIMATE MAXIMUM WATER INFILTRATION RATES FOR VARIOUS SOIL TYPES

Soil Type	Infiltration Rate[1] (in./hr)
Sand	2.0
Loamy sand	1.8
Sandy loam	1.5
Loam	1.0
Silt and clay loam	0.5
Clay	0.2

[1] Assumes a full crop cover. For bare soil, reduce the rate by half.

323

TABLE 7.10. BASIN IRRIGATION: APPROXIMATE AREA

Infiltration Rate of Soil (in./hr)	Quantity of Water to be Supplied	
	450 gal/min or 1 cu ft/sec	900 gal/min or 2 cu ft/sec
	Approximate Area (acre/basin)	
High (1.5 or more)	0.1	0.2
Medium (0.5–1.5)	0.2	0.4
Low (0.1–0.5)	0.5	1.0

TABLE 7.11. VOLUME OF WATER APPLIED FOR VARIOUS FLOW RATES AND TIME PERIODS

Flow Rate (gpm)	Volume (acre-in.) Applied			
	1 hr	8 hr	12 hr	24 hr
25	0.06	0.44	0.66	1.33
50	0.11	0.88	1.33	2.65
100	0.22	1.77	2.65	5.30
200	0.44	3.54	5.30	10.60
300	0.66	5.30	7.96	15.90
400	0.88	7.07	10.60	21.20
500	1.10	8.84	13.30	26.50
1,000	2.21	17.70	26.50	53.00
1,500	3.32	26.50	39.80	79.60
2,000	4.42	35.40	53.00	106.00

USE OF SIPHONS

Siphons of metal, plastic, or rubber can be used to carry water from a supply ditch to the area or furrow to be irrigated.

The inside diameter of the pipe and the head—the vertical distance from the surface of the water in the ditch to the surface of the water on the outlet side—determine the rate of flow.

When the outlet is not submerged, the head is measured to the center of the siphon outlet. You can determine how many gallons per minute are flowing through each siphon from the chart below.

> *Example:* You have a head of 4 in. and are using 2-in. siphons. Follow the 4-in. line across the chart until you reach the curve for 2-in. siphons. Move straight down to the scale at the bottom. You will find that you are putting on about 28 gal/min.

Figure 7.4. Method of measuring the head for water carried from a supply ditch to a furrow by means of a siphon. Adapted from University of California Division of Agricultural Science Leaflet 2956 (1977).

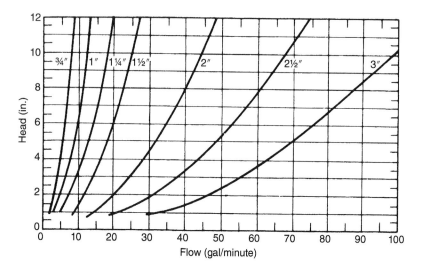

Figure 7.5 Chart for determining the flow of water through small siphons. Adapted from University of California Division of Agricultural Science Leaflet 2956 (1977). Also: E. C. Martin, *Measuring Water Flow in Surface Irrigation Ditches and Gated Pipe* (University of Arizona College Agriculture and Life Sciences, Arizona Water Series 31, 2011), https://extension.arizona. edu/pubs/measuring-water-flow-surface-irrigation-ditches-gated-pipe (accessed 11 Nov. 2021).

APPLICATION OF FERTILIZER IN WATER FOR FURROW IRRIGATION

There are certain limitations to the method of applying fertilizer solutions or soluble fertilizers in water supplied by furrow irrigation. You may not get uniform distribution of the fertilizer over the whole irrigated area. More of the dissolved material may enter the soil near the head than at the end of the furrow. You must know how long it will be necessary to run water to irrigate a certain area so as to meter the fertilizer solution properly. Soils vary considerably in their ability to absorb water.

Fertilizer solutions can be dripped from containers into the water as it flows. Devices are available that meter dry fertilizer materials into the irrigation water where they dissolve.

The rate of flow of dry soluble fertilizer or of fertilizer solutions into an irrigation head ditch can be calculated as follows:

$$\frac{\text{area to be irrigated}\left(\dfrac{\text{acres}}{\text{hr}}\right) \times \text{amount of nutrient wanted}\left(\dfrac{\text{lb}}{\text{acre}}\right)}{\text{nutrients in solution}\left(\dfrac{\text{lb}}{\text{gal}}\right)}$$

$$= \text{flow rate of fertilizer solution}\left(\frac{\text{gal}}{\text{hr}}\right)$$

$$\frac{\text{area to be irrigated}\,(\text{acres}) \times \text{amount of soluble fertilizer}\left(\dfrac{\text{lb}}{\dfrac{\text{acre or gal}}{\text{acre}}}\right)}{\text{time of irrigation}\,(\text{hr})}$$

$$= \text{flow rate of fertilizer solution}\left(\frac{\text{lb}}{\text{hr}} \text{ or } \frac{\text{gal}}{\text{hr}}\right)$$

Knowing the gallons of solution per hour to be added to the irrigation water, you can adjust the flow from the tank as directed by the following table (Table 7.12).

327

TABLE 7.12. RATE OF FLOW OF FERTILIZER SOLUTIONS

Amount of Solution Desired (gal/hr)	Approximate Time (sec) to Fill a 4-oz Container	Approximate Time (sec) to Fill an 8-oz Container
½	225	450
1	112	224
2	56	112
3	38	76
4	28	56
5	22	44
6	18	36
7	16	32
8	14	28
9	12	24
10	11	22
12	9	18
14	8	16
16	7	14
18	6	12
20	5.5	11

05

OVERHEAD IRRIGATION

Overhead irrigation generally refers to systems that sprinkle water over the top of a crop from sprinklers operated under pressure (Tables 7.13 – 7.21 and Figure 7.6). Examples include solid-set sprinklers, hose reel, lateral-move, and center-pivots.

TABLE 7.13. SPRINKLER IRRIGATION: APPROXIMATE APPLICATION RATE OF WATER

Infiltration Rate of Soil (in./hr)	Slope of Land (%)	
	0–5	5–12
	Approximate Application Rate (in./hr)	
High (1.5 or more)	1.0	0.75
Medium (0.5–1.5)	0.5	0.40
Low (0.1–0.5)	0.2	0.15

TABLE 7.14. ACREAGE COVERED BY MOVES OF PIPE OF VARIOUS LENGTHS

Lateral Move of Pipe (ft)	Length of Sprinkler Pipe (ft)	Area Covered Per Move (Acres)
20	2,640	1.21
20	1,320	0.61
20	660	0.30
20	330	0.15
30	2,640	1.82
30	1,320	0.91
30	660	0.46
30	330	0.23
40	2,640	2.42
40	1,320	1.21
40	660	0.61
40	330	0.30
50	2,640	3.03
50	1,320	1.52
50	660	0.76
50	330	0.38
60	2,640	3.64
60	1,320	1.82
60	660	0.91
60	330	0.46
80	2,640	4.85
80	1,320	2.42
80	660	1.21
80	330	0.61
100	2,640	6.06
100	1,320	3.03
100	660	1.52
100	330	0.76

TABLE 7.15. PRECIPITATION RATES FOR VARIOUS NOZZLE SIZES, PRESSURE, AND SPACINGS

Nozzle Size (in.)	Pressure (psi)	Discharge[1] (gpm)	Diameter of Spray[2] (ft)	Precipitation Rate at Spacings (in./hr)[1]		
				30 × 40 ft	30 × 45 ft	40 × 40 ft
$\frac{1}{16}$	45	0.76	60–72	0.061		
$\frac{1}{16}$	50	0.80	61–73	0.064		
$\frac{1}{16}$	55	0.85	62–74	0.068		
$\frac{1}{16}$	60	0.88	63–75	0.071		
$\frac{1}{16}$	65	0.93	64–76	0.075		
$\frac{5}{64}$	45	1.19	59–73	0.095	0.085	
$\frac{5}{64}$	50	1.25	62–72	0.100	0.089	
$\frac{5}{64}$	55	1.30	64–74	0.104	0.094	0.079
$\frac{5}{64}$	60	1.36	67–76	0.110	0.097	0.082

TABLE 7.15. PRECIPITATION RATES FOR VARIOUS NOZZLE SIZES, PRESSURE, AND SPACINGS *(Continued)*

Nozzle Size (in.)	Pressure (psi)	Discharge[1] (gpm)	Diameter of Spray[2] (ft)	Precipitation Rate at Spacings (in./hr)[1]		
				30×40 ft	30×45 ft	40×40 ft
$5/64$	65	1.45	68–77	0.116	0.103	0.087
$3/32$	45	1.72	68–76	0.138	0.123	0.103
$3/32$	50	1.80	69–77	0.145	0.128	0.108
$3/32$	55	1.88	70–78	0.151	0.134	0.113
$3/32$	60	1.98	71–79	0.159	0.141	0.119
$3/32$	65	2.08	72–80	0.167	0.148	0.125
$7/64$	45	2.32	71–78	0.186	0.165	0.140
$7/64$	50	2.44	72–80	0.196	0.174	0.147
$7/64$	55	2.56	74–81	0.205	0.182	0.154
$7/64$	60	2.69	76–82	0.216	0.192	0.161

7/64	65	2.79	77–83	0.224	0.199	0.168
1/8	45	3.04	76–82	0.244	0.217	0.183
1/8	50	3.22	78–82		0.230	0.193
1/8	55	3.39	79–83		0.242	0.204
1/8	60	3.55	80–84		0.253	0.213
1/8	65	3.70	81–85			0.222

Adapted from original publication: A. W. Marsh et al., "Solid Set Sprinklers for Starting Vegetable Crops," (University of California Division of Agricultural Science Leaflet 2265, 1977).

[1] Three-digit numbers are shown here only to indicate the progression as nozzle size and pressure increase.

[2] Range of diameters of spray for different makes and models of sprinklers.

TABLE 7.16. GUIDE FOR SELECTING SIZE OF ALUMINUM PIPE FOR SPRINKLER LATERAL LINES

Maximum Number of Sprinklers
to Use on Single Lateral Line

Sprinkler Discharge (gpm)	30-ft Sprinkler Spacing for Pipe Diameter (in.):			40-ft Sprinkler Spacing for Pipe Diameter (in.):		
	2	3	4	2	3	4
0.75	47	95	200	43	85	180
1.00	40	80	150	36	72	125
1.25	34	69	118	31	62	104
1.50	30	62	100	28	56	92
1.75	27	56	92	25	50	83
2.00	25	51	84	23	46	76
2.25	23	47	78	21	43	71
2.50	21	44	73	19	40	66
2.75	20	42	68	18	38	62
3.00	19	40	65	17	36	58
3.25	18	38	62	16	34	56
3.50	17	36	59	15	32	53
3.75	16	34	56	14	31	51
4.00	16	33	54	14	30	48

Adapted from original publication A. W. Marsh et al., "Solid Set Sprinklers for Starting Vegetable Crops," (University of California Division of Agricultural Science Leaflet 2265, 1977). Also, H. W. Otto and J. Meyer, "Family Farm Series Publications: *Vegetable Crop Production*" (University of California), http://www.sfc.ucdavis.edu/pubs/family_farm_series/veg/irrigating/irrigating.html.

TABLE 7.17. GUIDE TO MAIN-LINE PIPE SIZES[1]

Water Flow (gpm) for Pipe Diameter (in.):

Distance (ft)	200	400	600	800	1,000	1,200	1,400	1,600	1,800
200	3	4	5	5	6	6	6	7	7
400	4	5	5	6	6	7	7	8	8
600	4	5	6	7	7	7	8	8	8
800	4	5	6	7	7	8	8	8	10
1,000	5	6	6	7	8	8	8	10	10
1,200	5	6	7	7	8	8	10	10	10

Adapted from original publication: A. W. Marsh et al., "Solid Set Sprinklers for Starting Vegetable Crops," (University of California Division of Agricultural Science Leaflet 2265, 1977).

[1] Using aluminum pipe (C = 120) with pressure losses ranging from 5 to 15 psi, average about 10.

TABLE 7.18. CONTINUOUS POWER OUTPUT REQUIRED AT TRACTOR POWER TAKEOFF TO PUMP WATER

Flow (gpm)

Pressure[1] (psi)	Head[1] (ft)	100	200	300	400	500	600	700	800	1,000
				Horsepower Required[2]						
50	116	3.9	7.8	11.7	16	20	23	27	31	39
55	128	4.3	8.7	13	17	22	26	30	35	43
60	140	4.7	9.5	14	19	24	28	33	38	47
65	151	5.1	10	15	20	25	30	36	41	51
70	162	5.5	11	16	22	27	33	38	44	55
75	173	5.8	12	17	23	29	35	41	47	58
80	185	6.2	12	19	25	31	37	44	50	62

Adapted from original publication: A. W. Marsh et al., "Solid Set Sprinklers for Starting Vegetable Crops," (University of California Division of Agricultural Science Leaflet 2265, 1977).

[1] Including nozzle pressure, friction loss, and elevation lift.
[2] Pump assumed to operate at 75% efficiency.

TABLE 7.19. FLOW OF WATER REQUIRED TO OPERATE SOLID SET SPRINKLER SYSTEMS

Irrigation Rate (in./hr)	Area Irrigated Per Set (Acres)									
	4		8		12		16		20	
	gpm[1]	cfs[2]	gpm	cfs	gpm	cfs	gpm	cfs	gpm	cfs
0.06	108	0.5	217	0.5	326	1.0	435	1.0	543	1.5
0.08	145	0.5	290	1.0	435	1.0	580	1.5	725	2.0
0.10	181	0.5	362	1.0	543	1.5	724	2.0	905	2.5
0.12	217	0.5	435	1.0	652	1.5	870	2.0	1,086	2.5
0.15	271	1.0	543	1.5	815	2.0	1,086	2.5	1,360	3.5
0.20	362	1.0	724	2.0	1,086	2.5	1,448	2.5	1,810	4.5

Adapted from original publication: A. W. Marsh et al., "Solid Set Sprinklers for Starting Vegetable Crops," (University of California Division of Agricultural Science Leaflet 2265, 1977).

[1] Gallons per minute pumped into the sprinkler system to provide an average precipitation rate as shown. Pump must have this much or slightly greater capacity.

[2] Cubic feet per second—the flow of water to the next larger $\frac{1}{2}$ cfs that must be ordered from the water district, assuming that the district accepts orders only in increments of $\frac{1}{2}$ cfs. Actually, $\frac{1}{2}$ cfs = 225 gpm.

TABLE 7.20. AMOUNT OF FERTILIZER TO USE FOR EACH SETTING OF THE SPRINKLER LINE

Nutrient Per Setting of Sprinkler Line (lb):

Length of Line (ft)	Lateral Move of Line (ft)	10	20	30	40	50	60	70	80	90	100
		Nutrient Application Desired (lb/acre)									
330	40	3	6	9	12	15	18	21	24	27	30
	60	4	9	12	18	22	27	31	36	40	45
	80	6	12	18	24	30	36	42	48	54	60
660	40	6	12	18	24	30	36	42	48	54	60
	60	9	18	24	36	45	54	63	72	81	90
	80	12	24	36	48	60	72	84	96	108	120
990	40	9	18	24	36	45	54	63	72	81	90
	60	13	27	40	54	67	81	94	108	121	135
	80	18	36	54	72	90	108	126	144	162	180
1,320	40	12	24	36	48	60	72	84	96	108	120
	60	18	36	54	72	90	108	126	144	162	180
	80	24	48	72	96	120	144	168	192	216	240

TABLE 7.21. APPLICATION RATE RECOMMENDED FOR COLD PROTECTION UNDER DIFFERENT WIND AND TEMPERATURE CONDITIONS

	Wind Speed (mph)		
	0–1	2–4	5–8
Minimum Temperature Expected (°F)	(in./hr)		
27	0.10	0.10	0.10
26	0.10	0.10	0.14
24	0.10	0.16	0.30
22	0.12	0.24	0.50
20	0.16	0.30	0.60

Adapted originally from D. S. Harrison, J. F. Gerber, and R. E. Choate, *Sprinkler Irrigation for Cold Protection* (Florida Cooperative Extension Circular 348, 1974).

USING SPRINKLER SYSTEMS

Each irrigation system presents a separate engineering challenge. The advice of a competent irrigation engineer is essential. Many factors must be taken into consideration in developing a plan for the equipment:

Water supply available at period of greatest need
Distance from source of water to field to be irrigated
Height of field above water source and topography of the land
Type of soil (rate at which it absorbs water and its water-holding capacity)
Area to be irrigated
Desired frequency of irrigation
Quantity of water to be applied
Time of day in which application is to be made
Type of power available

Normal wind velocity and direction
Possible future expansion of the installation

Specific details of the plan must then include the following:

Size of power unit and pump to do the particular job
Pipe sizes and lengths for mains and laterals
Operating pressures of sprinklers
Size and spacing of sprinklers
Frictional losses in the system

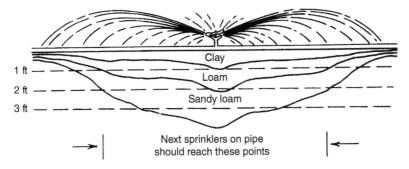

Figure 7.6. The diagram shows the approximate depth of penetration of available water from a 3-in. irrigation event on various classes of soil.

To avoid uneven water distribution, there should be enough distance between sprinklers to allow a 40% overlap in diameter of the area they are to cover.

CALCULATING RATES OF SPRINKLER APPLICATIONS

To determine the output per sprinkler needed to apply the desired rate of application:

$$\frac{\begin{array}{c}\text{distance between} \\ \text{sprinklers(ft)}\end{array} \times \begin{array}{c}\text{distance between} \\ \text{line settings(ft)}\end{array} \times \begin{array}{c}\text{precipitation} \\ \text{rate}\left(\dfrac{\text{in.}}{\text{hr}}\right)\end{array}}{96.3}$$

$$= \text{sprinkler rate}\left(\frac{\text{gal}}{\text{minute}}\right)$$

$$Example: \frac{30 \times 50 \times 0.4}{96.3} = \frac{6.23\,\text{gal}}{\text{minute per sprinkler}}$$

To determine the rate at which water is being applied:

$$\frac{\begin{array}{c}\text{sprinkler rate} \\ \left(\dfrac{\text{gal}}{\text{minute}}\right)\end{array} \times 96.3}{\begin{array}{c}\text{distance between} \\ \text{sprinklers(ft)}\end{array} \times \begin{array}{c}\text{distance between} \\ \text{line settings(ft)}\end{array}} = \text{precipitation rate } \text{in}/_{\text{hr}}$$

Manufacturer's specifications give the gallons per minute for each type of sprinkler at various pressures.

$$Example: \frac{10 \times 96.3}{40 \times 50} = 0.481 \frac{\text{in.}}{\text{hr}}$$

CENTER PIVOT OR LATERAL MOVE IRRIGATION SYSTEMS

Center pivot irrigation systems are commonly used in vegetable production especially for crops grown without plastic mulch. These crops would include potato, sweet corn, snap bean, and sweet potato. Center pivot irrigation systems can irrigate large expanses of crops and do so with automated movement in circular fashion about a center water supply point. Lateral move systems are similar to center pivots, but move across a field laterally rather than in circular fashion. Rate of movement governs the rate of water application. These systems can be adapted to apply crop fertilizers.

Further Reading

Center Pivots (University of Nebraska Institute of Agriculture and Natural Resources), https://water.unl.edu/article/agricultural-irrigation/center-pivots (accessed 11 Nov. 2021).

L. New and G. Fipps, (Center Pivot Irrigation. Texas Agricultural Extension Service. Publ. B-6096), https://irrigation.tamu.edu/files/2018/06/B-6096-Center-Pivot-Irrigation.pdf (accessed 11 Nov. 2021).

APPLYING FERTILIZER THROUGH A SPRINKLER SYSTEM

Anhydrous ammonia, aqua ammonia, and nitrogen solutions containing free ammonia should not be applied by sprinkler irrigation because of the excessive loss of the volatile ammonia. Ammonium nitrate, ammonium sulfate, calcium nitrate, sodium nitrate, urea-ammonium nitrate, and urea are all suitable materials for use through a sprinkler system. The water containing the ammonium salts should not have a reaction on the alkaline side of neutrality, or the loss of ammonia will be considerable.

It is best to put phosphorus fertilizers directly in the soil by a band application. Potash fertilizers can be used in sprinkler lines. However, a soil application ahead of or at planting time usually proves adequate and can be made efficiently at that time.

Manganese, boron, and copper can be applied through the sprinkler system. See Table 6.43, for possible rates of application.

The fertilizing material is dissolved in a tank of water. Calcium nitrate, ammonium sulfate, and ammonium nitrate dissolve completely. Commercial mixtures for injection are also available for delivery to the farm. The solution can then be introduced into the water line, either by suction or by pressure from a pump. See Table 6.17 for relative solubility of fertilizer materials.

Introduce the fertilizer into the line slowly, taking 10–20 min to complete the operation. After enough of the fertilizer solution has passed into the pipelines, shut the valve if suction by pump is used. This prevents unpriming the pump. Then run the system for 10–15 min to wash the fertilizer off the leaves. This also flushes out the lines, valves, and pump, if one has been used to force or suck the solution into the main line.

It is necessary to calculate the actual pounds of a fertilizing material that must be dissolved in the mixing tank to supply a certain number of pounds of the nutrient to the acre at each setting of the sprinkler line. This is done as follows. To apply 40 lb nitrogen to the acre when the sprinkler line is 660 ft long and will be moved 80 ft, if sodium nitrate is used, divide 48 (as

341

shown in Table 7.20) by 0.16 (the percentage of nitrogen in sodium nitrate). This equals 300 lb, which must be dissolved in the tank and applied at each setting of the pipe. Do the same with ammonium nitrate: Divide 48 by 0.33, which equals about 145 lb.

SPRINKLER IRRIGATION FOR COLD PROTECTION

Sprinklers are often used to protect vegetables from freezing (Table 7.21). Sprinkling provides cold protection because the latent heat of fusion is released when water changes from liquid to ice. When water is freezing, its temperature is near 32°F. The heat liberated as the water freezes maintains the temperature of the vegetable near 32°F even though the surroundings may be colder. As long as there is a mixture of both water and ice present, the temperature remains near 32°F. For all of the plant to be protected, it must be covered or encased in the freezing ice–water mixture. Enough water must be applied continuously so that the latent heat released compensates for the heat losses.

Further Reading

T. Coolong, E. Smith, and P. Knox, *Commercial Freeze Protection for Fruits and Vegetables* (University of Georgia Extension, 2017), https://extension. uga.edu/publications/detail.html?number=B1479&title=Commercial%20 Freeze%20Protection%20for%20Fruits%20and%20Vegetables (accessed 11 Nov. 2021).

M. Longstroth, *Using Sprinklers to Protect Plants from Spring Freezes.* (Michigan State University Extension, 2012), https://www.canr.msu.edu/ news/using_sprinklers_to_protect_plants_from_spring_freezes (accessed 11 Nov. 2021).

06
DRIP OR TRICKLE IRRIGATION

Drip or *trickle irrigation* refers to the frequent slow application of water directly to the root zone of plants (Tables 7.22 – 7.26 and Figure 7.7). Vegetables are usually irrigated by double-wall, thin-wall, or heavy-wall tubing to supply a uniform rate along the entire row. It is most often recommended that the tubing be placed 2 in. below the surface of the soil with emitters facing upward, near the row of plants. This helps keep pests from damaging the tubing and keeps the tubing from "snaking" when on the bed surface.

Pressure in the drip lines typically varies from 8 to 10 psi and about 12 psi in the submains. Length of the drip lines may be as long as 600 ft, but 200–250 ft is more common. Rate of water application is about $\frac{1}{4} - \frac{1}{2}$ gpm/100 ft of crop row. One acre of plants in rows 100 ft long and 4 ft apart use about 30 gpm water. Unless clear, sediment-free water is available, it is necessary to install a filter in the main line to prevent clogging of the small emitter pores in the drip lines.

Drip irrigation provides for considerable saving in water application, particularly during early plant growth. The aisles between rows remain dry because water is applied only next to plants in the row. Irrigation application efficiency is about 90% with drip irrigation.

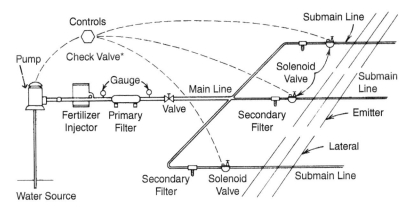

* A backflow preventer or vacuum breaker is required in some areas.

Figure 7.7.   Drip or trickle irrigation system components.

TABLE 7.22. VOLUME OF WATER TO APPLY (GAL) BY DRIP IRRIGATION PER 100 LINEAR FT BED FOR A GIVEN WETTED SOIL VOLUME, AVAILABLE WATER-HOLDING CAPACITY, AND AN ALLOWABLE DEPLETION OF 1/2[1]

Wetted Soil Volume Per 100 ft (Cubic ft)	Available Water-Holding Capacity (in. Water Per ft Soil)							
	0.25	0.50	0.75	1.00	1.25	1.50	1.75	2.00
	(gal Per 100 Linear Bed Feet)							
25	2.2	4.3	6.5	8.7	10.8	13.0	15.2	17.3
50	4.3	8.7	13.0	17.3	21.6	26.0	30.3	34.6
75	6.5	13.0	19.5	26.0	32.5	39.0	45.5	51.9
100	8.7	17.3	26.0	34.6	43.3	51.9	60.6	69.3
125	10.8	21.6	32.5	43.3	54.1	64.9	75.8	86.6
150	13.0	26.0	39.0	51.9	64.9	77.9	90.9	103.9
175	15.2	30.3	45.5	60.6	75.8	90.9	106.1	121.2
200	17.3	34.6	51.9	69.3	86.6	103.9	121.2	138.5
225	19.5	39.0	58.4	77.9	97.4	116.9	136.4	155.8
250	21.6	43.3	64.9	86.6	108.2	129.9	151.5	173.1
275	23.8	47.6	71.4	95.2	119.0	142.8	166.7	190.5
300	26.0	51.9	77.9	103.9	129.9	155.8	181.8	207.8
350	30.3	60.6	90.9	121.2	151.5	181.8	212.1	242.4
400	34.6	69.3	103.9	138.5	173.1	207.8	242.4	277.0
450	39.0	77.9	116.9	155.8	194.8	233.8	272.7	311.7
500	43.3	86.6	129.9	173.1	216.4	259.7	303.0	346.3
550	47.6	95.2	142.8	190.5	238.1	285.7	333.3	380.9
600	51.9	103.9	155.8	207.8	259.7	311.7	363.6	415.6
700	60.6	121.2	181.8	242.4	303.0	363.6	424.2	484.8
800	69.3	138.5	207.8	277.0	346.3	415.6	484.8	554.1
900	77.9	155.8	233.8	311.7	389.6	467.5	545.4	623.3

Adapted from original publication by G. A. Clark, C. D. Stanley, and A. G. Smajstrla, *Micro-irrigation on Mulched Bed Systems: Components, System Capacities, and Management* (Florida Cooperative Extension Service Bulletin 245, 2002), https://ufdc.ufl.edu/IR00004488/00001 (accessed 11 Nov. 2021).

[1] An irrigation application efficiency of 90% is assumed.

TABLE 7.23. **DISCHARGE PER GROSS ACRE (GPM/ACRE) FOR DRIP IRRIGATION BASED ON IRRIGATED LINEAR BED FEET AND EMITTER DISCHARGE**

Linear Bed Feet Crop Per Acre	Emitter Discharge (gpm/100 ft)						
	0.25	0.30	0.40	0.50	0.75	1.00	1.50
	(gal Per min/acre)						
3,000	7.5	9.0	12.0	15.0	22.5	30.0	45.0
3,500	8.8	10.5	14.0	17.5	26.3	35.0	52.5
4,000	10.0	12.0	16.0	20.0	30.0	40.0	60.0
4,500	11.3	13.5	18.0	22.5	33.8	45.0	67.5
5,000	12.5	15.0	20.0	25.0	37.5	50.0	75.0
5,500	13.8	16.5	22.0	27.5	41.3	55.0	82.5
6,000	15.0	18.0	24.0	30.0	45.0	60.0	90.0
6,500	16.3	19.5	26.0	32.5	48.8	65.0	97.5
7,000	17.5	21.0	28.0	35.0	52.5	70.0	105.0
7,500	18.8	22.5	30.0	37.5	56.3	75.0	112.5
8,000	20.0	24.0	32.0	40.0	60.0	80.0	120.0
8,500	21.3	25.5	34.0	42.5	63.8	85.0	127.5
9,000	22.5	27.0	36.0	45.0	67.5	90.0	135.0
9,500	23.8	28.5	38.0	47.5	71.3	95.0	142.5
10,000	25.0	30.0	40.0	50.0	75.0	100.0	150.0

Adapted from G. A. Clark, C. D. Stanley, and A. G. Smajstrla, *Micro-irrigation on Mulched Bed Systems: Components, System Capacities, and Management* (Florida Cooperative Extension Service Bulletin 245, 2002), https://ufdc.ufl.edu/IR00004488/00001.

TABLE 7.24. VOLUME OF WATER (GAL WATER PER ACRE PER MINUTE) DELIVERED UNDER VARIOUS BED SPACINGS WITH ONE TAPE LATERAL PER BED AND FOR SEVERAL EMITTER FLOW RATES

| Bed Spacing (in.) | Drip Tape Per Acre (ft) | Emitter Flow Rate (gal Per min Per 100 ft) | | | |
		0.50	0.40	0.30	0.25
		(gal Per acre/min)			
24	21,780	108.9	87.1	65.3	54.5
30	17,420	87.1	69.7	52.3	43.6
36	14,520	72.6	58.1	43.6	36.6
42	12,450	62.2	49.8	37.3	31.1
48	10,890	54.5	43.6	32.7	27.2
54	9,680	48.4	38.7	29.0	24.2
60	8,710	43.6	34.9	26.1	21.8
72	7,260	36.3	29.0	21.8	18.2
84	6,220	31.1	24.9	18.7	15.6
96	5,450	27.2	21.8	16.3	13.6
108	4,840	24.2	19.4	14.5	12.1
120	4,360	21.8	17.4	13.1	10.0

TABLE 7.25. VOLUME OF AVAILABLE WATER IN THE WETTED
CYLINDRICAL DISTRIBUTION PATTERN UNDER
A DRIP IRRIGATION LINE BASED ON THE
AVAILABLE WATER-HOLDING CAPACITY OF THE
SOIL

| Available Water (%) | Wetted Radius (in.)[1] | | | | |
	6	9	12	15	18
	(gal Available Water Per 100 Emitters)				
3	9	20	35	55	79
4	12	26	47	74	106
5	15	33	59	92	132
6	18	40	71	110	159
7	21	46	82	129	185
8	24	53	94	147	212
9	26	60	106	165	238
10	29	66	118	184	265
11	32	73	129	202	291
12	35	79	141	221	318
13	38	86	153	239	344
14	41	93	165	257	371
15	44	99	176	276	397

Adapted from original publication by G. A. Clark and A. G. Smajstrla, *Application Volumes and Wetted Patterns for Scheduling Drip Irrigation in Florida Vegetable Production* (Florida Cooperative Extension Service Circular 1041, 1993). https://ufdc.ufl.edu/UF00014456/00001.

[1] For a 1-ft depth of wetting.

TABLE 7.26. MAXIMUM APPLICATION TIMES FOR DRIP-IRRIGATED VEGETABLE PRODUCTION ON SANDY SOILS WITH VARIOUS WATER-HOLDING CAPACITIES

Available Water-Holding Capacity (in. Water Per in. Soil)	Tubing Flow Rate (gpm Per 100 ft)				
	0.2	0.3	0.4	0.5	0.6
	(Maximum min Per Application)[1]				
0.02	41	27	20	16	14
0.03	61	41	31	24	20
0.04	82	54	41	33	27
0.05	102	68	51	41	34
0.06	122	82	61	49	41
0.07	143	95	71	57	48
0.08	163	109	82	65	54
0.09	184	122	92	73	61
0.10	204	136	102	82	68
0.11	224	150	112	90	75
0.12	245	163	122	98	82

Adapted from original publication: C. D. Stanley and G. A. Clark, "Maximum Application Times for Drip-irrigated Vegetable Production as Influenced by Soil Type or Tubing Emission Characteristics," (Florida Cooperative Extension Service Drip Tip No. 9305, 1993).

[1] Assumes 10-in.-deep root zone and irrigation at 50% soil moisture depletion.

TREATING IRRIGATION SYSTEMS WITH CHLORINE

Chlorine can be used in irrigation systems to control the growth of algae and other microorganisms such as bacteria and fungi (Table 7.27). These organisms are found in surface and ground water and can proliferate with the nutrients present in the water inside the drip tube. Filtration alone cannot remove all of these contaminants. Hypochlorous acid, the agent responsible for controlling microorganisms in drip tubes, is more active under slightly acidic water conditions. Chlorine gas, solid (calcium hypochlorite), or liquid (sodium hypochlorite) are sources of chlorine; however, all forms might not be legal for injecting into irrigation systems. For example, only sodium hypochlorite is legal for use in Florida.

When sodium hypochlorite is injected, the pH of the water rises. The resulting chloride and sodium ions are not detrimental to crops at typical injection rates. Chlorine materials should be injected at a rate to provide 1–2 ppm free residual chlorine at the most distant part of the irrigation system.

In addition to controlling microorganisms, hypochlorous acid also reacts with iron in solution to oxidize the ferrous form to the ferric form, which precipitates as ferric hydroxide. If irrigation water contains iron, this reaction with injected chlorine should occur before the filter system so the precipitate can be removed.

TABLE 7.27. LIQUID CHLORINE (SODIUM HYPOCHLORITE) INJECTION

Treatment Level (ppm)	% Concentration of Chlorine in Stock			
	1	3	5.25	10
	gal/hr Injection Per 100 gal/min Irrigation Flow Rate			
1	0.54	0.18	0.10	0.05
2	1.1	0.36	0.21	0.14
3	1.6	0.54	0.31	0.16
4	2.2	0.72	0.41	0.22
5	2.7	0.90	0.51	0.27
6	3.3	1.1	0.62	0.32
8	4.4	1.5	0.82	0.43
10	5.5	1.8	1.0	0.54
15	8.3	2.8	1.5	0.81
20	11.0	3.7	2.1	1.1
30	16.5	5.5	3.1	1.6

Adapted from the original publication by G. Clark and A. Smajstrla, *Treating Irrigation Systems with Chlorine* (Florida Cooperative Extension Service Circular 1039, 2002), https://ufdc.ufl.edu/UF00014458/00001/5j.

METHODS OF INJECTING FERTILIZER AND OTHER CHEMICAL SOLUTIONS INTO IRRIGATION PIPELINES

Fertilizers and other chemicals can be injected into drip irrigation systems (Tables 7.28 – 7.33 and Figures 7.8 – 7.15). Nitrogen is the most commonly injected nutrient because it is needed in largest amounts and is the most likely nutrient to be leached. Spoon-feeding N is a good way to optimize efficiency for this nutrient. Typically, the injection program follows a "sigmoid" curve for amounts during the season, starting with small amounts and gradually building the rate of application in concert with plant growth. Rate of application can level off or decrease as the crop gets close to harvest. Most University Extension Services have recommendations for fertilizer management with drip irrigation.

Four principal methods are used to inject fertilizers and other solutions into drip irrigation systems: (1) pressure differential, (2) the venturi (vacuum), (3) centrifugal pumps, and (4) positive displacement pumps. It is essential that irrigation systems equipped with a chemical injection system have a vacuum breaker (anti-siphon device) and a backflow preventer (check valve) installed upstream from the injection point. The vacuum-breaking valve and backflow preventer prevent chemical contamination of the water source in case of a water pressure loss or power failure. Operators may need a license to chemigate in some states. Local backflow regulations should be consulted prior to chemigation to ensure compliance.

Further Reading

H. Bayabil, K. Migliaccio, J. Crane, T. Olczyk, and Q. Wang, *Regulations and Guidelines for Chemigation* (2020), https://edis.ifas.ufl.edu/publication/ae542 (accessed 11 Nov. 2021).

TABLE 7.28. **REQUIRED VOLUME (GAL) OF CHEMICAL MIXTURE TO PROVIDE A DESIRED LEVEL OF AN ACTIVE CHEMICAL FOR DIFFERENT CONCENTRATIONS (LB/GAL) OF THE CHEMICAL IN THE STOCK SOLUTION**

S_{mx}

Mass of Chemical (lb) Per gal Stock Solution

Weight of Chemical Desired (lb)	0.2	0.4	0.6	0.8	1.0	2.0	3.0	4.0
				(gal Stock Solution)				
20	100	50	33	25	20	10	7	5
40	200	100	67	50	40	20	13	10
60	300	150	100	75	60	30	20	15
80	400	200	133	100	80	40	27	20
100	500	250	167	125	100	50	33	25
150	750	375	250	188	150	75	50	38
200	1,000	500	333	250	200	100	67	50
250	1,250	625	417	313	250	125	83	63
300	1,500	750	500	375	300	150	100	75
350	1,750	875	583	438	350	175	117	88
400	2,000	1,000	667	500	400	200	133	100
450	2,250	1,125	750	563	450	225	150	113
500	2,500	1,250	833	625	500	250	167	125

Adapted from original publication by G. A. Clark, D. Z. Haman, and F. S. Zazueta, *Injection of Chemicals into Irrigation Systems: Rates, Volumes, and Injection Periods* (Florida Cooperative Extension Service Bulletin 250, 2002), https://edis.ifas.ufl.edu/publication/AE116.

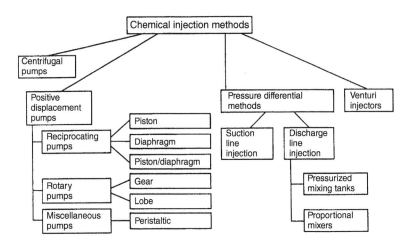

Figure 7.8. Classification of chemical injection methods for irrigation systems. D. Z. Haman and F. S. Zazueta, *Chemical Injection Methods for Irrigation* (Florida Cooperative Extension Service Circular 864, 2021), https://edis.ifas.ufl.edu/publication/wi004 (accessed 11 Nov. 2021).

TABLE 7.29. COMPARISON OF VARIOUS CHEMICAL INJECTION METHODS

Injector	Advantages	Disadvantages
	Centrifugal Pumps	
Centrifugal pump injector	Low cost. Can be adjusted while running.	Calibration depends on system pressure. Cannot accurately control low injection rates.
	Positive Displacement Pumps	
Piston pumps	High precision. Linear calibration. Very high pressure. Calibration independent of pressure.	High cost. May need to stop to adjust calibration. Chemical flow not continuous.
Diaphragm pumps	Adjust calibration while injecting. High chemical resistance.	Nonlinear calibration. Calibration depends on system pressure. Medium to high cost. Chemical flow not continuous.
Piston/diaphragm	High precision. Linear calibration. High chemical resistance. Very high pressure. Calibration independent of pressure.	High cost. May need to stop to adjust calibration.
	Rotary Pumps	
Gear pumps Lobe pumps	Injection rate can be adjusted when running.	Fluid pumped cannot be abrasive. Injection rate is dependent on system pressure. Continuity of chemical flow depends on number of lobes in a lobe pump.
	Miscellaneous	
Peristaltic pumps	High chemical resistance. Major adjustment can be made by changing tubing size. Injection rate can be adjusted when running.	Short tubing life expectancy. Injection rate dependent on system pressure. Low to medium injection pressure.

354

Method	Advantages	Disadvantages
Pressure Differential Methods		
Suction line injection	Very low cost. Injection can be adjusted while running.	Permitted only for surface water source and injection of fertilizer. Injection rate depends on main pump operation.
Pressurized mixing tanks	Low to medium cost. Easy operation. Total chemical volume controlled.	Variable chemical concentration. Cannot be calibrated accurately for constant injection rate.
Discharge Line Injection		
Proportional mixers	Low to medium cost. Calibrate while operating. Injection rates accurately controlled.	Pressure differential required. Volume to be injected is limited by the size of the injector. Frequent refills required.
Venturi Injectors		
Venturi injector	Low cost. Water powered. Simple to use. Calibrate while operating. No moving parts.	Pressure drop created in the system. Calibration depends on chemical level in the tank.
Combination Methods		
Proportional mixers/ venturi	Greater precision than proportional mixer or venturi alone.	Higher cost than proportional mixer or venturi alone.

Adapted from D. Haman, A. Smajstrla, and F. Zazueta, *Chemical Injection Methods for Irrigation* (Florida Cooperative Extension Service Circular 864, 2021), https://edis.ifas.ufl.edu/publication/wi004 (accessed 11 Nov. 2021).

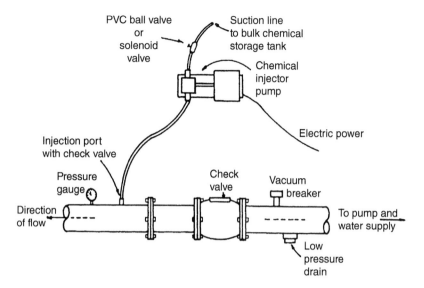

Figure 7.9. Single anti siphon device assembly. (Florida Cooperative Extension Service Bulletin 217, 1985).

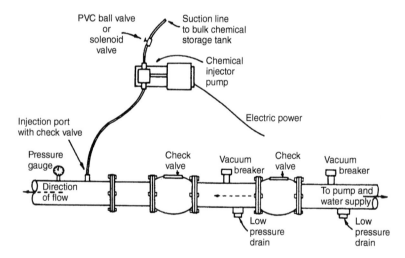

Figure 7.10. Double anti siphon device assembly. A. G. Smajstrla, D. S. Harrison, W. J. Becker, F. S. Zazueta, and D. Z. Haman, *Backflow Prevention Requirements for Florida Irrigation Systems* (Florida Cooperative Extension Service Bulletin 217, 1985).

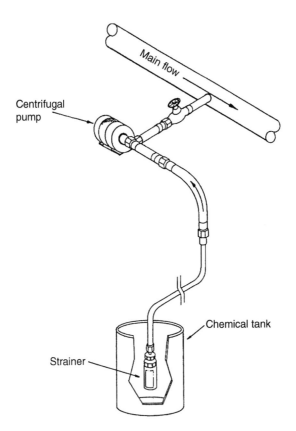

Figure 7.11. Centrifugal pump chemical injector. D. Z. Haman and F. S. Zazueta, *Chemical Injection Methods for Irrigation* (Florida Cooperative Extension Service Circular 864, 2021), https://edis.ifas.ufl.edu/publication/wi004 (accessed 11 Nov. 2021).

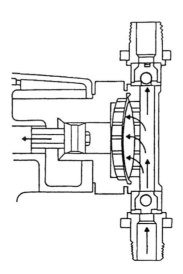

Figure 7.12. Diaphragm pump—suction stroke. (Florida Cooperative Extension Service Circular 864, 2021), https://edis.ifas.ufl.edu/publication/wi004 (accessed 11 Nov. 2021).

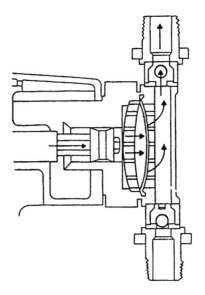

Figure 7.13. Diaphragm pump—discharge stroke. D. Z. Haman and F. S. Zazueta, *Chemical Injection Methods for Irrigation* (Florida Cooperative Extension Service Circular 864, 2021), https://edis.ifas.ufl.edu/publication/wi004 (accessed 11 Nov. 2021).

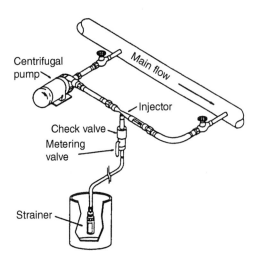

Figure 7.14. Venturi injector (Florida Cooperative Extension Service Circular 864, 2021), https://edis.ifas.ufl.edu/publication/ wi004 (accessed 11 Nov. 2021).

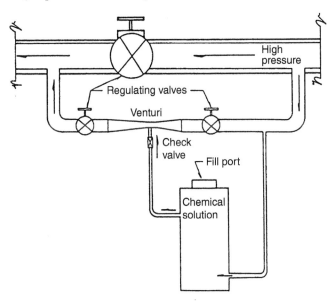

Figure 7.15. Proportional mixer/venturi. D. Z. Haman and F. S. Zazueta, *Chemical Injection Methods for Irrigation* (Florida Cooperative Extension Service Bulletin 864, 2021), https://edis.ifas.ufl.edu/publication/wi004 (accessed 11 Nov. 2021).

TABLE 7.30. WATER QUALITY GUIDELINES FOR IRRIGATION[1]

Situation	Degree of Problem		
	None	Increasing	Severe
Salinity			
EC (dS/m) or	Less than 0.75	0.75–3.0	More than 3.0
TDS (mg/L)	Less than 480	480–1,920	More than 1,920
Permeability			
Low EC (dS/m) or	More than 0.5	0.5–0	—
Low TDS (mg/L)	More than 320	320–0	—
SAR	Less than 6.0	6.0–9.0	More than 9.0
Toxicity of Specific Ions to Sensitive Crops			
ROOT ABSORPTION			
Sodium (evaluated by SAR)	SAR less than 3	3–9	More than 9
Chloride			
meq/L	Less than 2	2–10	More than 10.0
mg/L	Less than 70	70–345	More than 345
Boron (mg/L)	1.0	1.0–2.0	2.0–10.0
RELATED TO FOLIAR ABSORPTION (SPRINKLER IRRIGATED)			
Sodium			
meq/L	Less than 3.0	More than 3	—
mg/L	Less than 70	70	—
Chloride			
meq/L	Less than 3.0	More than 3	—
mg/L	Less than 100	100	—

TABLE 7.30. WATER QUALITY GUIDELINES FOR IRRIGATION[1]
(Continued)

	Degree of Problem		
Situation	None	Increasing	Severe
MISCELLANEOUS			
NH_4 and NO_3–N (mg/L)	Less than 5	5–30	More than 30
HCO3			—
meq/L	Less than 1.0	1.5–8.5	More than 8.5
mg/L	Less than 40	40–520	More than 520
pH	Normal range: 6.5–8.3	More than 8.3	—

Adapted from D. S. Farnham, R. F. Hasek, and J. L. Paul, "Water Quality," (University of California Division of Agricultural Science Leaflet 2995, 1985); B. Hanson, S. Gratham, and A. Fulton, *Agricultural Salinity and Drainage* (University of California Division of Agriculture and Natural Resources Publication 3375, 1999).

[1] Interpretation is related to type of problem and its severity but is modified by circumstances of soil, crop, and locality.

TABLE 7.31. MAXIMUM CONCENTRATIONS OF TRACE ELEMENTS IN IRRIGATION WATERS

Element	For Waters Used Continuously on All Soils (mg/L)	For Use Up to 20 Years on Fine-Textured Soils of pH 6.0–8.5 (mg/L)
Aluminum	5.0	20.0
Arsenic	0.10	2.0
Beryllium	0.10	0.50
Boron	0.75	2.0–10.0
Cadmium	0.01	0.05
Chromium	0.10	1.0
Cobalt	0.05	5.0

TABLE 7.31.   MAXIMUM CONCENTRATIONS OF TRACE ELEMENTS IN IRRIGATION WATERS (*Continued*)

Element	For Waters Used Continuously on All Soils (mg/L)	For Use Up to 20 Years on Fine-Textured Soils of pH 6.0–8.5 (mg/L)
Copper	0.20	5.0
Fluoride	1.0	15.0
Iron	5.0	20.0
Lead	5.0	10.0
Lithium	2.5	2.5
Manganese	0.20	10.0
Molybdenum	0.01	0.05[1]
Nickel	0.20	2.0
Selenium	0.02	0.02
Vanadium	0.10	1.0
Zinc	2.00	10.0

Adapted from D. S. Farnham, R. F Hasek, and J. L. Paul, "Water Quality," (University of California Division of Agricultural Science Leaflet 2995, 1985).

[1] Only for acidic, fine-textured soils or acidic soils with relatively high iron oxide contents.

TABLE 7.32.   ESTIMATED YIELD LOSS TO SALINITY OF IRRIGATION WATER

Crop	Electrical Conductivity of Water (mmho/cm or dS/m) for Following Yield Loss			
	0	10%	25%	50%
Bean	0.7	1.0	1.5	2.4
Carrot	0.7	1.1	1.9	3.1
Strawberry	0.7	0.9	1.2	1.7
Onion	0.8	1.2	1.8	2.9
Radish	0.8	1.3	2.1	3.4
Lettuce	0.9	1.4	2.1	3.4
Pepper	1.0	1.5	2.2	3.4

TABLE 7.32. ESTIMATED YIELD LOSS TO SALINITY OF IRRIGATION WATER (*Continued*)

Crop	Electrical Conductivity of Water (mmho/cm or dS/m) for Following Yield Loss			
	0	10%	25%	50%
Sweet potato	1.0	1.6	2.5	4.0
Sweet corn	1.1	1.7	2.5	3.9
Potato	1.1	1.7	2.5	3.9
Cabbage	1.2	1.9	2.9	4.6
Spinach	1.3	2.2	3.5	5.7
Cantaloupe	1.5	2.4	3.8	6.1
Cucumber	1.7	2.2	2.9	4.2
Tomato	1.7	2.3	3.4	5.0
Broccoli	1.9	2.6	3.7	5.5
Beet	2.7	3.4	4.5	6.4

Adapted from original publication by R. S. Ayers, *Journal of the Irrigation and Drainage Division* 103 (1977):135–154.Water Salinity and Plant Irrigation (Department of Primary Industries and Regional Development. Agriculture and Food. Government of Western Australia, 2019), https://www.agric.wa.gov.au/water-management/water-salinity-and-plant-irrigation (accessed 10 Aug. 2021).

TABLE 7.33. RELATIVE TOLERANCE OF VEGETABLE CROPS TO BORON IN IRRIGATION WATER[1]

10–15 ppm	4–6 ppm	2–4 ppm	1–2 ppm	0.5–1 ppm
Asparagus	Beet	Artichoke	Broccoli	Bean
	Parsley	Cabbage	Carrot	Garlic
	Tomato	Cantaloupe	Cucumber	Lima bean
		Cauliflower	Pea	Onion
		Celery	Pepper	
		Corn	Potato	
		Lettuce	Radish	
		Turnip		

Adapted from E. V. Mass, "Salt Tolerance of Plants," *Applied Agricultural Research* 1(1):12–26 (1986).

[1] Maximum concentrations of boron in soil water without yield reduction.

PART **8**

VEGETABLE PESTS AND PROBLEMS

Knott's Handbook for Vegetable Growers, Sixth Edition. George J. Hochmuth and Rebecca G. Sideman.
© 2023 John Wiley & Sons Ltd. Published 2023 by John Wiley & Sons Ltd.
Companion website: https://www.wiley.com/go/hochmuth/vegetablegrowers6e

01

INTEGRATED PEST MANAGEMENT

BASICS OF INTEGRATED PEST MANAGEMENT

Integrated Pest Management (IPM) attempts to make the most efficient use of the strategies available to control pest populations by taking action to prevent problems, suppress damage levels, and use chemical pesticides only where needed. Rather than seeking to eradicate all pests entirely, IPM strives to prevent their development or to suppress their population numbers below levels that would be economically damaging.

Integrated means that a broad, interdisciplinary approach is taken using scientific principles of crop protection, incorporating a variety of methods and tactics.

Pest includes insects, mites, nematodes, plant pathogens, weeds, and vertebrates that adversely affect crop quality and yield.

Management refers to the attempt to control pest populations in a planned, systematic way by keeping their numbers or damage within acceptable levels.

Depending on the situation and the specific pest, some combination of *prevention*, *suppression*, or *eradication* may be necessary. For predictable pests, *prevention* of crop damage may be possible by understanding the pest life cycle. *Suppression* of pests involves keeping pest populations below the level at which they cause economic damage. *Eradication* may be feasible when foreign pests are accidentally introduced but are not yet established in a location, but this is generally a very difficult (and unnecessary) goal for most common pest problems in outdoor agricultural settings.

The key steps of IPM include:

- *Identify key pests and beneficial organisms.*
- *Use preventive cultural practices* to minimize pest population development.
- *Monitor pest populations* by routinely sampling fields.
- *Set economic thresholds.* Economic thresholds are the level at which the pest population causes economically significant yield or quality losses. If these thresholds are known, pests are then controlled only when the pest population threatens acceptable levels of quality and yield.

- *Make action decisions.* In some cases, cultural or chemical interventions may be necessary to reduce the crop threat, whereas in other cases, a decision will be made to wait, and continue monitoring.
- *Evaluation and follow-up* continuously, to make corrections, assess success, and identify future possibilities for improvement.

The following tools may be part of an IPM approach:

- *Choose resistant crop varieties* to protect against key pests.
- *Use natural enemies* to regulate pest populations using biological controls.
- *Use pheromone* (sex lure) traps to lure male insects to help monitor (and/or reduce) populations.
- *Avoid peak pest populations* by modifying planting times or using crop rotation, or by using physical barriers that exclude pests.
- *Use cultural practices that encourage plant health and discourage pests.* This can include building soil fertility to encourage healthy crops, minimizing flooding, irrigating appropriately, and maintaining row and plant spacing to encourage air flow.
- *Use pesticides that will be effective and specific to the pest of interest.* In an effective IPM program, pesticides are applied to manage a particular pest, and are chosen so as to have minimum impact on people and the environment. Pesticides are applied only when a pest population is large enough to threaten acceptable levels of yield and quality and no other economic control measure is available.
- *Use calibrated and effective pesticide application equipment,* so that the correct amounts of materials are applied to the right place to maximize control.

INTEGRATED PEST MANAGEMENT IN FIELD VS. PROTECTED CULTURE

The pest complex, or the group of diseases, insects, and weeds that attack a particular crop, varies with environment. Inside protected culture (screenhouses, greenhouses, or high tunnels), certain pests are limited, and others are encouraged. For example, certain insects may be physically excluded through screens, whereas others (e.g. two-spotted spider mite, broad mite, and thrips) may thrive in protected environments. Similarly, while fungal pathogens that require leaf wetness may be prevented in protected culture, those that do well in high humidity conditions (e.g. botrytis and powdery mildew) can thrive. While IPM principles and tools are the same in field and protected culture, several factors should be considered when designing an IPM plan for a protected environment:

- Consider the use of insect-exclusion materials like nettings to prevent pests from entering the greenhouse. Note that adding nettings may require alternative approaches to increase ventilation.
- In a closed system, inundative releases or other biological control strategies are often more feasible than in open field situations. The use of sentinel plants (that reveal pests before they attack crop plants) and banker plants (that host predators and parasites of pest insects) may be particularly useful in closed or semi-closed environments.
- There are some limitations on the chemicals that may be used inside high tunnels and greenhouses. Residues may last longer when not exposed to the environment, and worker exposure can be greater inside a structure. Make sure to read labels carefully to be sure that use on crops in protected cultivation is legal.
- Sanitation is key. Use clean planting material, pathogen-free growing media, clean tools and equipment, and keep the area surrounding protected growing areas free from diseased and infested material.

Resources for Integrated Pest Management

- UC IPM Online-Statewide Integrated Pest Management Program, http://www.ipm.ucdavis.edu/.
- The Pennsylvania Integrated Pest Management Program, http://paipm.cas.psu.edu/.
- North Central Integrated Pest Management Center, https://www.ncipmc.org/.
- Northeastern IPM Center, http://northeastipm.org/.
- The Southern Integrated Pest Management Center, https://southernipm.org/.
- New York State Integrated Pest Management Program. Cornell University, http://www.nysipm.cornell.edu/.
- IPM Florida, http://ipm.ifas.ufl.edu/.

02

PEST MANAGEMENT IN ORGANIC PRODUCTION SYSTEMS

Organic management of diseases, insects, nematodes, weeds, and other pests employs an approach similar to that used in IPM. A diverse set of cultural tactics are used first to manage pests, with the use of pesticides as a last resort. A thorough understanding of the biology and ecology of pests is essential. Many of the cultural practices described throughout this section (crop rotation, resistant varieties, scouting, avoidance, exclusion, etc.) are foundations of organic pest management programs.

In general, any pesticides permitted in certified organic systems are of natural origin, and synthetic ones are prohibited, but there are exceptions. In the United States, the National List of Allowed and Prohibited Substances (https://www.ams.usda.gov/rules-regulations/organic/national-list) lists synthetic materials that are allowed, and natural materials that are prohibited, in organic crop and livestock production. The Organic Materials Review Institute also compiles regular lists of products and materials allowed for organic use (https://www.omri.org/omri-lists).

A particular challenge when designing an organic pest management plan is to find accurate efficacy data for the many natural and biological pest management products that are now commercially available. Some excellent resources that provide some of these data, and guidance in preparing an organic pest management plan, are listed below.

SELECTED RESOURCES FOR PEST MANAGEMENT IN ORGANIC SYSTEMS

B. Caldwell, E. Rosen, E. Sideman, A. Shelton, and C. Smart, *Resource Guide for Organic Insect and Disease Management,* 2nd ed. (Cornell University, 2013), http://web.pppmb.cals.cornell.edu/resourceguide/ (accessed 13 Sept. 2021).

A. Shelton, *Biological Control: A Guide to Natural Enemies in North America*, https://biocontrol.entomology.cornell.edu/index.php (accessed 13 Sept. 2021).

J. Masabni, F. Dainello, and P. Lillard, *Organic Vegetable Production Guide* (Texas A&M Publication B-6237), https://aggie-horticulture.tamu.edu/vegetable/guides/organic-vegetable-production-guide/ (accessed 13 Sept. 2021).

Sustainable Agriculture Research and Education, https://www.sare.org/resources/organic-production/ (accessed 13 Sept 2021).

eOrganic, The Organic Agriculture Community of Practice for the Extension Foundation, http://eorganic.org/ (accessed 13 Sept. 2021).

A. Seaman, *Organic Production and IPM Guide for Various Vegetable Crops* (Cornell University, 2016).

Peas, https://ecommons.cornell.edu/handle/1813/42896.

Spinach, https://ecommons.cornell.edu/handle/1813/42898

Snap Beans, https://ecommons.cornell.edu/handle/1813/42891

Carrots, https://ecommons.cornell.edu/handle/1813/42892

Cole Crops, https://ecommons.cornell.edu/handle/1813/42893

Lettuce, https://ecommons.cornell.edu/handle/1813/42895

Cucumbers and Squash, https://ecommons.cornell.edu/handle/1813/42894

Potato, https://ecommons.cornell.edu/handle/1813/42897

03
SOIL SOLARIZATION

BASICS OF SOIL SOLARIZATION

Soil solarization is a nonchemical control method that is particularly effective in areas that have high temperatures and long days for the required 4–6 weeks. In the northern hemisphere, this generally means that solarization is done during the summer months in preparation for a fall crop or for a crop in the following spring.

Soil solarization captures radiant heat energy from the sun, thereby causing physical, chemical, and biological changes in the soil. Transparent polyethylene plastic placed on moist soil during the hot summer months increases soil temperatures to levels lethal to many soil-borne plant pathogens, weed seeds, and seedlings (including parasitic seed plants), nematodes, and some soil-residing mites. Soil solarization also improves plant nutrition by increasing the availability of nitrogen and other essential nutrients.

Time of Year. Highest soil temperatures are obtained when the days are long, air temperatures are high, the sky is clear, and there is no wind.

Plastic Color. Clear polyethylene should be used, not black plastic. Transparent plastic results in greater transmission of solar energy to the soil. Polyethylene should be UV stabilized so the tarp does not degrade during the solarization period.

Plastic Thickness. Polyethylene 1 mil thick is the most efficient and economical for soil heating. However, it is easier to rip or puncture and is less able to withstand high winds than thicker plastic. Users in windy areas may prefer to use plastic 1 1/2 to 2 mils thick. Thick transparent plastic (4–6 mils) reflects more solar energy than does thinner plastic (1–2 mils) and results in slightly lower temperatures.

Preparation of the Soil. It is important that the area to be treated is level and free of weeds, plants, debris, and large clods that would raise the plastic off the ground. Maximum soil heating occurs when the plastic is close to the soil; therefore, air pockets caused by large clods or deep furrows should be avoided. Soil should be tilled as deeply as possible and have moisture at field capacity.

Partial vs. Complete Soil Coverage. Polyethylene tarps may be applied in strips (a minimum of 2–3 ft wide) over the planting bed or as continuous large pieces of sheeting held in place by soil. In some cases, strip coverage may be more practical and economical than full soil coverage.

Soil Moisture. Soil must be moist (field capacity) for maximum effect because moisture not only makes organisms more sensitive to heat, but it also conducts heat faster and deeper into the soil.

Duration of Soil Coverage. Killing of pathogens and pests is related to time and temperature of exposure. The longer the soil is heated, the deeper the control. Although some pest organisms are killed within days, 4–6 weeks of treatment in full sun during the summer is usually best. The upper limit for temperature is about 115°F for vascular plants, 130°F for nematodes, 140°F for fungi, and 160–212°F for bacteria.

BIOSOLARIZATION, ANAEROBIC SOIL DISINFESTATION, AND TARPING

Variations in the theme of solarization have increased in popularity in recent years. These include alternative strategies that can be effective in cooler regions, where solar heating is insufficient to kill pathogens and weeds under clear plastic.

Biosolarization. The addition of organic materials to the soil prior to solarization has been shown, in some cases, to increase the efficacy of solarization. This is thought to be because both heat and microbial decomposition are at work, attacking pests and pathogens while retaining some beneficial biological organisms.

Anaerobic Soil Disinfestation (ASD). ASD is very similar to biosolarization; some use the terms interchangeably. ASD involves incorporating organic amendments such as wheat bran or molasses to the soil, irrigating to field capacity, and then covering the soil with a gas-impermeable polyethylene sheet to create anaerobic conditions. The polyethylene sheet in this system may be clear or black. This practice has been shown to reduce the presence of pathogens, nematodes, weeds, and other pests. This is an area of current research, and more information will inform concrete research-based recommendations in the near future.

Tarping. Tarping uses opaque coverings in place of the clear covers used in solarization, and the primary goal is weed control. Blocking light is the primary mechanism by which this works, and it therefore takes longer than solarization (often 6 weeks). In smaller-scale agriculture in the Northeastern United States, there has been a recent interest in tarping (sometimes called *occultation*), where growers utilize recycled silage tarps, and reuse them several times as they prepare different cropping areas.

Original material adapted from G. S. Pullman, J. E. DeVay, C. L Elmore, and W. H. Hart, *Soil Solarization* (California Cooperative Extension Leaflet 21377, 1984), and from D. O. Chellemi, *Soil Solarization for Management of Soilborne Pests* (Florida Cooperative Extension Fact Sheet PPP51, 1995).

R. Kruger and R. McSorley, *Solarization for Pest Management in Florida* (University of Florida, 2018), https://edis.ifas.ufl.edu/publication/IN824.

The Soil Solarization Home (Hebrew University of Jerusalem, 2021), https://plantpathology.agri.huji.ac.il/jaacov-katan/solarization.

J. J. Stapleton, C. A. Wilen, and R. H. Molinar, *Soil Solarization for Gardens and Landscapes* (University of California, 2019), http://ipm.ucanr.edu/PMG/PESTNOTES/pn74145.html.

A. Hagan and W. Gazaway, *Soil Solarization for the Control of Nematodes and Soilborne Diseases* (Auburn University, 2018), https://www.aces.edu/wp-content/uploads/2018/09/ANR-0713_SoilSolarization_030918.pdf.

J. J. Stapleton, R. M. Dahlquist-Willard, Y. Achmon, M. N. Marshall, J. S. VandeGheynst, and C. W. Simmons, *Advances in Biosolarization Technology to Improve Soil Health and Organic Control of Soilborne Pests* (2016), https://eorganic.info/sites/eorganic.info/files/u27/1.1.2-Stapleton-Biosolarization-Final.pdf.

04

PESTICIDE-USE PRECAUTIONS

GENERAL SUGGESTIONS FOR PESTICIDE SAFETY

All chemicals are potentially hazardous and should be used carefully. Follow exactly the directions, precautions, and limitations given on the container label. Store all chemicals in a safe place where children, pets, and livestock cannot reach them. Do not reuse pesticide containers. Avoid inhaling fumes and dust from pesticides. Avoid spilling chemicals; if they are accidentally spilled, remove contaminated clothing and thoroughly wash the skin with soap and water immediately.

Observe the following rules:

1. Avoid drift from the application area to adjacent areas occupied by other crops, humans, or livestock, or to bodies of water.
2. Wear goggles, an approved respirator, and neoprene gloves when loading or mixing pesticides. Aerial applicators should be loaded by a ground crew.
3. Pour chemicals at a level well below the face to avoid splashing or spilling onto the face or eyes.
4. Have plenty of soap and water on hand to wash contaminated skin in the event of spilling.
5. Change clothing and bathe after the job is completed.
6. Know the insecticide, the symptoms of overexposure to it, and a physician who can be called quickly. If symptoms appear (contracted pupils, blurred vision, nausea, severe headache, dizziness), stop operations at once and contact a physician.

DESCRIPTION OF THE WORKER PROTECTION STANDARD

The worker protection standard (WPS) is a set of regulations issued by the U.S. Environmental Protection Agency designed to protect agricultural workers and pesticide handlers from exposure to pesticides.

Important terms and definitions:

Agricultural Worker: Those who perform tasks related to the growing and harvesting plants on farms or in greenhouses, nurseries and forests. Workers include anyone employed for any type of compensations (including self-employed).

Personal Protective Equipment (PPE): Clothing and equipment, such as goggles, gloves, boots, aprons, coveralls and respirators, that provide protection from exposure to pesticides.

Pesticide Handler: Those who mix, load, transfer or apply pesticides; or do other tasks that bring them into direct contact with pesticides.

Restricted Entry Interval (REI): The time immediately after a pesticide application when entry into the treated area is limited. Lengths of restricted entry intervals range between 4 and 72 hours. The exact amount of time is product-specific and is indicated on the pesticide label.

The WPS applies to all agricultural employers whose employees are involved in the production of agricultural plants on a farm, forest, nursery, or greenhouse. The employers include owners or managers of farms, forests, nurseries, or greenhouses as well as commercial (custom) applicators and crop advisors who provide services for the production of agricultural plants on these sites, and to employers of researchers who help grow and harvest plants. The WPS requires that specific protections be provided to workers and pesticide handlers to prevent pesticide exposure. The agricultural employer is responsible for providing those protections to employees.

Under the WPS, employers are required to do the following:

- Do not retaliate against a worker or a handler.
- *Provide Pesticide Safety Training.* Employers must provide annual pesticide safety training for pesticide handlers and agricultural workers before they begin work in an area where a pesticide has been used or an REI has been in effect in the past 30 days. Employers must keep records of this training for 2 years.
- *Provide Access to Information at a Central Location.* This must contain:
  1. An approved EPA safety poster showing how to keep pesticides from a person's body and how to clean up any contact with a pesticide.
  2. Name, address, and telephone number of the nearest medical care facility.
  3. Information about each pesticide application, including description of treated area, product name, EPA registration number of product, active ingredient of pesticide, time and date of application, and the restricted entry interval (REI) for the pesticide. Safety data sheets (SDSs) for all applied materials must also be kept for 30 days after REI expires.
- Application information and SDS must be kept for 2 years from end of REI and made available upon request.
- *Provide Decontamination Areas and Supplies, including:*

1. Water for washing and eye flushing
2. Soap
3. Single-use towels
4. Water for whole-body wash (pesticide handler sites only)
5. Clean coveralls (pesticide handler sites only)

- Provide emergency assistance by making transportation available to a medical care facility in case of a pesticide injury or poisoning and providing information about the pesticide(s) to which the person may have been exposed.
- *Maintain Restricted Entry Intervals.* Pesticides have restricted entry intervals, a period after a pesticide application during which employers must keep everyone, except appropriately trained and equipped handlers, out of areas being treated with pesticides. Workers must be notified about applications and pesticide-treated areas and not to enter during the REI by providing oral warnings or by posting warning signs.

In addition to the duties listed above for all employers, employers of pesticide handlers are required to do the following:

- Implement restrictions during applications by ensuring that pesticides applied do not contact workers or other people. Also, handlers must suspend an application if workers or other people are in the application exclusion zone.
- Monitor handlers working with toxic pesticides.
- Provide specific instructions for handlers.
- Provide access to labeling information for handlers.
- Take steps to ensure equipment safety.
- *Personal Protective Equipment.* Personal protective equipment (PPE) must be provided by the employer to all pesticide handlers.
  1. Employers are responsible for providing clean PPE for each day. All PPE must be clean, in operating condition, used correctly, inspected each day before use, and repaired or replaced as needed.
  2. Respirators must fit correctly and respirator-purifying elements must be replaced at recommended intervals. Employers must provide respirator and fit testing, training, and medical evaluation according to OSHA standards. Records must be kept to document completion of first test, training, and medical evaluation.

3. Handlers must be provided clean, pesticide-free areas to store personal clothing and PPE.

4. Employers must provide instructions for people who clean PPE.

5. Employers must dispose of contaminated PPE.

Adapted from EPA, *Agricultural Worker Protection Standard*. Available at: https://www.epa.gov/ pesticide-worker-safety/agricultural-worker-protection-standard-wps (accessed 13 Sept. 2021).

ADDITIONAL INFORMATION ON SAFETY AND RULES AFFECTING PESTICIDE USE

Licenses and Permits

Farmers who use pesticides may require pesticide applicator licenses or permits according to state and federal law. It is important to check with your state to determine what is required in your state. In general:

- Farmers who apply restricted use pesticides need a private applicator license or permit.
- Workers who help someone who is licensed or certified to apply restricted-use materials may also need a license to assist.
- Farmers who use only general use pesticides may also require licenses or permits; requirements vary from state to state.
- In most states, commercial (for hire) applicators must follow rules that are more restrictive than those of private applicators.

The EPA Worker Protection Standards (WPS) must still be followed regardless of whether a pesticide license or certification is required.

Record Keeping

Federal Pesticide Recordkeeping Program, which requires all certified private pesticide applicators to keep records of their use of federally restricted use pesticides (RUP) for a period of 2 years. Under this law, all certified private pesticide applicators who have no requirement through State regulations to maintain RUP records must comply with the Federal pesticide recordkeeping regulations. The following must be recorded: Brand or product name, EPA registration number, Total quantity applied, Date, Location, Crop/Commodity/Site treated, Size of area treated, Name of applicator, and Certification number of applicator.

A template record-keeping forms can be found at:
https://www.ams.usda.gov/sites/default/files/media/Pesticide%20
Record%20Form.pdf

Emergency Planning and Community Right-to-Know Act (EPCRA)

Authorized by the Superfund Amendments and Reauthorization Act
(SARA), this act sets rules requiring anyone who stores extremely
hazardous materials to inform the proper authorities, in case of accidental
release. Each state has related statutes. Lists of chemicals regulated, and
reportable quantities, are available at:
   https://www.epa.gov/epcra/epcra-section-304#chemicals

Endangered Species Act

The U. S. E.P.A. has determined threshold pesticide application rates that
may affect listed species. This information is presented on the pesticide label.
   https://www.epa.gov/endangered-species

PESTICIDE HAZARDS TO HONEYBEES

Honeybees and other bees are necessary for pollination of vegetables in the
gourd family—cantaloupe and other melons, cucumber, pumpkin, squash,
and watermelon. Bees and other pollinating insects are necessary for all the
insect-pollinated vegetables grown for seed production. Some pesticides are
extremely toxic to bees and other pollinating insects, so certain precautions
are necessary to avoid injury to them.

*Recommendations for Vegetable Growers and Pesticide Applicators to
protect bees:*

1. Participate actively in areawide integrated crop management
   programs.
2. Follow pesticide label directions and recommendations.
3. Apply hazardous chemicals in late afternoon, night, or early morning
   (generally 6 P.M.–7 A.M.) when honeybees are not actively foraging.
   Evening applications are generally somewhat safer than morning
   applications.
4. Use pesticides that are relatively nonhazardous to bees whenever this
   is consistent with other pest-management strategies. Choose the least
   hazardous pesticide formulations or tank mixes.

5. Become familiar with bee foraging behavior and types of pesticide applications that are hazardous to bees.
6. Know the location of all apiaries in the vicinity of fields to be sprayed.
7. Avoid drift, overspray, and dumping of toxic materials in noncultivated areas.
8. Survey pest populations and be aware of current treatment thresholds in order to avoid unnecessary pesticide use.
9. Determine if bees are foraging in target area so protective measures can be taken.

TOXICITY OF CHEMICALS USED IN PEST CONTROL

The danger in handling pesticides does not depend exclusively on toxicity values (see Table 8.1 for EPA accepted categories of pesticide toxicity). Hazard is a function of both toxicity and the amount and type of exposure. Some chemicals are very hazardous from dermal (skin) exposure as well as oral (ingestion). Although inhalation values are not given, this type of exposure is similar to ingestion. A compound may be highly toxic but present little hazard to the applicator if the precautions are followed carefully.

TABLE 8.1. CATEGORIES OF PESTICIDE TOXICITY[1]

		LD50 Value (mg/kg)	
Categories	Signal Word	Oral	Dermal
I	Danger–Poison	0–50	0–200
II	Warning	50–500	200–2,000
III	Caution	500–5,000	2,000–20,000
IV	None[2]	5,000	20,000

Resources on toxicity of pesticides: W. Hock, *Toxicity of Pesticides* (Pennsylvania State University, 2016), https://extension.psu.edu/toxicity-of-pesticide (accessed 13 Sept. 2021).

[1] EPA accepted categories.

[2] No signal word required based on acute toxicity; however, products in this category usually display "Caution."

Toxicity values are expressed as acute oral LD_{50} in terms of milligrams of the substance per kilogram (mg/kg) of test animal body weight required to kill 50% of the population. The acute dermal LD_{50} is also expressed in mg/kg. These acute values are for a single exposure and not for repeated exposures such as may occur in the field. Rats are used to obtain the oral LD_{50}, and the test animals used to obtain the dermal values are usually rabbits.

PESTICIDE FORMULATIONS

Several formulations of many pesticides are available commercially. Some are emulsifiable concentrates, flowables, wettable powders, dusts, and granules. After any pesticide recommendation, one of these formulations is suggested; however, unless stated to the contrary, equivalent rates of another formulation or concentration of that pesticide can be used. In most cases, pesticide experts suggest that sprays rather than dusts be applied to control pests of vegetables. This is because sprays have produced better control and resulted in less drift.

PREVENTING SPRAY DRIFT

Pesticides sprayed onto soil or crops may be subject to movement or drift away from the target due mostly to wind. Drift may lead to risks to nearby people and wildlife, damage to nontarget plants, and pollution of surface water or groundwater.

Factors That Affect Drift

1. *Droplet size*. Smaller droplets can drift longer distances. Droplet size can be managed by selecting the proper nozzle type and size, adjusting the spray pressure, and by increasing the viscosity of the spray mixture.
2. *Equipment adjustments*. Routine calibration of spraying equipment and general maintenance should be practiced. Equipment can be fitted with hoods or shields to reduce drift away from the sprayed area.
3. *Weather conditions*. Spray applicators must be aware of wind speed and direction, relative humidity, temperature, and atmospheric stability at time of spraying.

Spraying Checklist to Minimize Drift

1. Do not spray on windy days (>12 mph).
2. Avoid spraying on extremely hot and dry days.
3. Use minimum required pressure.
4. Select correct nozzle size and spray pattern.
5. Keep the boom as close as possible to target.
6. Install hoods or shields on spray boom.
7. Leave unsprayed border of 50–100 ft near water supplies, wetlands, neighbors, and nontarget crops.
8. Spray when wind direction is favorable for keeping drift off of nontarget areas.

05

PESTICIDE APPLICATION AND EQUIPMENT

ESTIMATION OF CROP AREA

To calculate approximately the acreage of a crop in the field, multiply the length of the field by the number of rows or beds. Divide by the factor for spacing of beds.

Examples: Field 726 ft long with 75 rows 48 in. apart.

$$\frac{726 \times 75}{10,890} = 5 \text{ acres}$$

Field 500 ft long with 150 beds on 40-in. centers.

$$\frac{500 \times 150}{13,068} = 5.74 \text{ acres}$$

TABLE 8.2. FACTORS USED IN CALCULATING TREATED CROP AREA

Row or Bed Spacing (in.)	Factor
12	43,560
18	29,040
24	21,780
30	17,424
36	14,520
40	13,068
42	12,445
48	10,890
60	8,712
72	7,260
84	6,223

TABLE 8.3. DISTANCE TRAVELED AT VARIOUS TRACTOR SPEEDS

mph	ft/min	mph	ft/min
1.0	88	3.1	273
1.1	97	3.2	282
1.2	106	3.3	291
1.3	114	3.4	299
1.4	123	3.5	308
1.5	132	3.6	317
1.6	141	3.7	325
1.7	150	3.8	334
1.8	158	3.9	343
1.9	167	4.0	352
2.0	176	4.1	361
2.1	185	4.2	370
2.2	194	4.3	378
2.3	202	4.4	387
2.4	211	4.5	396
2.5	220	4.6	405
2.6	229	4.7	414
2.7	237	4.8	422
2.8	246	4.9	431
2.9	255	5.0	440
3.0	264		

CALCULATIONS OF SPEED OF EQUIPMENT AND AREA WORKED

To review the actual performance of a tractor, determine the number of seconds required to travel a certain distance. Then use the formula

$$\text{Speed(mph)} = \frac{\text{distance traveled(ft)} \times 0.682}{\text{time to cover distance(sec)}}$$

or the formula

$$\text{Speed(mph)} = \frac{\text{distance traveled(ft)}}{\text{time to cover distance(sec)} \times 1.47}$$

Another method is to walk beside the machine counting the number of normal paces (2.93 ft) covered in 20 seconds. Move decimal point one place. Result equals tractor speed (mph).

Example: 15 paces/20 sec = 1.5 mph.

The working width of an implement multiplied by mph equals the number of acres covered in 10 hr. This includes an allowance of 17.5% for turning at the ends of the field. By moving the decimal point one place, which is equivalent to dividing by 10, the result is the acreage covered in 1 hr.

Example: A sprayer with a 20-ft boom is operating at 3.5 mph. Thus, 20 × 3.5 = 70 acres/10 hr or 7 acres/hr.

TABLE 8.4. APPROXIMATE TIME REQUIRED TO WORK AN ACRE[1]

Rate (mph):	1	2	3	4	5	10
Rate (ft/min):	88	176	264	352	440	880

Effective Working Width of Equipment (in.)	Approximate Time Required (min/acre)					
18	440	220	147	110	88	44
24	330	165	110	83	66	33
36	220	110	73	55	44	22
40	198	99	66	50	40	20
42	189	95	63	47	38	19
48	165	83	55	41	33	17
60	132	66	44	33	26	13
72	110	55	37	28	22	11
80	99	50	33	25	20	10
84	94	47	31	24	19	9
96	83	42	28	21	17	8
108	73	37	24	19	15	7
120	66	33	22	17	13	6
240	33	17	11	8	7	3
360	22	11	7	6	4	2

[1] These figures have been calculated on the basis of 75% field efficiency to allow for turning and other lost time.

USE OF PESTICIDE APPLICATION EQUIPMENT

Ground Application

Boom-Type Sprayers
High-pressure, high-volume sprayers have been used for row-crop pest control for many years. Now a trend exists toward the use of sprayers that utilize lower volumes and pressures.

Airblast-Type Sprayers
Airblast sprayers are used in the vegetable industry to control insects and diseases. Correct operation of an airblast sprayer is more critical than for a boom-type sprayer. Do not operate an airblast sprayer under high wind conditions. Wind speed below 5 mph is preferable unless it becomes necessary to apply the pesticide for timely control measures, but drift and nearby crops must be considered. Do not overextend the coverage of the machine. Considerable visible mist from the machine moves into the atmosphere and does not deposit on the plant. If in doubt, use black plastic indicator sheets in the rows to determine deposit and coverage before a pest problem appears as evidence. Use correct gallonage and pressures to obtain proper droplet size to ensure uniform coverage across the effective swath width. Adjust the vanes and nozzles on the sprayer unit to give best coverage. Vane adjustment must occur in the field, depending on terrain, wind, and crop. Cross drives in the field allow the material to be blown down the rows instead of across them and help to give better coverage in some crops, such as staked or trellised tomato.

Air-boom Sprayers
These sprayers combine the airblast spray with the boom spray delivery characteristics. Air-boom sprayers are becoming popular with vegetable producers in an effort to achieve high levels of spray coverage with minimal quantity of pesticide.

Electrostatic Sprayers
These sprayers create an electrical field through which the spray droplet moves. Charged spray droplets are deposited more effectively onto plant surfaces and less drift results.

Aerial Application

Spraying should not be done when wind is more than 6 mph. A slight crosswind during spraying is advantageous in equalizing the distribution of the spray within the swath and between swaths. Proper nozzle angle and arrangements along the boom are critical and necessary to obtain proper distribution at ground level. Use black plastic indicator sheets in the rows to determine deposit and coverage patterns. Cover a swath no wider than is reasonable for the aircraft and boom being used. Fields of irregular shape or topography and ones bounded by woods, power lines, or other flight hazards should not be sprayed by aircraft.

CALIBRATION OF FIELD SPRAYERS

Width of Boom

The boom coverage is equal to the number of nozzles multiplied by the space between two nozzles.

Ground Speed (mph)

Careful control of ground speed is very important for accurate spray application. Select a gear and throttle setting to maintain constant speed. A speed of 2–3 mph is desirable. From a "running start," mark off the beginning and ending of a 30-sec run. The distance traveled in this 30-sec period divided by 44 will equal the speed in miles per hour.

Sprayer Discharge (gallons per minute—gpm)

Run the sprayer at a certain pressure, and catch the discharge from each nozzle for a known length of time. Collect all the discharge and measure the total volume. Divide this volume by the time in minutes to determine discharge in gallons per minute.

Before Calibrating

1. Thoroughly clean all nozzles, screens, etc., to ensure proper operation.
2. Check to be sure that all nozzles are the same.
3. Check the spray patterns of all nozzles for uniformity. Check the volume of delivery by placing similar containers under each nozzle. Replace nozzles that do not have uniform patterns or do not fill containers at the same rate.
4. Select an operating speed. Note the tachometer reading or mark the throttle setting. When spraying, be sure to use the same speed as used for calibrating.
5. Select an operating pressure. Adjust to desired pressure (pounds per square inch-psi) while pump is operating at normal speed and water is actually flowing through the nozzles. This pressure should be the same during calibration and field spraying.

Calibration (Jar Method)

Either a special calibration jar or a home-made one can be used. If you buy one, carefully follow the manufacturer's instructions.

Make accurate speed and pressure readings and jar measurements. Make several checks.

Any 1-quart or larger container, such as a jar or measuring cup, if calibrated in fluid ounces, can easily be used in the following manner:

1. Measure a course on the same type of surface (sod, plowed, etc.) and same type of terrain (hilly, level, etc.) as that to be sprayed, according to nozzle spacing as follows:

TABLE 8.5. COURSE LENGTH SELECTED BASED ON NOZZLE SPACING

Nozzle spacing (in.)	16	20	24	28	32	36	40
Course length (ft)	255	204	170	146	127	113	102

2. Time the seconds it takes the sprayer to cover the measured distance at the desired speed.
3. With the sprayer standing still, operate at selected pressure and pump speed. Catch the water from several nozzles for the number of seconds measured in Step 2.
4. Determine the average output per nozzle in ounces. The ounces per nozzle equal the gallons per acre applied for one nozzle per spacing.

Calibration (Boom or Airblast Sprayer)

1. Fill sprayer with water.
2. Spray a measured area (width of area covered × distance traveled) at constant speed and pressure selected from manufacturer's information.
3. Measure amount of water necessary to refill tank (gallons used).
4. Multiply gallons used by 43,560 and divide by the number of square feet in area sprayed. This gives gallons per acre.

$$\text{Gal / acre} = \frac{\text{gal used} \times 43,560}{\text{area sprayed (sq ft)}}$$

5. Add correct amount of spray material to tank to give the recommended rate per acre.

EXAMPLE

Assume: 10 gal of water used to spray an area 660 ft long and 20 ft wide
Tank size: 100 gal
Spray material: 2 lb (actual)/acre
Calculation:

$$\frac{\text{Gal used} \times 43,560}{\text{area sprayed(sq ft)}} = \frac{10 \times 43,560}{660 \times 20} = 33 \, \text{gal/acre}$$

$$\frac{\text{Tank capacity}}{\text{gal/acre}} \quad \frac{100(\text{tank size})}{33} = 3.03 \text{ acres sprayed per tank}$$

3.03 × 2 (lb/acre) = 6.06 lb material per tank

If 80% material is used:

$$\frac{6.06}{0.8} = 7.57 \, \text{lb material needed per tank to give 2 lb/acre rate}$$

Adapted from *2020/2021 Mid-Atlantic Commercial Vegetable Production Recommendations* (2021),
https://extension.umd.edu/resource/2020-2021-mid-atlantic-commercial-vegetable-production-
recommendations (accessed Oct. 2021).

CALIBRATION OF GRANULAR APPLICATORS

Sales of granular fertilizer, herbicides, insecticides, etc., for application
through granular application equipment have been on the increase.

Application rates of granular application equipment are affected by
several factors: gate openings or settings, ground speed of the applicator,
shape and size of granular material, and roughness of the ground.

Broadcast Application

1. From the label, determine the application rate.
2. From the operator's manual, set dial or feed gate to apply desired rate.
3. On a level surface, fill hopper to a given level and mark this level.
4. Measure test area—length of run will depend on size of equipment.
   It need not be one long run but can be multiple runs at shorter
   distances.
5. Apply material to measured area, operating at the speed applicator
   will travel during application.

6. Weigh amount of material required to refill hopper to the marked level.

7. Determine application rate:

Area covered =

$$\frac{\text{number of runs} \times \text{length of run} \times \text{width of application}}{43,560}$$

Application Rate =

$$\frac{\text{amount applied (pounds to refill hopper)}}{\text{area covered}}$$

Note. Width of application is width of the spreader for drop or gravity spreaders. For spinner applicators, it is the working width (distance between runs). Check operator's manual for recommendations, generally one-half to three-fourths of overall width spread.

Example

Assume: 50 lb/acre rate
Test run: 200 ft
Four runs made
Application width: 12 ft
11.5 lb to refill hopper

Calculation:

$$\text{Area covered} = \frac{4 \times 200 \times 12}{43,560} = 0.22 \, \text{acre}$$

$$\text{Application Rate} = \frac{11.5}{0.22} = 52.27 \, \text{lb}$$

8. If application rate is not correct, adjust feed gate opening and recheck.

Band Application

1. From the label, determine application rate.
2. From the operator's manual, determine applicator setting and adjust accordingly.
3. Fill hopper half full.

4. Operate applicator until all units are feeding.
5. Stop applicator; remove feed tubes at hopper.
6. Attach paper or plastic bag over hopper openings.
7. Operate applicator over measured distance at the speed equipment will be operated.
8. Weigh and record amount delivered from each hopper. (Compare to check that all hoppers deliver the same amount.)
9. Calculate application rate:

$$\text{Area covered in bands} = \frac{\text{length of run} \times \text{band width} \times \text{number of bands}}{43,560}$$

Application Rate

$$\frac{\text{Amount applied in bands} = \text{total amount collected}}{\text{area covered in bands}}$$

Changing from Broadcast to Band Application

$$\frac{\text{Band width in inches}}{\text{row spacing in inches}} \times \frac{\text{broadcast}}{\text{rate}}_{\text{per acre}} = \text{amount needed per acre}$$

Adapted from *2020/2021 Mid-Atlantic Commercial Vegetable Production Recommendations* (2021), https://extension.umd.edu/resource/2020-2021-mid-atlantic-commercial-vegetable-production-recommendations (accessed Oct. 2021).

CALIBRATION OF AERIAL SPRAY EQUIPMENT

Calibration

$$\text{Acres covered} = \frac{\text{length of swath}(\text{miles}) \times \text{width}(\text{ft})}{8.25}$$

$$\text{Acres/min} = \frac{2 \times \text{swath width} \times \text{mph}}{1,000}$$

$$\text{GPM} = \frac{2 \times \text{swath width} \times \text{mph} \times \text{gal / acre}}{1,000}$$

Adapted from O. C. Turnquist et al., *Weed, Insect, and Disease Control Guide for Commercial Vegetable Growers, Minnesota Agricultural Extension Service Special Report* 5 (1978).

CALIBRATION OF DUSTERS

Select a convenient distance that multiplied by the width covered by the duster, both expressed in feet, equals a convenient fraction of an acre. With the hopper filled to a marked level, operate the duster at this distance. Take a known weight of dust in a bag or other container and refill hopper to the marked level. Weigh the dust remaining in the container. The difference is the quantity of dust applied to the fraction of an acre covered.

Example:

Distance duster is operated × width covered by duster = area dusted
108 ft x 10 ft. = 1089 ft2
1089 ft2/43560 = 1/40 acre

If it takes 1 lb of dust to refill the hopper, the rate of application is 40 lb/acre.

MORE INFORMATION ON CALIBRATION OF SPRAYERS

Pesticide Equipment Calibration, http://edis.ifas.ufl.edu/
TOPIC_Calibration.
Sprayers 101, https://sprayers101.com/airblast101/.
Proper Calibration and Operation of Backpack and Hand Can Sprayers, https://ohioline.osu.edu/factsheet/fabe-531.
D. Overhults, *Applicator Training Manual for Aerial Application of Pesticides* (University of Kentucky), http://www.uky.edu/Agriculture/PAT/Cat11/Cat11.htm.

SPRAY EQUIVALENTS AND CONVERSIONS

Pesticide containers give directions usually in terms of pounds or gallons of material in 100 gal of water. The following tables make the conversion for smaller quantities of spray solution easy.

TABLE 8.6. SOLID EQUIVALENT TABLE

100 gal	25 gal	5 gal	1 gal
4 oz	1 oz	$3/16$ oz	$1/2$ oz
8 oz	2 oz	$3/8$ oz	1 tsp
1 lb	4 oz	$7/8$ oz	2 tsp
2 lb	8 oz	$1\frac{3}{4}$ oz	4 tsp
3 lb	12 oz	$2\frac{3}{8}$ oz	2 tbsp
4 lb	1 lb	$3\frac{1}{4}$ oz	2 tbsp + 2 tsp

TABLE 8.7. LIQUID EQUIVALENT TABLE

100 gal	25 gal	5 gal	1 gal
1 gal	1 qt	$6\frac{1}{2}$ oz	$1\frac{1}{4}$ oz
2 qt	1 pt	$3\frac{1}{4}$ oz	$5/8$ oz
1 qt	$1/2$ pt	$1^{9/16}$ oz	$5/16$ oz
$1\frac{1}{2}$ pt	6 oz	$1\frac{1}{4}$ oz	$1/4$ oz
1 pt	4 oz	$7/8$ oz	$3/16$ oz
8 oz	2 oz	$7/16$ oz	$1/2$ tsp
4 oz	1 oz	$1/4$ oz	$1/4$ tsp

TABLE 8.8. DILUTION OF LIQUID PESTICIDES TO VARIOUS CONCENTRATIONS

Dilution	1 gal	3 gal	5 gal
1 : 100	2 tbsp + 2 tsp	$1/2$ cup	$3/4$ cup + 5 tsp
1 : 200	4 tsp	$1/4$ cup	$6\frac{1}{2}$ tbsp
1 : 800	1 tsp	1 tbsp	1 tbsp + 2 tsp
1 : 1,000	$3/4$ tsp	$2\frac{1}{2}$ tsp	1 tbsp + 1 tsp

TABLE 8.9. PESTICIDE DILUTION CHART

Amount of Formulation Necessary to Obtain Various Amounts of
Active Ingredients

Insecticide Formulation	Amount of Formulation (At Left) Needed to Obtain the Following Amounts of Active Ingredients			
	¼ lb	½ lb	¾ lb	1 lb
1% dust	25	50	75	100
2% dust	12½	25	37½	50
5% dust	5	10	15	20
10% dust	2½	5	7½	10
15% wettable powder	1²/₃ lb	3¹/₃ lb	5 lb	6²/₃ lb
25% wettable powder	1 lb	2 lb	3 lb	4 lb
40% wettable powder	⅝ lb	1¼ lb	1⅞ lb	2½ lb
50% wettable powder	½ lb	1 lb	1½ lb	2 lb
73% wettable powder	¹/₃ lb	²/₃ lb	1 lb	1¹/₃ lb
23–25% liquid concentrate (2 lb active ingredient per gallon)	1 pt	1 qt	3 pt	2 qt
42–46% liquid concentrate (4 lb active ingredient per gallon)	½ pt	1 pt	1½ pt	1 qt
60–65% liquid concentrate (6 lb active ingredient per gallon)	¹/₃ pt	²/₃ pt	1 pt	1¹/₃ pt
72–78% liquid concentrate (8 lb active ingredient per gallon)	¼ pt	½ pt	¾ pt	1 pt

TABLE 8.10. PESTICIDE APPLICATION RATES FOR SMALL CROP PLANTINGS

Distance Between Rows (ft)	Amount (gal/acre)	Amount (Per 100 ft of Row)	Length of Row Covered (ft/gal)
1	75	22 oz	581
	100	30 oz	435
	125	1 qt, 5 oz	348
	150	1 qt, 12 oz	290
	175	1 qt, 20 oz	249
	200	1 qt, 27 oz	218
2	75	1 qt, 12 oz	290
	100	1 qt, 27 oz	218
	125	2 qt, 10 oz	174
	150	2 qt, 24 oz	145
	175	3 qt, 7 oz	124
	200	3 qt, 21 oz	109
3	75	2 qt, 2 oz	194
	100	2 qt, 24 oz	145
	125	3 qt, 14 oz	116
	150	4 qt, 4 oz	97
	175	4 qt, 26 oz	83
	200	5 qt, 16 oz	73

GUIDELINES FOR EFFECTIVE PEST CONTROL

Often, failure to control a pest is blamed on the pesticide when frequently the cause lies elsewhere. Some common reasons for failure include:

1. Delaying applications until pests are already well established.
2. Applying insufficient gallonage or using clogged or poorly arranged nozzles.
3. Selecting the wrong pesticide.

The following points are suggested for more effective pest control:

1. *Inspect fields regularly.* Frequent examinations (at least twice per week) help determine the proper timing of the next pesticide

application. Often less toxic and safer-to-handle chemicals are effective when pests are small in size and population.

2. *Control pests according to economic thresholds or schedule.* Economic thresholds are intended to reflect the population size that will cause economic damage and thus would warrant the cost of treatment.

3. *Weather conditions.* Spray only when wind velocity is less than 10 mph. Dust only when it is perfectly calm. Do not spray when sensitive plants are wilted during the heat of the day. If possible, make applications when ideal weather conditions prevail. Read label instructions to determine the ideal weather conditions for a particular material.

4. *Strive for adequate coverage of plants.* The principal reason aphids, mites, cabbage loopers, and diseases are serious pests is that they occur beneath leaves, where they are protected from spray deposit or dust particles. Improved control can be achieved by adding and arranging nozzles so that the application is directed toward the plants from the sides as well as from the tops. In some cases, nozzles should be arranged so that the application is directed beneath the leaves. As the season progresses, plant size increases, and so does the need for increased spray gallonage to ensure adequate coverage. Applying sprays with sufficient spray volume and pressure is important. Sprays from high-volume, high-pressure rigs (airblast) should be applied at 40–100 gal/acre at approximately 400 psi pressure. Sprays from low-volume, low-pressure rigs (boom type) should be applied at 50–100 gal/acre at approximately 100–300 psi pressure.

5. *Select the proper pesticide.* Know the pests to be controlled and choose the recommended pesticide and rate of application. For certain pests that are extremely difficult to control or are resistant, it may be important to rotate between pesticides with different modes of action, indicated by IRAC or FRAC (insecticide or fungicide resistance action committee) group.

6. *Pesticide compatibility.* To determine if two pesticides are compatible, use the following "jar test" before you tank mix pesticides or pesticides and fluid fertilizers:

   a. Add 1 pint of water or fertilizer solution to a clean quart jar. Then add the pesticide to the water or fertilizer solution in the same proportion as used in the field.

   b. To a second clean quart jar, add 1 pint of water or fertilizer solution. Then add ½ teaspoon of an adjuvant to keep the mixture emulsified. Finally, add the pesticide to the water-adjuvant or fertilizer-adjuvant in the same proportion as to be used in the field.

c. Close both jars tightly and mix thoroughly by inverting 10 times. Inspect the mixtures immediately and again after standing for 30 min. If the mix in either jar remains uniform for 30 min, the combination can be used. If either mixture separates but readily remixes, constant agitation is required. If nondispersible oil, sludge, or clumps of solids form, do not use the mixture.

7. *Calibrate application equipment.* Periodic calibration of sprayers, dusters, and granule distributors is necessary to ensure accurate delivery rates of pesticides per acre.

8. *Select correct sprayer tips.* The selection of proper sprayer tips for use with various pesticides is very important. Flat fan-spray tips are designed for preemergence and postemergence application of herbicides. They can also be used with insecticides, fertilizers, and other pesticides. Flat fan spray tips produce a tapered-edge spray pattern for uniform coverage where patterns overlap. Spray nozzles with even flat-spray tips (often designated E) are designed for band spraying where uniform distribution is desired over a zone 8–14 in. wide; they are generally used for herbicides.

Flood-type nozzle tips are generally used for complete fertilizer or liquid nitrogen, and are sometimes for spraying herbicides onto the soil surface prior to incorporation. They are less suited for spraying postemergence herbicides or for applying fungicides or insecticides to plant foliage. Spray at the maximum pressure recommended for the nozzle.

Wide-spray angle tips with full or hollow cone patterns are usually used for fungicides and insecticides. They are used at higher water volume and spray pressures than are commonly recommended for herbicide application with flat fan or flood-type nozzle tips.

9. *pH and pesticides.* Some materials carry a label cautioning the user against mixing the pesticide with alkaline materials because they undergo a chemical reaction known as "alkaline hydrolysis." This reaction occurs when the pesticide is mixed in water with a pH greater than 7. Check the pH of the mixing water. If acidification is necessary, there are several commercial nutrient buffer materials available on the market.

Adapted from *2020/2021 Mid-Atlantic Commercial Vegetable Production Recommendations* (2021), https://extension.umd.edu/resource/2020-2021-mid-atlantic-commercial-vegetable-production-recommendations (accessed Oct. 2021).

SPRAY ADJUVANTS OR ADDITIVES

Adjuvants are chemicals that, when added to a liquid spray, make it mix, wet, spread, stick, or penetrate better. Water is almost a universal diluent for pesticide sprays. However, water is not compatible with oily pesticides, and an *emulsifier* may be needed in order to obtain good mixing. Furthermore, water from sprays often remains as large droplets on leaf surfaces. A *wetting agent* lowers the interfacial tension between the spray droplet and the leaf surface and thus moistens the leaf. *Spreaders* are closely related to wetters and help to build a deposit on the leaf and improve weatherability. *Stickers* cause pesticides to adhere to the sprayed surface and are often called spray-stickers. They are oily and serve to increase the amounts of suspended solids held on the leaves or fruits by holding the particles in a resinlike film. *Extenders* form a sticky, elastic film that holds the pesticide on the leaves and thus reduces the rate of loss caused by sunlight and rainfall.

There are a number of adjuvants on the market. Read the label not only for dosages, but also for crop uses and compatibilities, because some adjuvants must not be used with certain pesticides. Although many formulations of pesticides contain adequate adjuvants, some do require additions on certain crops, especially cabbage, cauliflower, onion, and pepper.

Spray adjuvants for use with herbicides often serve a function distinctly different from that of adjuvants used with insecticides and fungicides. For example, adjuvants such as oils used with atrazine greatly improve penetration of the chemical into crop and weed leaves, rather than just give more uniform coverage. Do not use any adjuvant with herbicides unless there are specific recommendations for its use. Plant damage or even crop residues can result from using an adjuvant that is not recommended.

Resources

B. Young, *Compendium of Herbicide Adjuvants*, 7th ed. (Southern Illinois University, 2016), http://siu-weeds.com/adjuvants/index-adj.html.

06

VEGETABLE SEED TREATMENTS

Various vegetable seed treatments prevent early infection by seed-borne pathogens, protect the seed from infection by soil microorganisms, and guard against a poor crop stand or crop failure caused by attacks on seeds by soil insects. Commercial seed is often supplied with the appropriate treatment.

Two general categories of vegetable seed treatments are used. Eradication treatments kill disease-causing agents on or within the seed, whereas protective treatments are applied to the surface of the seed to protect against seed decay, damping-off, and soil insects. Hot-water treatment is the principal means of eradication, and chemical treatments usually serve as protectants. Follow time-temperature directions precisely for hot-water treatment and label directions for chemical treatment. When insecticides are used, seeds should also be treated with a fungicide.

Hot-Water Treatment

For a detailed discussion of hot-water treatment to eradicate to seed-borne plant pathogens, see Part 2.

Chemical Seed Treatments

Chemicals such as sodium hypochlorite (bleach) can also be used as an eradication treatment for certain bacterial pathogens that are present on the seed surface, for example:

1. *Tomato Bacterial Canker*. Soak seeds in 1.05% sodium hypochlorite solution for 20–40 min or 5% hydrochloric acid for 5–10 hr, rinse, and dry.
2. *Tomato Bacterial Spot*. Soak seeds in 1.3% sodium hypochlorite for 1 min, rinse, and dry.
3. *Pepper Bacterial Spot*. Soak seeds in 1.3% sodium hypochlorite for 1 min, rinse, and dry.

There are several commonly used fungicides for treating vegetable seed. These fungicides protect against fungal attack during the germination process, often resulting in more uniform plant stands. Seed protectants are most effective under cool germination conditions in a greenhouse or field, when germination is likely to be slow, and where germinating seeds might be exposed to disease-causing organisms. There is some recent evidence suggesting that fungicidal coatings can negatively affect beneficial fungal organisms in the soil that form relationships with germinating seeds, and that common insecticide seed treatments can also have negative effects on nontarget organisms.

Fungicidal seed treatments are applied as a dust or slurry and when dry, can be dusty. To reduce dust, the fungicide-treated seed can be covered by a thin polymer film and many seed companies offer film coated seeds. Large-seeded vegetables may require treatment with a labeled insecticide as well as a fungicide. Always follow label directions when pesticides are used.

DO NOT USE CHEMICALLY TREATED SEED FOR FOOD OR FEED

Originally adapted from A. F. Sherf and A. A. MacNab, *Vegetable Diseases and Their Control* (New York, Wiley, 1986).

Additional References for Seed Treatment

- M. McGrath, *Treatments for Managing Bacterial Pathogens in Vegetable Seed* (2005), http://vegetablemdonline.ppath.cornell.edu/NewsArticles/All_BactSeed.htm.
- S. Miller and M. Lewis Ivey, *Hot Water and Chlorine Treatment of Vegetable Seeds to Eradicate Bacterial Plant Pathogens* (Cooperative Extension Service, Ohio State University), https://u.osu.edu/vegprolab/grafting-publications/hot-water-and-chlorine-treatment-of-vegetable-seeds-to-eradicate-bacterial-plant-pathogens/.
- J. R. Lamichhane, M. P. You, V. Laudinot, M. J. Barbetti, J.-N. Aubertot, Revisiting sustainability of fungicide seed treatments for field crops. *Plant Disease* 104:610–623 (2020).

ORGANIC SEED TREATMENTS

Priming, pelleting, and hot water treatment are all strategies to improve seedling health that are consistent with organic production. The seed treatment technology for organic seeds is increasing in scope, and includes identifying seed coatings and treatments to increase germination uniformity and adaptation to mechanized seeding.

Plant-based oils and extracts have in some cases shown promise for suppressing damping off and other diseases, but efficacy data remain very limited. Compounds of interest include oils derived from thyme, cinnamon, clove, lemongrass, oregano, savory, and garlic.

There are several commercially available biological seed treatments that have potential to increase seedling health, for example: Kodiak (*Bacillus subtilis*), Mycostop (*Streptomyces grieseoviridis*), SoilGard (*Gliocladium virens*), T-22 Planter Box (*Trichoderma harzianum*), and Actinovate (*Streptomyces lydicus*). Efficacy data have been mixed, and additional research in this area is needed.

With the increasing interest in organic vegetables, the need for information about organic seed production, sources, and treatment is greater. Seed quality and vigor are important aspects for high-quality organic seeds. The seed industry is working to provide vegetable seeds that meet the requirements for organic crop production. Likewise, the seed treatment technology for organic seeds is increasing in scope, for example seed coating and treatment to increase germination uniformity and adaptation to mechanized seeding, among other needs.

For additional information, see:

E. Gatch, *Organic Seed Treatments and Coatings* (Washington State University), https://eorganic.org/node/749.

07
NEMATODES

PLANT PARASITIC NEMATODES

Nematodes are unsegmented round worms that range in size from microscopic to many inches long. Some nematodes, usually those that are microscopic or barely visible without magnification, attack vegetable crops and cause maladies, restrict yields, or in severe cases, lead to total crop failure. Many types of nematodes are known to infest the roots and aboveground plant parts of vegetable crops. Their common names are usually descriptive of the affected plant part and the resulting injury.

TABLE 8.11. COMMONLY OBSERVED NEMATODES IN VEGETABLE CROPS

Common Name	Scientific Name
Awl nematode	*Dolichodorus* spp.
Bud and leaf nematode	*Aphelenchoides* Spp.
Cyst nematode	*Heterodera* spp.
Dagger nematode	*Xiphinema* spp.
Lance nematode	*Hoplolaimus spp.*
Root-lesion nematode	*Pratylenchus* spp.
Root-knot nematode	*Meloidogyne* spp.
Spiral nematode	*Helicotylenchus* spp. and *Scutellonema* spp.
Sting nematode	*Belonolaimus* spp.
Stubby-root nematode	*Trichodorus* spp.
Stunt nematode	*Tylenchorhynchus* spp.

Nematodes are the most troublesome in areas with mild winters where soils are not subject to freezing and thawing. Management practices and chemical control are both required to keep nematode numbers low enough to permit normal plant growth where populations are not kept in check naturally by severe winters.

The first and most obvious control for nematodes is avoiding their introduction into uninfected fields or areas. This may be done by quarantine over large geographical areas or by means of good sanitation in smaller areas. A soil sample for a nematode assay through the County Extension Service can provide information on which nematodes are present and their population levels. This information is valuable for planning a nematode management program.

Once nematodes have been introduced into a field, several management practices will help to control them: rotating with crops that a particular species of nematode does not attack, frequent disking during hot weather, and alternating flooding and drying cycles.

If soil management practices are not possible or are ineffective, chemicals (nematicides) may have to be used to control nematodes. Some fumigants are effective against soil-borne disease, insects, and weed seeds—these are termed multipurpose soil fumigants. Growers should select a chemical for use against the primary problem to be controlled and use it according to label directions.

TABLE 8.12. PLANT PARASITIC NEMATODES KNOWN TO BE OF ECONOMIC IMPORTANCE TO VEGETABLES

Nematode	Bean and Peas	Carrot	Celery	Crucifers	Cucurbits	Leaf Crops	Okra	Onion	Potato	Sweet Corn	Sweet Potato	Tomato	Pepper	Eggplant
Root knot	X	X	X	X	X	X	X	X	X		X	X	X	X
Sting	X	X	X	X	X	X	X	X	X			X	X	X
Stubby root	X	X	X	X		X	X	X	X			X	X	X
Root lesion										X				
Cyst				X										
Awl	X	X												
Stunt										X				
Lance										X				
Spiral										X				
Ring														
Dagger														
Bud and leaf														
Reniform	X										X			

From J. Noling, *Nematodes and Their Management* (2009), https://www.semanticscholar.org/paper/Nematodes-and-Their-Management-Noling/2d306034 1ed08e4c 067af3d9540813637bb391fa. For more information on nematode management, see the *Florida Nematode Management Guide*, https://edis.ifas.ufl.edu/entity/topic/ nematode_management.

MANAGEMENT TECHNIQUES FOR CONTROLLING NEMATODES

1. *Crop rotation.* Exposing a nematode population to an unsuitable host crop is an effective means of reducing nematodes in a field. Cover crops should be established rapidly and kept weed-free. There are many cover crop options but growers need to consider the species of nematode in question and how long the alternate crop is needed. Also, certain cover crops may be effective as nonhosts for nematodes, but may not fit into a particular cropping sequence.

2. *Fallowing.* Practicing clean fallowing in the intercropping season is probably the single-most effective nonchemical means to reducing nematode populations. Clean discing of the field must be practiced frequently to keep weeds controlled because certain nematodes can survive of weeds.

3. *Plant resistance.* Certain varieties have genetic resistance to nematode damage. Where possible, these varieties should be selected when other horticultural traits in these varieties are acceptable. There are nematode resistant varieties in tomato and pepper.

4. *Soil amendments.* There are certain soil amendments, such as compost, manures, cover crops, chitin, and other materials that have been shown to reduce nematode populations. The age of the material, amount applied, and level of incorporation have effects on performance.

5. *Flooding.* Extended periods of flooding can reduce nematode populations. This practice should be practiced only where flooding is approved by environmental agencies. Alternating periods of flooding and drying seems to be more effective than a single flooding event.

6. *Soil solarization.* Nematodes can be killed by elevated heat created in the soil by covering the soil with clear polyethylene film for extended periods of time. Solarization works best under clear, hot, and dry climates. More information on solarization can be found in Part 8, section 3.

7. *Crop management.* Growers should rapidly destroy and till crops that are infested with nematodes. Irrigation water should not drain from infested fields to noninfested fields and equipment should be cleaned between infested and clean fields.

8. *Chemical control.* There are fumigant and nonfumigant nematicides approved for use against nematodes. Not all vegetable crops have

405

nematicides recommended for use. Effectiveness of the treatment depends on many factors including timing, placement in the soil, soil moisture, soil temperature, soil type, and presence of plastic mulch. Growers should refer to their state Extension recommendations for the approved chemicals for particular crops.

08

DISEASES

GENERAL DISEASE CONTROL PROGRAMS

Diseases of vegetable crops are caused by fungi, bacteria, viruses, and mycoplasmas. For a disease to occur, organisms must be transported to a susceptible host plant. This may be done by infected seeds or plant material, contaminated soil, wind, water, animals (including humans), or insects. Suitable environmental conditions must be present for the organism to infect and thrive on the crop plant. Effective disease control requires knowledge of the disease life cycle, time of likely infection, agent of distribution, plant part affected, and the symptoms produced by the disease.

Crop Rotation: Root-infecting diseases are the usual targets of crop rotation, although rotation can help reduce innocula of leaf- and stem-infecting organisms. Land availability and costs can make rotation challenging, but a well-planned rotation program is still a very important part of an effective disease control program.

Site Selection: Consider using fields that are free of volunteer crops and perimeter weeds that may harbor disease organisms. If aerial applications of fungicides are to be used, try to select fields that are geometrically adapted to serial spraying (long and wide), are away from homes, and have no bordering trees or power lines.

Deep Plowing: Use tillage equipment such as plows to completely bury plant debris to fully decompose plant material and kill disease organisms.

Weed Control: Certain weeds, particularly ones botanically related to the crop, may harbor disease agents that could move to the crop, especially viruses. Also, weeds within the crop could harbor diseases and by their physical presence could interfere with deposition of fungicides on the crop. Volunteer plants from previous crops should be carefully controlled in any nearby fallow fields.

Resistant Varieties: Where possible growers should choose varieties that carry genetic resistance to disease. Varieties with disease resistance will require less pesticide application.

Seed Protection: Seeds can be treated with fungicides to offer some degree of protection of the young seedling against disease attack. Seeds planted in warm soil will germinate fast and possibly outgrow disease development.

Healthy Transplants: Growers should always purchase or grow disease-free transplants. Growers should contract with good transplant growers and should inspect their transplants before having them shipped to the farm. Paying a little extra to a reputable transplant producer is good insurance.

Be Observant in the Field: The fields should be periodically checked for disease development by walking the field and inspecting the plants up close, not from behind the windshield. Have any suspicious situations diagnosed by a competent disease diagnosis laboratory.

Foliar Fungicides: Plant disease outbreaks sometimes can be prevented or minimized by timely use of fungicides. For some diseases, it is essential to have a preventative protectant fungicide program in place. For successful fungicide control, growers should consider proper chemical selection, use well-calibrated sprayers, use correct application rate, and follow all safety recommendations for spray application.

Integrated approach: Successful vegetable growers use a combination, or the entire set of strategies listed above. Routine implementation of combinations of strategies is needed where a grower desires to reduce the use of fungicides on vegetable crops.

COMMON VEGETABLE DISEASES

Some of the more common vegetable diseases are described below (Table 8.13). Consult your local Extension Service for recommendations for specific fungicide information, since products and use recommendations change frequently. When using fungicides, read the label and carefully follow the instructions. Do not exceed maximum rates, observe the interval between application and harvest, and apply only to those crops on the label. Make a record of the product used, trade name, concentration of the fungicide, dilution, rate applied per acre, and dates of application. Follow local recommendations for efficacy, and follow all directions on the label.

TABLE 8.13. DISEASE CONTROL FOR VEGETABLES

Crop	Disease	Description	Control
Asparagus	Fusarium root rot	Damping-off of seedlings. Yellowing, stunting, or wilting of the growing stalks; vascular bundle discoloration. Crown death gives fields a spotty appearance	Use disease-free crowns. Select fields where asparagus has not grown for 8 years. Use preplant fungicide crown dip
	Rust	Reddish or black pustules on stems and foliage	Cut and burn diseased tops. Use resistant varieties. Use approved fungicides
Bean	Anthracnose	Brown or black sunken spots with pink centers on pods, dark red or black cankers on stems and leaf veins	Use disease-free seed and rotate crops every 2 years. Plow stubble. Do not cultivate when plants are wet. Use approved fungicides
	Bacterial blight	Large, dry, brown spots on leaves, often encircled by yellow border; water-soaked spots on pods; reddish cankers on stems. Plants may be girdled	Use disease-free seed. Do not cultivate when plants are wet. Use 3-year rotation. Use approved fungicides
	Mosaic (several)	Mottled (light and dark green) and curled leaves; stunting, reduced yields	Use mosaic-resistant varieties. Control weeds in areas adjacent to field. Control insect vector (aphid, white fly) with insecticides
	Powdery mildew	Faint, slightly discolored spots appear first on leaves, later on stems and pods, from which white powdery spots develop and may cover the entire plant	Use approved fungicides

TABLE 8.13. DISEASE CONTROL FOR VEGETABLES (*Continued*)

Crop	Disease	Description	Control
	Rust	Red to black pustules on leaves; leaves yellow and drop	Use approved fungicides
	Seed rot	Seed or seedling decay, which results in poor stands. Occurs most commonly in cold, wet soils	Crop rotation. Treat seed with approved fungicides
	White mold	Water-soaked spots on plants. White, cottony masses on pods	Use approved fungicides
Beet	Cercospora	Numerous light tan to brown spots with reddish to dark brown borders on leaves	Use a long rotation. Use approved fungicides
	Damping-off	Seed decay in soil; young seedlings collapse and die	Avoid wet soils, rotate crops. Treat seed with approved fungicides
	Downy mildew	Lighter than normal leaf spots on upper surface and white mildew areas on lower side. Roots, leaves, flowers, and seed balls distorted on stecklings	Use approved fungicides
Broccoli, Brussels sprouts, cabbage, cauliflower, kale, kohlrabi	Alternaria leaf spot	Damping-off of seedlings. Small, circular yellow areas that enlarge in concentric circles, and become black and sooty	Use approved fungicides
	Black leg	Sunken areas on stem near ground line resulting in girdling; gray spots speckled with black dots on leaves and stems	Use hot-water-treated seed and long rotation. Sanitation

Crop	Disease	Symptoms	Control
	Black rot	Yellowing and browning of the foliage; blackened veins; stems show blackened ring when cross-sectioned	Use hot-water-treated seed and long rotation. Do not work wet fields. Sanitation
	Club root	Yellow leaves or green leaves that wilt on hot days; large, irregular swellings or "clubs" on roots	Start plants in new, steamed or fumigated plant beds. Adjust soil pH to 7.2 with hydrated lime before planting. Use approved fungicides
	Downy mildew	Begins as slight yellowing on upper side of leaves; white mildew on lower side; spots enlarge until plant dies	Use approved fungicides
	Fusarium yellows	Yellowish-green leaves; stunted plants; lower leaves drop	Use yellows-resistant varieties
Cantaloupe (see Vine crops)			
Carrot	Alternaria leaf blight	Small, brown to black, irregular spots with yellow margins may enlarge to infect the entire top	Use approved fungicides
	Cercospora leaf blight	Small, necrotic spots that may enlarge and infect the entire top	Use approved fungicides
	Aster yellows	Purpling of tops; yellowed young leaves at center of crown followed by bushiness due to excessive petiole formation. Roots become woody and form numerous adventitious roots	Control leafhopper vector with insecticides

TABLE 8.13. DISEASE CONTROL FOR VEGETABLES (*Continued*)

Crop	Disease	Description	Control
Celery	Aster yellows	Yellowed leaves; stunting; tissues brittle and bitter in taste	Use resistant varieties. Control leafhopper vector with insecticides. Control weeds in adjacent areas
	Bacterial blight	Bright-yellow leaf spots, center turns brown and a yellow halo appears with enlargement	Seedbed sanitation. Copper compounds
	Early blight	Dead, ash gray, velvety areas on leaves	Use approved fungicides
	Late blight	Yellow spots on old leaves and stalks that turn dark gray speckled with black dots	Use approved fungicides
	Mosaic	Dwarfed plants with narrow, gray, or mottled leaves	Control weeds in adjacent areas. Control aphid vector with insecticides
	Pink rot	Water-soaked spots; white- to pink-colored cottony growth at base of stalk leads to rotting	Crop rotation. Flooding for 4–8 weeks. Use approved fungicides
Cucumber (*see* Vine crops)			
Eggplant	Anthracnose	Sunken, tan fruit lesions	Use disease-free seed. Use approved fungicides
	Phomopsis blight	Young plants blacken and die; older plants have brown spots on leaves and fruit covered with brownish-black pustules	Use resistant varieties. Use approved fungicides
	Verticillium wilt	Slow wilting; browning between leaf veins; stunting	Fumigate soil with approved fumigants. Use verticillium-tolerant varieties. Use long rotation

412

Crop	Disease	Symptoms	Control
Endive, escarole, lettuce	Aster yellows	Center leaves bleached, dwarfed, curled or twisted. Heads do not form; young plants particularly affected	Control leafhopper vector with insecticides
	Big vein	Leaves with light green, enlarged veins developing into yellow, crinkled leaves; stunting; delayed maturity	Avoid cold, wet soils. Use tolerant varieties. Crop rotation
	Bottom rot	Damage begins at base of plants; blades of leaves rot first, then the midrib but the main stem is hardly affected	Avoid wet, poorly drained areas. Plant on raised beds. Practice 3-year rotation. Use approved fungicides
	Downy mildew	Light green spots on upper side of leaves; lesions enlarge and white mycelium appears on opposite side of spots; browning and dwarfing of plant	Use approved fungicides
	Drop	Wilting of outer leaves; watery decay on stems and old leaves	Crop rotation. Deep plowing. Raised beds. Use approved fungicides
	Mosaic	Mottling (yellow and green), ruffling, or distortion of leaves; plants have unthrifty appearance	Use virus-free MTO seed. Plant away from old lettuce beds. Control weeds. Control aphid vector with insecticide
	Tipburn	Edges of tender leaves brown and die; may interfere with growth; most severe on head lettuce	Use tolerant varieties. Prevent stress by providing good growing conditions
Lima bean	Downy mildew	Purpling and distortion of leaf veins; white downy mold on pods; blackened beans	Use resistant varieties and disease-free seed. Use approved fungicides
Okra	Southern blight	Mass of pinkish fungus bodies around base of plant; sudden loss of leaves	Crop rotation. Deep plowing of plant stubble
	Verticillium wilt	Stunting; chlorosis; shedding of leaves	Crop rotation. Avoid planting where disease previously present

413

TABLE 8.13. DISEASE CONTROL FOR VEGETABLES (*Continued*)

Crop	Disease	Description	Control
Onion	Blight (blast)	Papery spots on leaves; browning and death of upper portion of leaves; delayed maturity	Use approved fungicides
	Downy mildew	Begins as pale-green spot near tip of leaf; purple mold found when moisture present; infected leaves olive-green to black	Use approved fungicides
	Neck rot	Soft, brownish tissue around neck; scales around neck are dry, and black sclerotia may form. Essentially a dry rot if soft rot bacteria not present	Undercut, and windrow plants until inside neck tissues are dry before storage. Cure at 93–95°F for 5 days
	Pink rot	Plants are affected from seedling stage onward throughout life cycle. Affected roots turn pink, shrivel, and die	Avoid infected soils. Use tolerant varieties
	Purple blotch	Small, white sunken lesions with purple centers enlarge to girdle leaf or seed stem. Leaves and stems fall over 3–4 weeks after infection in severe cases. Bulb rot at and after harvest	Use approved fungicides
	Smut	Black spots on leaves; cracks develop on side of spot revealing black, sooty powder within	Crop rotation. Use approved fungicides

Crop	Disease	Symptoms	Control
Parsnip	Canker	Brown discoloration near shoulder or crown of root	Ridge soil over shoulders
Pea	Leaf blight	Leaves and petioles turn yellow and then brown. Entire plant may be killed	Practice 2-year rotation, use well-drained soil with pH 7.0
	Powdery mildew	White, powdery mold on leaves, stems, and pods, affected areas become brown and necrotic	Use disease-free seed and resistant varieties. Use approved fungicides
	Root rot	Rotted and yellowish-brown or black stems (below ground) and roots; outer layers of root along off leaving a central core	Early planting and 3-year rotation. Do not double crop with bean. Seed treatment
	Virus	Several viruses affect pea causing mottling, distortion of leaves, rosetting, chlorosis, or necrosis	Use resistant varieties. Control aphid vector with insecticides
Pepper	Wilt	Yellowing leaves; dwarfing, browning of xylem; wilting	Early planting and 3-year rotation. Use resistant varieties
	Anthracnose	Dark, round spots with black specks on fruits	Use approved fungicides
	Bacterial leaf spot	Yellowish-green spots on young leaves; raised, brown spots on undersides of older leaves; brown, cracked, rough spots on fruit; old leaves turn yellow	Use disease-free seed, hot-water-treated seed. Use approved bactericides. Use resistant varieties
Potato	Mosaic	Mottled (yellowed and green) and curled leaves; fruits yellow or show green ring spots; stunted; reduced yields	Use resistant varieties. Control insect vectors (particularly aphids) and weed hosts. Stylet oil
	Early blight	Dark brown spots on leaves; foliage injured; reduced yields	Bury all cull potatoes. Use approved fungicides

TABLE 8.13. DISEASE CONTROL FOR VEGETABLES (Continued)

Crop	Disease	Description	Control
	Late blight	Dark, then necrotic area on leaves and stem; infected tubers rot in storage. Disease is favored by moist conditions	Bury cull piles, Use approved fungicides
	Rhizoctonia	Necrotic spots, girdling and death of sprouts before or shortly after emergence. Brown to black raised spots on mature tubers	Avoid deep planting to encourage early emergence. Use disease-free seed. Use approved fungicides
	Scab	Rough, scabby, raised or pitted lesions on tubers	Crop rotation. Use resistant varieties. Maintain soil pH about 5.3
	Virus	A large number of viruses infect potato causing leaf mottling, distortion, and dwarfing. Some viruses cause irregularly shaped or necrotic area in tubers	Use certified seed. Control aphid and leafhopper vectors with insecticides
Radish	Downy mildew	Internal discoloration of root crown tissue. Outer surface may become dark and rough at the soil line	Select clean, well-drained soils. Use approved fungicides
	Fusarium wilt	Young plants yellow and die rapidly in warm weather. Stunting, unilateral leaf yellowing; vascular discoloration of fleshy roots	Use tolerant varieties. Avoid infected soil
Rhubarb	Crown rot	Wilting of leaf blades; browning at base of leaf stalk leading to decay	Plant in well-drained soil

Crop	Disease	Symptoms	Control
	Leaf spot	Tiny, greenish-yellow spots (resembling mosaic) on upper side of leaf, eventually browning and forming a white spot surrounded by a red band; these spots may drop out to give a shot-hole appearance	Use approved fungicides
Rutabaga, turnip	Alternaria	Small, circular, yellow areas that enlarge in concentric circles and become a black sooty color. Roots may become infested in storage	Use hot-water-treated seed. Use approved fungicides
	Anthracnose	Small, water-soaked spots on all above ground parts, which become light colored and may drop out, Small, sunken, dry spots on turnip roots, which are subject to secondary decay	Use approved fungicides
	Club root	Tumorlike swellings on taproot. Main root may be distorted. Diseased roots decay prematurely	Avoid soil previously infected with club root. Adjust acid soil to pH 7.3 by liming
	Downy mildew	Small, purplish, irregular spots on leaves, stems, and seedpods, which produce fluffy white growth. Desiccation of roots in storage	Use approved fungicides
	Mosaic virus	Stunted plants having ruffled leaves. Infected roots store poorly	Destroy volunteer plants. Control aphid vector with insecticides
Southern pea	Fusarium wilt	Yellowed leaves; wilted plants; interior of stems lemon yellow	Avoid infested soil

TABLE 8.13. DISEASE CONTROL FOR VEGETABLES (Continued)

Crop	Disease	Description	Control
Spinach	Blight (CMV)	Yellowed and curled leaves; stunted plants; reduced yields	Use tolerant varieties. Control aphid vector with insecticides
	Downy mildew	Yellow spots on upper surface of leaves; downy or violet-gray mold on undersides	Use resistant varieties. Use approved fungicides
Squash (see Vine crops)			
Strawberry	Anthracnose	Spotting and girdling of stolens and petioles, crown rot, fruit rot, and a black leaf spot; commonly occurs in southeastern United States	Use disease-free plants and resistant varieties. Use approved fungicides
	Gray mold	Rot on green or ripe fruit, beginning at calyx or contact with infected fruit; affected area supports white or gray mycelium	Use less susceptible varieties. Use approved fungicides
	Leaf scorch	Numerous irregular, purplish blotches with brown centers; entire leaves dry up and appear scorched	Use disease-free plants and resistant varieties. Renew perennial plantings frequently. Use approved fungicides
	Leaf spot	Indefinite-shaped spots with brown, gray, or white centers and purple borders	Use disease-free plants and resistant varieties. Use approved fungicides
	Powdery mildew	Characteristic white mycelium on leaves, flower, and fruit	Use resistant varieties. Use approved fungicides

Crop	Disease	Symptoms	Control
	Red stele	Stunted plants having roots with red stele that is seen when root is cut lengthwise	Improve drainage and avoid compaction of soil. Use disease-free plants and resistant varieties
Sweet corn	Verticillium	Marginal and interveinal necrosis of outer leaves, inner leaves remain green	Preplant soil fumigation. Use resistant varieties
	Bacterial blight	Dwarfing; premature tassels die; yellow bacterial slime oozes from wet stalks; stem dries and dies	Use resistant varieties. Control corn flea beetle with insecticides
	Leaf blight	Canoe-shaped spots on leaves	Use resistant varieties. Use approved fungicides
	Maize dwarf mosaic	Stunting; mottling of new leaves in whorl and poor ear fill at the base	Use tolerant varieties. Plant tolerant varieties around susceptible ones. Control aphid vector with insecticides
	Seed rot	Seed decays in soils	Use seed treated with approved fungicides
	Smut	Large, smooth, which galls, or outgrowths on ears, tassels, and nodes; covering drives and breaks open to release black, powdery, or greasy spores	Use tolerant varieties. Control corn borers with insecticides
Sweet potato	Black rot	Black depressions on sweet potato; black cankers on underground stem parts	Select disease-free potato seed. Rotate crops and planting beds. Use vine cuttings for propagation rather than slips

TABLE 8.13. DISEASE CONTROL FOR VEGETABLES (Continued)

Crop	Disease	Description	Control
	Internal cork	Dark brown to black, hard, corky lesions in flesh developing in storage at high temperature. Yellow spots with purple borders on new growth of leaves	Select disease-free seed potatoes
	Pox	Plants dwarfed; only one or two vines produced; leaves thin and pale green; soil rot pits on roots	Use disease-free stock and clean planting beds. Sulfur to lower soil pH to 5.2
	Scurf	Brown to black discoloration of root; uniform rusting of root surface	Rotation of crops and beds. Use disease-free stock. Use vine cuttings rather than slips
	Stem rot	Yellowing between veins; vines wilt; stems darken inside and may split	Select disease-free seed potatoes. Rotate fields and plant beds
Tomato	Anthracnose	Begins with circular, sunken spots on fruit; as spots enlarge, center becomes dark and fruit rots	Use approved fungicides
	Bacterial canker	Wilting, rolling, and browning of leaves; pith may discolor or disappear; fruit displays bird's-eye spots	Use hot-water-treated seed. Avoid planting in infected fields for 3 years
	Bacterial spot	Young lesions on fruit appear as dark, raised spots; older lesions blacken and appear sunken with brown centers; leaves brown and dry	Use hot-water-treated seed. Use approved bactericides

420

Disease	Symptoms	Control
Early blight	Dark brown spots on leaves; brown cankers on stems; girdling; dark, leathery, decayed areas at stem end of fruit	Use approved fungicides
Late blight	Dark, water-soaked spots on leaves; white fungus on undersides of leaves; withering of leaves; water-soaked spots on fruit turn brown. Disease is favored by moist conditions	Use approved fungicides
Fusarium wilt	Yellowing and wilting of lower, older leaves; disease affects whole plant eventually	Use resistant varieties
Gray leaf spot	Symptoms appear first in seedlings. Small brown to black spots on leaves, which enlarge and have shiny gray centers. The centers may drop out to give shotgun appearance. Oldest leaves affected first	Use resistant varieties. Use approved fungicides
Leaf mold	Chlorotic spots on upper side of oldest leaves appear in humid weather. Underside of leaf spot may have green mold. Spots may merge until entire leaf is affected. Disease advances to younger leaves	Use resistant varieties. Stake and prune to provide air movement. Use approved fungicides
Mosaic	Mottling (yellow and green) and roughening of leaves; dwarfing; reduced yields; russeting of fruit	Avoid contact by smokers. Control aphid vector with insecticides. Stylet oil may be effective.
Tomato spotted wilt virus	Brown spots, some circular on youngest leaves, stunted plant. Fruits misshapen often with circular brown rings, a diagnostic characteristic of this disease.	Use a combination of resistant varieties, highly reflective mulch to repel the silverleaf white fly, and plant activators

TABLE 8.13. DISEASE CONTROL FOR VEGETABLES (Continued)

Crop	Disease	Description	Control
	Verticillium wilt	Differs from fusarium wilt by appearance of disease on all branches at the same time; yellow areas on leaves become brown; midday wilting; dropping of leaves beginning at bottom	Use resistant varieties
Vine crops: cantaloupe, cucumber, pumpkin, squash, watermelon	Alternaria leaf spot	Circular spots showing concentric rings as they enlarge, appear first on oldest leaves	Field sanitation. Use disease-free seed. Use approved fungicides
	Angular leaf spot	Irregular, angular, water-soaked spots on leaves which later turn gray and die. Dead tissue may tear away leaving holes. Nearly circular fruit spots, which become white	Use tolerant varieties. Use approved bactericides
	Anthracnose	Reddish-black spots on leaves; elongated tan cankers on stems; fruits have sunken spots with flesh-colored ooze in center, later turning black	Use tolerant varieties. Use approved fungicides
	Bacterial wilt	Vines wilt and die; stem sap produces strings; no yellowing occurs	Control striped cucumber beetles with insecticides. Remove wilting plants from field

Disease	Symptoms	Control
Black rot (squash and pumpkin only)	Water-soaked areas appear on rinds of fruit in storage. Brown or black infected tissue rapidly invades entire plant	Use disease-free seed. Crop rotation. Cure fruit for storage at 85°F for 2 weeks, store at 50–55°F. Use approved fungicides
Downy mildew	Angular, yellow spots on older leaves; purple fungus on undersides of leaves when moisture present; leaves wither, die; fruit may be dwarfed with poor flavor	Use tolerant varieties. Use approved fungicides
Fusarium wilt	Stunting and yellowing of vine; water-soaked streak on one side of vine turning yellow eventually cracks and oozes sap	Use resistant varieties. Avoid infected soils
Gummy stem blight	Lesions may occur on stems, leaves, and fruit from which a reddish gummy exudate may ooze	Use disease-free seed. Rotate crops. Use approved fungicides
Mosaic (several)	Mottling (yellow and green) and curling of leaves; mottled and warty fruit; reduced yields; burning and dwarfing of entire plant	Control striped cucumber beetle or aphid with insecticides. Use resistant varieties. Destroy surrounding perennial weeds
Powdery mildew	White, powdery growth on upper leaf surface and petioles; wilting of foliage	Use tolerant varieties. Use approved fungicides
Cucumber scab	Water-soaked spots on leaves turning white; sunken cavity on fruit later covered by grayish-olive fungus; fruit destroyed by soft rot	Use resistant varieties. Use approved fungicides
Squash silverleaf	Silvering or white coloration to leaves. Associated with silverleaf whitefly feeding, disorder is worse in fall crops in southern U. S.	Little control, except to avoid planting under high whitefly populations and use tolerant varieties

RESOURCES WITH DISEASE PHOTOGRAPHS AND DIAGNOSTIC INFORMATION

Effective disease management requires accurate identification, and a thorough knowledge of the pathogen's life cycle. There are many excellent websites and apps that provide good photographs and information that can assist with identification of plant diseases. Although these resources may help in identifying pathogens, growers should consult a knowledgeable expert to provide confirmation of the identification before any control strategy is implemented. Most state Extension systems offer diagnostic services on a county- or state-wide basis, and they are likely to be very familiar with the most significant local pests. Providing a digital photo or physical sample can help provide confirmation of any preliminary identification.

California, https://vric.ucdavis.edu/veg_info_topic/diseases.htm.

New England, https://ag.umass.edu/vegetable/publications/guides/ northeast-vegetable-strawberry-pest-identification-guide.

New York, http://vegetablemdonline.ppath.cornell.edu/PhotoPages/ PhotoGallery.htm.

Washington, http://mtvernon.wsu.edu/path_team/diseasegallery.htm.

In addition, the American Phytopathological Society (APS) publishes detailed compendia of diseases and pests on a crop-by-crop basis (https:// apsjournals.apsnet.org/series/compendia). For growers who specialize in a few crops, or larger growers, these valuable resources may be a worthwhile investment. They contain many excellent photos and information on managing disease, insect, and other pests.

SOME INSECTS THAT ATTACK VEGETABLES

Most vegetable crops are attacked by insects at one time or another in the crop growth cycle. Several common insect pests of vegetable crops are listed in Table 8.14. Growers should strive to minimize insect problems by employing cultural methods aimed at reducing insect populations. These tactics include using resistant varieties, reflective mulches, crop rotation for soil-borne insects, destruction of weed hosts, stylet oils, insect repellants, trap crops, floating row covers, or other tactics. Sometimes insecticides must be used in an integrated approach to insect control when the economic threshold insect population has been reached. When using insecticides, read the label and carefully follow the instructions. Do not exceed maximum rates given; observe the interval between application and harvest, and apply only to crops on the label. Make a record of the product used, trade name, formulation, dilution, rate applied per acre, and dates of application. Read and follow all label precautions to protect the applicator and workers from insecticide injury, and environment from contamination. Follow local recommendations for efficacy, and follow all directions on the label.

TABLE 8.14.   INSECTS THAT ATTACK VEGETABLES

Crop	Insect	Description
Artichoke	Aphid	Small, green, pink, or black soft-body insects that rapidly reproduce to large populations. Damage results from sucking plant sap; indirectly from virus transmission to crop plants
	Plume moth	Small wormlike larvae blemish bracts and may destroy the base of the bract
Asparagus	Beetle and 12-spotted beetle	Metallic blue or black beetles (¼ in.) with yellowish wing markings and reddish, narrow head. Larvae are humpbacked, slate gray. Both feed on shoots and foliage
	Cutworm	Dull-colored moths lay eggs in the soil producing dark-colored smooth worms, 1–2 in. long, which characteristically curl up when disturbed. May feed belowground, or aboveground at night
Bean	Aphid	*See* Artichoke
	Corn earworm	Gray-brown moth (1½ in.) with dark wing tips deposits eggs, especially on fresh corn silk. Brown, green, or pink larvae (2 in.) feed on silk, kernels, and foliage
	Leafhopper	Green wedge-shaped, soft bodies (⅛ in.). When present in large numbers, sucking of plant sap causes plant distortion or burned appearance. Secondary damage results from transmission of yellows disease

TABLE 8.14.   INSECTS THAT ATTACK VEGETABLES (*Continued*)

Crop	Insect	Description
	Mexican bean beetle	Copper-colored beetle (¼ in.) with 16 black spots on its back. Orange to yellow spiny larva (⅓ in.) Beetle and larvae feeding on leaf undersides cause a lacework appearance
	Seed corn maggot	Grayish-brown flies (⅕ in.) deposit eggs in the soil near plants. Cream-colored, wedge-shaped maggots (¼ in.) tunnel into seeds, potato seed pieces, and sprouts
	Spider mite	Reddish, yellow, or greenish, tiny, eight-legged spiders that suck plant sap from leaf undersides causing distortion. Fine webs may be visible when mites are present in large numbers. Mites are not true insects
	Spotted-cucumber beetle	Yellowish, elongated beetle (¼ in.) with 11 or 12 black spots on its back. Leaf-feeding may destroy young plants when present in large numbers. Transmits bacterial wilt of cucurbits
	Striped-cucumber beetle	Yellow (⅕ in.) with three black stripes on its back; feeds on leaves. White larvae (⅓ in.) feed on roots and stems. Transmits bacterial wilt of cucurbits
	Tarnish plant bug	Brownish, flattened, oval bugs (¼ in.) with a clear triangular marking at the rear. Bugs damage plants by sucking plant sap
Beet	Aphid	*See* Artichoke

TABLE 8.14.   INSECTS THAT ATTACK VEGETABLES (*Continued*)

Crop	Insect	Description
	Flea beetle	Small (⅙ in.) variable-colored, usually dark beetles, often present in large numbers in the early part of the growing season. Feeding results in numerous small holes, giving a shotgun appearance. Indirect damage results from diseases transmitted
	Leaf miner	Tiny, black and yellow adults. Yellowish-white maggotlike larvae tunnel within leaves and cause white or translucent, irregularly damaged areas
	Webworm	Yellow to green worm (1¼ in.) with a black stripe and numerous black spots on its back
Broccoli, Brussels sprouts, cabbage, cauliflower, kale, kohlrabi	Aphid	*See* Artichoke
	Flea beetle	*See* Beet
	Harlequin cabbage bug	Black, shield-shaped bug (⅜ in.) with red or yellow markings
	Cabbage maggot	Housefly-like adult lays eggs in the soil at the base of plants. Yellowish, legless maggot (¼–⅓ in.) tunnels into roots and lower stem
	Cabbage looper	A brownish moth (1½ in.) that lays eggs on upper leaf surfaces. Resulting worms (1½ in.) are green with thin white lines. Easily identified by their looping movement
	Diamondback moth	Small, slender, gray, or brown moths. The folded wings of male moths show three diamond markings. Small (⅓ in.) larvae with distinctive V at rear, wiggle when disturbed

TABLE 8.14.   INSECTS THAT ATTACK VEGETABLES (*Continued*)

Crop	Insect	Description
	Imported cabbage worm	White butterflies with black wing spots lay eggs on undersides of leaves. Resulting worms (1¼ in.) are sleek, velvety, green
Cantaloupe (*See* Vine crops)		
Carrot	Leafhopper	*See* Bean
	Rust fly	Shiny, dark fly with a yellow head; lays eggs in the soil at the base of plants. Yellowish-white, legless maggot tunnel into roots
Celery	Aphid	*See* Beet
	Leaf miner	Adults are small, shiny, black flies with a bright yellow spot-on upper thorax. Eggs are laid within the leaf. Larvae mine between upper and lower leaf surfaces
	Spider mite	*See* Bean
	Tarnished plant bug	*See* Bean
	Loopers and worms	*See* Broccoli, etc.
Cucumber (*See* Vine crops)		
Eggplant	Aphid	*See* Artichoke
	Colorado potato beetle	Oval beetle (⅜ in.) with 10 yellow and 10 black stripes lays yellow eggs on undersides of leaves. Brick-red, humpbacked larvae (½ in.) have black spots. Beetles and larvae are destructive leaf feeders
	Flea beetle	*See* Beet
	Leaf miner	*See* Beet
	Spider mite	*See* Bean

TABLE 8.14.   INSECTS THAT ATTACK VEGETABLES (*Continued*)

Crop	Insect	Description
Endive, escarole, lettuce	Aphid	*See* Artichoke
	Flea beetle	*See* Beet
	Leafhopper	*See* Bean
	Leaf miner	*See* Beet
	Looper	*See* Broccoli
Mustard greens	Aphid	*See* Artichoke
	Worms	*See* Broccoli, etc.
Okra	Aphid	*See* Artichoke
	Green stinkbug	Large, flattened, shield-shaped, bright green bugs; various-sized nymphs with reddish markings.
Onion	Maggot	Slender, gray flies (¼ in.) lay eggs in soil. Small (⅓ in.) maggots bore into stems and bulbs
	Thrips	Yellow or brown winged or wingless tiny (½₅ in.) insect damages plant by sucking plant sap causing white areas or brown leaf tips
Parsnip	Carrot rust fly	*See* Carrot
Pea	Aphid	*See* Artichoke
	Seed maggot	Housefly-like gray adults lay eggs that develop into maggots (¼ in.) with sharply pointed heads
	Weevil	Brown-colored adults marked by white, black, or gray (1/5 in.) lay eggs on young pods. Larvae are small and whitish with a brown head and mouth. Adults feed on blossoms. May infect seed before harvest and remain in hibernation during storage
Pepper	Aphid	*See* Artichoke
	Corn borer	*See* Sweet corn
	Flea beetle	*See* Beet
	Leaf miner	*See* Beet

TABLE 8.14. INSECTS THAT ATTACK VEGETABLES (*Continued*)

Crop	Insect	Description
	Maggot	Housefly-sized adults have yellow stripes on body and brown stripes on wings. Larvae are typical maggots with pointed heads
	Weevil	Black-colored, gray or yellow marked, snout beetle, with the snout about one-half the length of the body. Grayish-white larvae are legless and have a pale brown head. Both adults and larvae feed on buds and pods; adults also feed on foliage
Potato	Aphid	*See* Artichoke
	Colorado potato beetle	*See* Eggplant
	Cutworm	*See* Asparagus
	Flea beetle	*See* Beet
	Leafhopper	*See* Bean
	Leaf miner	*See* Beet
	Tuberworm	Small, narrow-winged, grayish-brown moths (½ in.) lay eggs on foliage and exposed tubers in evening. Purplish or green caterpillars (¾ in.) with brown heads burrow into exposed tubers in the field or in storage
	Wireworm	Adults are dark-colored, elongated beetles (click beetles). Yellowish, tough-bodied, segmented larvae feed on roots and tunnel through fleshy roots and tubers
Radish	Maggot	*See* Broccoli, etc.
Rhubarb	Curculio	Yellow-dusted snout beetle that damages plants by puncturing stems

431

TABLE 8.14. INSECTS THAT ATTACK VEGETABLES (*Continued*)

Crop	Insect	Description
Rutabaga, turnip	Flea beetle	*See* Beet
	Maggot	*See* Broccoli, etc.
Squash (*See* Vine crops)		
Southern pea	Curculio	Black, humpbacked snout beetle. Eats small holes in pods and peas. Larvae are white with yellowish head and no legs
	Leafhopper	*See* Bean
	Leaf miner	*See* Beet
Spinach	Aphid	*See* Artichoke
	Leaf miner	*See* Beet
Strawberry	Aphid	*See* Artichoke
	Mites	Several mite species attack strawberry *See* Bean
	Tarnished plant bug	*See* Bean
	Thrips	*See* Onion
	Weevils	Several weevil species attack strawberry
	Worms	Several worm species attack strawberry
Sweet corn	Armyworms	Moths (1½ in.) with dark gray front wings and light-colored hind wings lay eggs on leaf undersides. Tan, green, or black worms (1¼ in.) feed on plant leaves and corn ears
	Earworm	*See* Bean
	European corn borer	Pale, yellowish moths (1 in.) with dark bands lay eggs on undersides of leaves. Caterpillars hatch, feed on leaves briefly, and tunnel into stalk and to the ear
	Flea beetle	*See* Beet

TABLE 8.14.   INSECTS THAT ATTACK VEGETABLES (*Continued*)

Crop	Insect	Description
	Japanese beetle	Shiny, metallic green with coppery-brown wing covers, oval beetles (½ in.). Severe leaf feeding results in a lacework appearance. Larvae are grubs that feed on grass roots
	Seed-corn maggot	*See* Bean
	Stalk borer	Grayish moths (1 in.) lay eggs on weeds. Small, white, brown-striped caterpillars hatch and tunnel into weed and crop stalks. Most damage is usually at edges of fields
Sweet potato	Flea beetle	*See* Beet
	Weevil	Blue-black and red adult (¼ in.) feeds on leaves and stems, grublike larva tunnels into roots in the field and storage
	Wireworm	*See* Potato
Tomato	Aphid	*See* Artichoke
	Colorado potato beetle	*See* Eggplant
	Corn earworm (tomato fruitworm)	*See* Bean
	Flea beetle	*See* Beet
	Fruit fly	Small, dark-colored flies usually associated with overripe or decaying vegetables
	Hornworm	Large (4–5 in.) moths lay eggs that develop into large (3–4 in.) green fleshy worms with prominent white lines on sides and a distinct horn at the rear. Voracious leaf feeders

TABLE 8.14. INSECTS THAT ATTACK VEGETABLES (*Continued*)

Crop	Insect	Description
	Leaf miner	*See* Beet
	Pinworm	Tiny yellow, gray, or green, purple-spotted, brown-headed caterpillars cause small fruit lesions, mostly near calyx. Presence detected by large white blotches near folded leaves
	Mite	*See* Bean
	Stink bug	*See* Okra
	White fly	Small, white flies that move when disturbed
Vine crops: cantaloupe, cucumber, pumpkin squash, watermelon	Aphid	*See* Artichoke
	Cucumber beetle (spotted or striped)	
	Leafhopper	*See* Bean
	Leaf miner	*See* Beet
	Mite	*See* Bean
	Pickleworm	White moths (1 in.), later become greenish with black spots, with brown heads and brown-tipped wings with white centers, and a conspicuous brush at the tip of the body, lay eggs on foliage. Brown-headed, white, later becoming greenish with black spots. Larvae (¾ in.) feed on blossoms, leaves, and fruit
	Squash bug	Brownish, flat stinkbug (⅝ in.). Nymphs (⅜ in.) are gray to green. Plant damage is due to sucking of plant sap

TABLE 8.14. INSECTS THAT ATTACK VEGETABLES (*Continued*)

Crop	Insect	Description
	Squash vine borer	Black, metallic moth (1½ in.) with transparent hind wings and abdomen ringed with red and black; lays eggs at the base of the plant. White caterpillars bore into the stem and tunnel throughout
	White fly	*See* Tomato

IDENTIFICATION OF VEGETABLE INSECTS

Effective insect management requires accurate identification, and a thorough knowledge of the insect's habits and life cycle. There are many excellent websites and apps that provide good photographs and information that can assist with identification of insect pests. Although these resources may help in identifying pests, growers should consult a knowledgeable insect expert to provide confirmation of the identification before any control strategy is implemented. Most state Extension systems offer diagnostic services on a county- or state-wide basis, and they are likely to be very familiar with the most significant local pests. Providing a digital photo or physical sample can help provide confirmation of any preliminary identification.

SOME USEFUL RESOURCES FOR INSECT IDENTIFICATION

California, http://www.ipm.ucdavis.edu/PMG/crops-agriculture.html. http://www.ipm.ucdavis.edu/PCA/pcapath.html#SPECIFIC
Florida, https://edis.ifas.ufl.edu/entity/topic/vegetable_pest_insects.
New England, https://ag.umass.edu/vegetable/publications/guides/northeast-vegetable-strawberry-pest-identification-guide.
Utah, https://extension.usu.edu/pests/files/pubs/Vegetable-Pest-of-Utah-ID-Guide.pdf.
Iowa, https://bugguide.net/node/view/15740.
North Carolina, http://www.ces.ncsu.edu/chatham/ag/SustAg/insectlinks.html.

10
ABIOTIC DISORDERS

Many vegetable crops experience disorders that are caused by nonliving, or abiotic, factors. Several common disorders are provided in Table 8.15. Disorders are frequently caused by adverse weather conditions, deficiencies or excesses of plant nutrients or water, or mechanical injuries, and in many cases the causes are unknown. Symptoms vary and can be confused with problems caused by living factors.

TABLE 8.15. ABIOTIC DISORDERS OF VEGETABLE CROPS

Crop	Disorder	Description	Cause/Control
Asparagus	Autotoxicity	Plants replanted into prior asparagus plantings stunted	Rotate several years
	Cold injury	Spears curved and purple	Protect from frost, shelter from wind
Beet, chard, spinach	Heart rot	Young leaves stunted; roots may have internal necrosis	Boron deficiency
Bean	Baldhead	Plants with broken or dead growing points	Mechanical damage to the seed
	Sunscald	Small water-soaked spots on the exposed plant parts, may form large necrotic or russetted areas, flowers and pods may abort	Intense sunlight, especially after periods of high humidity and cloud cover
Carrot	Growth cracks	Vertical cracks, esp. on larger roots	Fluctuating soil moisture

TABLE 8.15. ABIOTIC DISORDERS OF VEGETABLE CROPS (*Continued*)

Crop	Disorder	Description	Cause/Control
	Heat canker	Collapse/death of seedlings, or shriveling at top of root	Excessive heating of the soil surface, prevent by irrigation, increasing plant density, or earlier planting
Celery, celeriac	Blackheart	Younger leaves wilt, turn black, growing point may die	Poor calcium uptake in the plant, fluctuating moisture conditions
	Chlorosis	Leaves yellow, starting on oldest leaves	Manganese or magnesium deficiency, sometimes promoted by heavy rains
	Cracked stem	Brown, cross-wise cracks on petioles of ribs, petioles bend and can dry out	Boron deficiency, may also be associated with high levels of potassium and nitrogen
	Spongy petiole	Hollow petioles that can easily be crushed, interior appears spongy	Uneven growth conditions and stresses, including low nitrogen and potassium, irregular supply of water
	Stringiness	Petioles have thickened, fibrous strands around the outer edge	Low nitrogen availability, and wind stress

TABLE 8.15. ABIOTIC DISORDERS OF VEGETABLE CROPS (*Continued*)

Crop	Disorder	Description	Cause/Control
Broccoli, cauliflower, cabbage, radish, rutabaga, turnip, Brussels sprouts	Buttoning	Young plants bolt prematurely, and form heads when still very small plants	Early-season stresses including high or low temperature
	Black speck (cauliflower)	Black necrotic specks on branches or flower stalks in the interior of the curd	Unknown, possibly a nutrient deficiency
	Brown bead (broccoli)	Floral buds turn tan or brown and become easily detached	Usually associated with rapid growth during periods of high temperature following periods of abundant rainfall
	Growth cracks (rutabaga, turnip, cabbage)	Cracks extend from the neck region down the side of the root	Occurs during periods of rapid growth, especially when heavy rainfall or irrigation follows a dry spell
	Hollow stem (broccoli, cauliflower)	Internal cracks within the stem, usually not visible externally	Unknown, possibly boron deficiency, calcium and nitrogen maybe play a role

TABLE 8.15.   ABIOTIC DISORDERS OF VEGETABLE CROPS (*Continued*)

Crop	Disorder	Description	Cause/Control
	Tipburn (Chinese cabbage, Brussels sprout, cabbage, cauliflower)	Inner leaves are affected, with no external symptoms	Inadequate transport of calcium to rapidly growing tissues, a dry spell after a period of abundant moisture may aggravate
	Black midrib, black speck, gray speck, necrotic speck, and vein streaking	All storage disorders of cabbage, symptoms include spots on leaves and midribs, and may be confused	Causes are not known, but certain nutrient deficiencies, storage conditions, and other factors variably promote or reduce
Cucumber, melon, pumpkin, squash, zucchini	Blossom end rot	Young fruit show decay at the blossom end, and fail to develop	Inadequate transport of calcium to rapidly growing tissues, period of rapid growth, or stresses caused by heavy fruit set can promote
	Cold injury (cucumber)	All plant tissues can be bleached, water-soaked, pitted	Temperatures below 50°F can cause chilling injury, delay planting until warm enough

439

TABLE 8.15. ABIOTIC DISORDERS OF VEGETABLE CROPS (*Continued*)

Crop	Disorder	Description	Cause/Control
	Cold injury (squashes)	Certain leaves become brilliant yellow, usually no lasting damage	Varieties with the "B" gene for precocious yellow fruit pigmentation are susceptible to this phenomenon
Lettuce	Tipburn	Necrotic spots on leaf margins, or entire leaf edge brown	Calcium deficiency in the growing tissues of the inner leaves, can be reduced by maintaining high humidity, reducing nitrogen to limit growth, maintaining adequate irrigation, and maintaining high Ca levels
	Pink rib	Pink discoloration at the base of the midveins of leaves	Unknown cause, aggravated by bruising, tight packing, and high storage temperature
	Russet spot	Tan-colored pits on the outer leaves, may be along the midrib or the blades	Caused primarily by ethylene injury during storage
Onion, garlic, leek, shallot, chives	Sunscald	Injured tissues shrivel, and the plant wilts and withers	Seed as early as possible to avoid high temperatures on young sensitive plants

TABLE 8.15. **ABIOTIC DISORDERS OF VEGETABLE CROPS** (*Continued*)

Crop	Disorder	Description	Cause/Control
	Ozone injury	Starts as flecks on leaves, coalesce to form whitish spots with diffuse margins, leaves may become yellowish and die back from tips	High ozone levels, which usually occur with warm, hazy, humid weather when air pollution levels are high
	Tipburn	Tips of oldest leaves turn yellow and then white, may extend down the entire leaf	Unknown causes, various stresses including drought and saturated soil are possible causes
	Translucent scale	Grayish, watery, translucent appearance of one or more layers of scales, often only affecting the second and third scales	Most often seen on Spanish onion, cause unknown, connected with storage conditions, excessive relative humidity could predispose onions to this problem
	Sprout inhibitor injury	Spongy, soft bulbs resulting from spaces between the rings of the bulb	Caused by application of maleic hydrazide as a sprout inhibitor for storage onions too early in the season, while onions are still producing new leaves

TABLE 8.15. ABIOTIC DISORDERS OF VEGETABLE CROPS
(*Continued*)

Crop	Disorder	Description	Cause/Control
Pea	Cold injury	Death of growing tips, roughened edges or leathery young leaves, water-soaked lesions, white lesions on pods	Caused by temperatures dropping below 32°F, cold-hardened plants are less susceptible
	Water congestion	Water-soaked spots on leaves that eventually become necrotic	Caused by high humidity, high temperature and high soil moisture, especially on clay or muck soil
Potato	Blackheart	Intense, blackish, irregular discoloration in the center of tubers. Initially firm, becomes soft and watery	Limited oxygen in storage, especially at temperatures above 60°F, or in fields that were excessively wet before harvest
	Growth cracks	Cracks may be thumbnail-shaped or longitudinal, and may heal over if damage occurred early in the season	Rough handling, mechanical pressure, or impact during harvest, or periods of excessively rapid growth. High humidity storage can minimize thumbnail cracking

TABLE 8.15. ABIOTIC DISORDERS OF VEGETABLE CROPS
(*Continued*)

Crop	Disorder	Description	Cause/Control
	Hollow heart	Brown areas, becoming tan- to brown-walled angular cavities in the centers of tubers, no external symptoms	Conditions that favor rapid tuber enlargement (moisture after a dry period, unbalanced fertilization), soils with low organic matter and potassium deficiency are also associated with this disorder
	Jelly end rot	Flesh at the stem end of the potato becomes glassy and jelly-like, and then dries up	Conditions that interfere with starch deposition, such as fluctuating soil moisture, are often responsible
	Secondary tubers	Tubers form secondary, bead-like tubers but no shoots	Storing tubers above 60°F or planting tubers into cold, dry soil can induce this disorder
Tomato, pepper	Blossom end rot	Light brown patches that become black sunken areas near the blossom end of the fruit, may be accompanied by an internal rot	Localized calcium deficiency in the fruit, promoted by rapid plant growth, damage to the roots, high soil salinity, and fluctuating soil moisture

TABLE 8.15. **ABIOTIC DISORDERS OF VEGETABLE CROPS**
(*Continued*)

Crop	Disorder	Description	Cause/Control
	Catfacing (tomato)	Bands of cork-like scar tissue cover the end of the fruit, more common in heirloom large-fruited varieties	Abnormal development of tissue at the junction of the style and ovary, temperatures below 60°F promote this disorder
	Cracking (tomato)	Radial or concentric cracks on the fruit	Variations in soil moisture, periods of rapid growth when relative humidity and air temperatures are high
	Internal white tissue (tomato)	Firm, white areas in the vascular tissue and outer walls of fruit, no external symptoms	High temperatures during the ripening period, exacerbated by potassium deficiency
	Leaf roll	Beginning with older leaves, the edges roll upwards and leaves become leathery	Appears following hot, dry growing conditions
	Zippering	Thin scars extending partially or fully along the side of the fruit from calyx to blossom end	Usually caused by floral parts attached to newly forming fruit, forming a scar as the fruit expands

TABLE 8.15.   ABIOTIC DISORDERS OF VEGETABLE CROPS (*Continued*)

Crop	Disorder	Description	Cause/Control
	Yellow shoulder (tomato)	Shoulders of the fruit become firm and yellow, never ripening to red	High temperatures and sun exposure during the ripening period, exacerbated by potassium deficiency, "uniform green" varieties are more resistant
	Puffiness	Fruit soft, light-weight and bloated, some locules filled only with gel	Unfavorable weather conditions negatively affect pollination
Sweet potato	Chilling injury	Surface pitting and, in some cases, internal tissue breakdown, becoming spongy and watery	Storage below 50°F can result in chilling injury
	Cracking	Longitudinal cracks and/or splits in roots	Fluctuations in soil moisture

Adapted from R. J. Howard, J. A. Garland and W. L. Seaman (eds.), *Diseases and Pests of Vegetable Crops in Canada* (The Canadian Phytopathological Society and the Entomological Society of Canada, 1994); J. B. Jones, T. A. Zitter, T. M. Momol and S. A. Miller (eds.), *Compendium of Tomato Diseases and Pests*, 2nd ed. (APS Press, 2014); R. M. Davis, K. Pernezny, and J. C. Broome, *Tomato Health Management* (APS Press, 2012); and C. A. Clark, D. M. Ferrin, T. P. Smith, and G. J. Holmes, *Compendium of Sweetpotato Diseases, Pests and Disorders*, 2nd ed. (APS Press, 2013).

AIR POLLUTION DAMAGE TO VEGETABLE CROPS

Plant damage by pollutants depends on meteorological factors leading to air stagnation, the presence of a pollution source, and the susceptibility of the plants. Among the pollutants that affect vegetable crops are sulfur dioxide (SO_2), ozone (O_3), peroxyacetyl nitrate (PAN), chlorine (Cl_2), and ammonia (NH_3). Volatile herbicides can also move through the air and affect vegetable crops. The following symptoms are most likely to be observed on vegetables produced near heavily urbanized areas or industrialized areas, particularly where weather conditions frequently lead to air stagnation.

Sulfur Dioxide: SO_2 causes acute and chronic plant injury. Acute injury is characterized by dead tissue between the veins or on leaf margins. The dead tissue may be bleached, ivory, tan, orange, red, reddish-brown, or brown, depending on the plant species, time of year, and weather. Chronic injury is marked by brownish-red, turgid, or bleached white areas on the leaf blade. Young leaves rarely display damage, whereas fully expanded leaves are very sensitive.

Ozone: Common symptoms of O_3 injury are very small irregularly shaped spots that are dark brown to black (stipplelike) or light tan to white (flecklike) on the upper leaf surface. Very young and old leaves are normally resistant to ozone. Recently matured leaves are most susceptible. Injury is usually more pronounced at the leaf tip and along the margins. With severe damage, symptoms may extend to the lower leaf surface.

Peroxyacetyl Nitrate: Typically, PAN affects the under-leaf surface of newly matured leaves and causes bronzing, glazing, or silvering on the lower surface of sensitive leaf areas. The leaf apex of broad-leaved plants becomes sensitive to PAN approximately five days after leaf emergence. About four leaves on a shoot are sensitive at any one time. PAN toxicity is specific for tissue in a particular stage of development. Only with successive exposure to PAN will the entire leaf develop injury. Injury may consist of bronzing or glazing with little or no tissue collapse on the upper leaf surface. Pale green to white stipplelike areas may appear on upper and lower leaf surfaces. Complete tissue collapse in a diffuse band across the leaf is helpful in identifying PAN injury.

Chlorine: Injury from chlorine is usually of an acute type and is similar in pattern to sulfur dioxide injury. Foliar necrosis and bleaching are common. Necrosis is marginal in some species but scattered in others either between or along veins. Lettuce plants exhibit necrotic injury on the margins of outer leaves, which often extends in solid areas toward the center and base of the leaf. Inner leaves remain unmarked.

Ammonia: Field injury from NH_3 has been primarily due to accidental spillage. Slight amounts of the gas produce color changes in the pigments of vegetable skin. The dry outer scales of red onion may become greenish or black, whereas scales of yellow or brown onion may turn dark brown.

Hydrochloric Acid Gas: HCl causes an acid-type burn. The usual acute response is a bleaching of tissue. Leaves of lettuce, endive, and escarole exhibit a tip burn that progresses toward the center of the leaf and soon dries out. Tomato plants develop interveinal bronzing.

Original material from *Commercial Vegetable Production Recommendations* (Maryland Agricultural Extension Service EB-236, 1986), and *Bulletin* 137 (2005). Additional Resource: H. Griffiths, *Effects of Air Pollution on Agricultural Crops* (Ontario Ministry of Agriculture, Food, and Rural Affairs, 2003). http://www.omafra.gov.on.ca/english/crops/facts/01-015.htm.

REACTION OF VEGETABLE CROPS TO AIR POLLUTANTS

Vegetable crops may be injured acutely following exposure to high concentrations of various atmospheric pollutants. Prolonged exposure to lower concentrations may also result in chronic plant damage. Injury appears progressively as leaf chlorosis (yellowing), necrosis (death), and perhaps restricted growth and yields. On occasion, plants may be killed, but usually not until they have suffered persistent injury. Symptoms of air pollution damage vary with the individual crops and plant age, specific pollutant, concentration, duration of exposure, and environmental conditions. The sensitivity of vegetable crops to several air pollutants are summarized in Table 8.16. Note that the terms "sensitive," "intermediate" and "tolerant" are relative, and the reactions of specific crops can vary widely.

TABLE 8.16. SENSITIVITY OF VEGETABLES TO SELECTED AIR POLLUTANTS

Pollutant	Sensitive	Intermediate	Tolerant
Ozone	Bean	Carrot	Beet
	Broccoli	Endive	Cucumber
	Onion	Parsley	Lettuce
	Potato	Parsnip	
	Radish	Turnip	
	Spinach		
	Sweet corn		
	Tomato		
Sulfur dioxide	Bean	Cabbage	Cucumber
	Beet	Pea	Onion
	Broccoli	Tomato	Sweet corn
	Brussels sprouts		
	Carrot		
	Endive		
	Lettuce		
	Okra		
	Pepper		
	Pumpkin		
	Radish		
	Rhubarb		
	Spinach		
	Squash		
	Sweet potato		
	Swiss chard		
	Turnip		
Fluoride	Sweet corn		Asparagus
			Squash
			Tomato
Nitrogen dioxide	Lettuce		Asparagus
			Bean
PAN	Bean	Carrot	Broccoli
	Beet		Cabbage
	Celery		Cauliflower
	Endive		Cucumber
	Lettuce		Onion
	Mustard		Radish
	Pepper		Squash

TABLE 8.16. **SENSITIVITY OF VEGETABLES TO SELECTED AIR POLLUTANTS** (*Continued*)

Pollutant	Sensitive	Intermediate	Tolerant
	Spinach		
	Sweet corn		
	Swiss chard		
	Tomato		
Ethylene	Bean	Carrot	Beet
	Cucumber	Squash	Cabbage
	Pea		Endive
	Southern pea		Onion
	Sweet potato		Radish
	Tomato		
2,4-D	Tomato	Potato	Bean
			Cabbage
			Eggplant
			Rhubarb
Dicamba	Tomato	Watermelon	
	Pepper	Cucumber	
	Eggplant	Summer Squash	
	Soybean		
	Bean		
	Sweet potato		
Chlorine	Mustard	Bean	Eggplant
	Onion	Cucumber	Pepper
	Radish	Southern pea	
	Sweet corn	Squash	
		Tomato	
Ammonia	Mustard	Tomato	
Mercury vapor	Bean	Tomato	
Hydrogen sulfide	Cucumber	Pepper	Mustard
	Radish		
	Tomato		

Adapted from J. S. Jacobson and A. C. Hill (eds.), *Recognition of Air Pollution Injury to Vegetation* (Pittsburgh, PA: Air Pollution Control Association, 1970); M. Treshow, *Environment and Plant Response* (New York: McGraw-Hill, 1970); H. Griffiths, *Effects of Air Pollution on Agricultural Crops. Ontario Ministry of Agriculture, Food, and Rural Affairs* (2003), http://www.omafra.gov.on.ca/english/crops/facts/01-015.htm; and M. Wasacz and T. Besancon, *Specialty Crops Injury Caused by Dicamba Herbicide Drift* (Rutgers Cooperative Extension, 2020), https://plant-pest-advisory.rutgers.edu/specialty-crops-injury-caused-by-dicamba-herbicide-drift/.

12
WILDLIFE CONTROL

Wildlife can be serious pests on farms, especially those fields near urban areas where habitat has been taken for development. Controlling wildlife on farms is difficult as many techniques are met with limited success. One point to be made is to consider local regulations regarding the use of various control measures. Trapping, shooting, and other means may need special permits. In some cases, such as trapping (especially relocation), a certified wildlife specialist may be needed.

1. Deer

Repellants. May be effective for low-density deer populations. Apply before damage is expected, when no precipitation is expected, and when temperatures are between 40 and 80°F.

Fencing. Woven wire fences are the most effective and should be 8–10 ft tall. Electric fences may act as a deterrent. Some growers have success with a 5- or 6-foot high-tensile electric fence even though deer might be able to jump fences this tall.

2. Raccoons

Many states have laws controlling the manner in which these pests can be removed. Usually, trapping is the only means of ridding a field of raccoons. Crops can be protected with a double-strand electric fence with wires at 5 and 10 in. above the ground.

3. Birds

Exclusion. Bird-proof netting can be used to protect high-value crops.

Sound devices. Some success has been reported with recorded distress calls. Other sound devices such as propane guns might be effective for short periods. Use of these devices should be random and with a range of sound frequency and intervals.

Visual devices. Eye-spot balloons have been used with some success against grackles, blue jays, crows, and starlings and might be the control method of choice for urban farms. Reflective tape has been used with variable success and is very labor intensive to install.

4. *Mice and voles*

Habitat control. Remove any possible hiding or nesting sites near the field. Sometimes rodents will nest underneath polyethylene mulch not applied tightly to the ground and in thick windbreaks.

Traps and baits. Strategically placed traps and bait stations can be used to reduce rodent populations.

Transplanting. For some particularly attractive crops, such as cucurbits, the seed is a favorite food and the seed is often removed from the ground soon after planting. One option to reduce stand losses would be transplanting instead of direct seeding.

PART **9**

WEED MANAGEMENT

Knott's Handbook for Vegetable Growers, Sixth Edition. George J. Hochmuth
and Rebecca G. Sideman.
© 2023 John Wiley & Sons Ltd. Published 2023 by John Wiley & Sons Ltd.
Companion website: https://www.wiley.com/go/hochmuth/vegetablegrowers6e

01

WEED MANAGEMENT STRATEGIES

Weeds reduce yield and quality of vegetables through direct competition for light, moisture, and nutrients as well as by interference with harvest operations. Early season competition is most critical, and a major emphasis on control should be made during this period. For example, common amaranth reduces yields of lettuce, watermelon, and muskmelon at least 20% if allowed to compete with these crops for only the first 3 weeks of growth. Weeds can be controlled, but this requires good management practices in all phases of production. Because there are many kinds of weeds, with much variation in growth habit, they obviously cannot be managed by a single method. The incorporation of several of the following management practices into vegetable strategies increases the effectiveness for controlling weeds.

Crop Competition

An often-overlooked tool in reducing weed competition is to establish a good crop stand in which plants emerge and rapidly shade the ground. The plant that emerges first and grows the most rapidly is the plant with the competitive advantage. Utilization of good production management practices such as fertility, well-adapted varieties, proper water control (irrigation and drainage), and establishment of adequate plant populations is very helpful in reducing weed competition. Everything possible should be done to ensure that crops, not weeds, have the competitive advantage.

Crop Rotation

If the same crop is planted in the same field year after year, usually some weed or weeds are favored by the cultural practices and herbicides used on that crop. By rotating to other crops, the cultural practices and herbicide programs are changed. This often reduces the population of specific weeds that were tolerant in the previous cropping rotation. Care should be taken, however, to not replant vegetables in soil treated with a nonregistered herbicide. Crop injury as well as vegetables containing illegal residues may result. Check the labels for plant-back limitations before herbicide application and planting rotational crops.

Mechanical Control

Mechanical control includes field preparation by plowing or disking, cultivation, mowing, hoeing, and hand pulling of weeds. Mechanical control practices are among the oldest weed management techniques. Weed control is a primary reason for preparing land for crops planted in rows. Seedbed preparation by plowing or disking exposes many weed seeds to variations in light, temperature, and moisture. For some weeds, this process breaks weed seed dormancy, leading to early season control with herbicides or additional cultivation. Cultivate only deep enough in the row to achieve weed control; deep cultivation may prune crop roots, bring weed seeds to the surface, and disturb the soil previously treated with an herbicide. Follow the same precautions between rows. When weeds can be controlled without cultivation, there is no advantage to the practice. In fact, there may be disadvantages, such as drying out the soil surface, bringing weed seeds to the surface, and disturbing the root system of the crop.

Mulching

The use of polyethylene mulch increases yield and earliness of vegetables. The proper injection of fumigants under the mulch controls nematodes, soil insects, soil-borne diseases, and weed seeds. Mulches, especially black mulch, act as a barrier to the growth of many weeds. Nutsedge, however, is one weed that will grow through the mulch. Nutsedge is difficult to manage and needs several approaches to control, including herbicides.

Prevention

Preventing weeds from infesting or reinfesting a field should always be considered. Weed seeds may enter a field in several ways. Seeds may be distributed by wind, water, machinery, in cover crop seed, and other means. Fence rows and ditch banks are often neglected when controlling weeds in the crop. Seeds produced in these areas may move into the field. Weeds in these areas can also harbor insects and diseases (especially viruses) that may move onto the crop. It is also important to clean equipment before entering fields or when moving from a field with a high weed infestation to a relatively clean field. Nutsedge tubers especially are moved easily on disks, cultivators, and other equipment.

Herbicides

Properly selected herbicides are effective tools for weed control. Herbicides may be classified several ways depending on how they are applied and their mode of action in or on the plant. Generally, herbicides are either soil applied or foliage applied. They may be selective or nonselective, and they may be either contact or translocated through the plant. For example, paraquat is a foliage-applied, contact, nonselective herbicide, whereas atrazine usually is described as a soil-applied, translocated, selective herbicide.

Foliage-applied herbicides may be applied to leaves, stems, and shoots of plants. Herbicides that kill only those parts of the plants they touch are *contact herbicides*. Those herbicides that are taken into the plant and moved throughout it are *translocated herbicides*. Paraquat is a contact herbicide, whereas glyphosate (Roundup) or sethoxydim (Poast) are translocated herbicides. For foliage-applied herbicides to be effective, they must enter the plant. Good coverage is very important. Most foliage-applied herbicides require either the addition of a specified surfactant or a specified formulation to be used for best control.

Soil-applied herbicides are either applied to the surface of the soil or incorporated into the soil. Surface-applied herbicides require rainfall or irrigation shortly after application for best results. Lack of moisture often results in poor weed control. Incorporated herbicides are not dependent on rainfall or irrigation and generally give more consistent and wider-spectrum control. They do, however, require more time and equipment for incorporation. Herbicides that specify incorporation into the soil improve the contact of the herbicide with the weed seed and/or minimize the loss of the herbicide by volatilization or photodecomposition from the surface of the soil. Some herbicides, if not incorporated, may be lost from the soil surface. Although most soil-applied herbicides must be moved into the soil to be effective, the depth of incorporation into the soil can be used to achieve selectivity. For example, if a crop seed is planted 2 in. deep in the soil and the herbicide is incorporated by irrigation only in the top 1 in., where most of the problem weed seeds are found, the crop roots will not come in contact with the herbicide. If too much irrigation or rain moves the herbicide down into the crop seed zone or if the herbicide is incorporated mechanically too deep, crop injury may result.

Adapted from P. Dittmar, N. Boyd, and R. Kanissery, "Weed Management Principles in Commercial Vegetable Production," (Florida Cooperative Extension Service, Serv, 2018). https://edis.ifas.ufl.edu/pdf/CV/CV11300.pdf (accessed 23 Aug. 2021).

02
WEED IDENTIFICATION

Accurate identification of the particular weed species is the first step to controlling the problem. Several university and cooperative extension websites offer information about and assistance with identifying weeds. Some sources offer information across crops and commodities, and many have very good photos of weed species. There are weed identification apps for your smart phone, but be sure you are downloading an app with weed identification for weeds in your region. As good as these websites and apps are for assisting in weed identification, the grower is encouraged to obtain confirmation of identification from a knowledgeable weed expert before implementing a weed control strategy. The following websites, among many others, offer photos and guides to the identification of weeds (accessed 23 Aug. 2021).

California, www.ipm.ucdavis.edu/PMG/weeds_common.html.
Illinois, http://web.aces.uiuc.edu/weedid.
Minnesota, https://extension.umn.edu/weed-management/weed-identification.
Weed identification. Weed Science Society of America, https://wssa.net/wssa/weed/weed-identification/.
Plant and weed identification. Michigan State University Extension, https://www.canr.msu.edu/pestid/resources/plant-and-weed-identification/index.

03
NOXIOUS WEEDS

Noxious weeds are plant species so injurious to agricultural crop interests that they are regulated or controlled by federal and/or state laws. Propagation, growing, or sales of these plants may be controlled. Some states divide noxious weeds into "prohibited" species, which may not be grown or sold, and "restricted" species, which may occur in the state and are considered nuisances or of economic concern for agriculture. States have different lists of plants considered noxious, and there is a regularly-updated federal noxious weed list. The Animal and Plant Health Inspection Service (APHIS) of the USDA has a website with current information about management of noxious weeds.

The Noxious Weeds Program Home Page is found at https://www.aphis. usda.gov/aphis/ourfocus/planthealth/plant-pest-and-disease-programs/pests-and-diseases/sa_weeds/sa_noxious_weeds_program (accessed 23 Aug. 2021).

04

WEED CONTROL IN ORGANIC FARMING

Weeds can be a serious threat to vegetable production in organic systems. Weed control is one of the costliest activities in successful organic vegetable production. Some of the main organic weed control strategies include:

Rotate crops.

Use cover crops to compete with weeds in the noncrop season.

Be knowledgeable about potential weed contamination of manures, composts, and organic soil amendments.

Employ mechanical and manual control through cultivation, hoeing, mowing, hand weeding, etc.

Clean equipment to minimize transfer of weed propagules from one field to another.

Control weeds at the crop perimeter.

Completely till crop and weeds immediately after final harvest.

Plan for fallow periods with mechanical destruction of weeds.

Use the stale seedbed technique when appropriate.

Use only approved weed control materials.

Encourage crop competition due to optimum crop vigor, correct plant spacing, or shading of weeds.

Use approved soil mulching practices to smother weed seedlings around crops or in crop alleys.

Practice precise placement of fertilizers and irrigation water to minimize availability to weeds in walkways and row middles.

Websites offering information on weed control practices in organic vegetable production include:

- Weed Management Menu—Sustainable Farming Connection, http://www.ibiblio.org/farming-connection/weeds/home.htm.
- National Sustainable Agriculture Information Service, http://www.attra.org.
- G. Boyhan, R. Westerfield, and S. Stone, *Growing Vegetables Organically,* Bull. 1011 (University of Georgia, 2017), https://secure.caes.uga.edu/extension/publications/files/pdf/B%201011_6.PDF (accessed 2 Dec. 2021).

- M. Schonbeck, *Twelve Steps Toward Ecological Weed Management in Organic Vegetables* (2019), https://eorganic.org/node/2320 (accessed 2 Dec. 2021).
- M. Schonbeck, *A Systems Approach to Organic Weed Management* (2021), https://eorganic.org/menu/ (accessed 2 Dec. 2021).

05

COVER CROPS AND ROTATION IN WEED MANAGEMENT

Vegetable growers can take advantage of certain cover crops to help control weeds in vegetable production systems and crop rotation systems. Cover crops compete with weeds, reducing the growth and weed seed production capacity of weeds. Cover crops help build soil organic matter and can lead to more vigorous vegetable crops that compete more effectively with weeds. Rotation introduces weed populations to different crops with different weed control options and helps keep herbicide-resistant weed populations from becoming established.

For detailed information about the use of cover crops and rotation for weed management in vegetable production:

C. Mohler and S. E. Johnson (eds.), *Crop Rotation on Organic Farms: A Planning Manual* (Sustainable Agriculture Research & Education Program, NRAES-177, 2020), https://www.sare.org/wp-content/uploads/Crop-Rotation-on-Organic-Farms.pdf (accessed 2 Dec. 2021).

WEED CONTROL WITH HERBICIDES

Chemical weed control minimizes labor and is effective if used with care. The following precautions should be observed:

1. Do not use an herbicide unless the label states that it is registered for that particular crop. Be sure to use as directed by the manufacturer.
2. Use herbicides so that no excessive residues remain on the harvested product, which may otherwise be confiscated. Residue tolerances are established by the U.S. Environmental Protection Agency (EPA).
3. Note that some herbicides kill only certain weeds.
4. Make certain the soil is sufficiently moist for effective action of preemergence sprays. Do not expect good results in dry soil.
5. Keep in mind that postemergence herbicides are most effective when conditions favor rapid weed germination and growth.
6. Avoid using too much herbicide. Overdoses can injure the vegetable crop. Few crops, if any, are entirely resistant.
7. Use less herbicide on light sandy soils than on heavy clay soils. Muck soils require somewhat greater rates than do heavy mineral soils.
8. When using wettable powders, be certain the liquid in the tank is agitated constantly as spraying proceeds.
9. Use a boom and nozzle arrangement that fans out the material close to the ground in order to avoid drift.
10. Thoroughly clean spray tank after use.

CLEANING SPRAYERS AFTER APPLYING HERBICIDES

Sprayers must be kept clean to avoid injury to the crop on which they are to be used for applying insecticides or fungicides as well as to prevent possible deterioration of the sprayers after use of certain materials.

1. Rinse all parts of sprayer with water before and after any special cleaning operation is undertaken.
2. If in doubt about the effectiveness of water alone to clean the herbicide from the tank, pump, boom, hoses, and nozzles, use a cleaner. In some cases, it is desirable to use activated carbon to reduce contamination.

3. Fill the tank with water. Use one of the following materials for each 100 gal water: 5 lb paint cleaner (trisodium phosphate), 1 gal household ammonia, or 5 lb Sal soda.

4. If hot water is used, let the solution stand in the tank for 18 hr. If cold water is used, leave it for 36 hr. Pump the solution through the sprayer.

5. Rinse the tank and parts several times with clear water.

6. If copper has been used in the sprayer before a weed control operation is performed, put 1 gal vinegar in 100 gal water and let the solution stay in the sprayer for 2 hr. Drain the solution and rinse thoroughly. Copper interferes with the effectiveness of some herbicides.

DETERMINING RATES OF APPLICATION OF WEED CONTROL MATERIALS

Commercially available herbicide formulations differ in their content of the active ingredient. The label indicates the amount of the active ingredient (lb/gal). Refer to this amount in Table 9.1 to determine how much of the formulation you need in order to supply the recommended amount of the active ingredient per acre.

TABLE 9.1. HERBICIDE DILUTION TABLE: QUANTITY OF LIQUID CONCENTRATES TO USE TO GIVE DESIRED DOSAGE OF ACTIVE CHEMICAL

	Active Ingredient Needed (lb/acre):						
Active Ingredient Content of Liquid Concentrate (lb/gal)	0.125	0.25	0.50	1	2	3	4
	Liquid Concentrate to Use (pint/acre)						
1	1.0	2.0	4.0	8.0	16.0	24.0	32.0
1½	0.67	1.3	2.6	5.3	10.6	16.0	21.3
2	0.50	1.0	2.0	4.0	8.0	12.0	16.0
3	0.34	0.67	1.3	2.7	5.3	8.0	10.7
4	0.25	0.50	1.0	2.0	4.0	6.0	8.0
5	0.20	0.40	0.80	1.6	3.2	4.8	6.4
6	0.17	0.34	0.67	1.3	2.6	4.0	5.3
7	0.14	0.30	0.60	1.1	2.3	3.4	4.6
8	0.125	0.25	0.50	1.0	2.0	3.0	4.0
9	0.11	0.22	0.45	0.9	1.8	2.7	3.6
10	0.10	0.20	0.40	0.8	1.6	2.4	3.2

Adapted from Spraying Systems Co., Catalog 36, Wheaton, Ill. (1978).

SUGGESTED CHEMICAL WEED CONTROL PRACTICES

State recommendations for herbicides vary because the effect of herbicides is influenced by growing area, soil type, temperature, and soil moisture. Growers should consult local authorities for specific recommendations. The EPA has established residue tolerances for those herbicides that may leave injurious residues in or on a harvested vegetable and has approved certain materials, rates, and methods of application. Laws regarding vegetation and herbicides are constantly changing. Growers and commercial applicators should not use a chemical on a crop for which the compound is not registered. Herbicides should be used exactly as stated on the label regardless of information presented here. Growers are advised to give special attention to plant-back restrictions.

New Technologies

Many scientists are working on new weed control technologies for reducing the amounts of herbicides required, reducing labor, improved placement of herbicides, and increased effectiveness of herbicides. For example, directed spray technologies using artificial vison and robotics can apply the spray only when a weed is identified. This helps control weeds in between plants in a row.

Robotic mechanical weeders have been around for a while and are being perfected to help growers thin plant hills or cultivate in specific ways to remove plants in the row. Unmanned aerial vehicles (UAVs) can help fly fields to see where the greatest weed pressure can be found.

WEED CONTROL RECOMMENDATIONS

The Cooperative Extension Service in each state publishes recommendations for weed control practices. We present below some websites containing recommendations for weed control in vegetable crops. The list is not exhaustive, and these websites are presented for information purposes only. Because weed control recommendations, especially recommended herbicides, may differ from state to state and year to year, growers are encouraged to consult the proper experts in their state for the latest information about weed control.

SELECTED WEBSITES FOR VEGETABLE WEED CONTROL RECOMMENDATIONS

- A. S. Culpepper, "Commercial Vegetables: Weed Control," https://site. extension.uga.edu/brooksag/files/2015/03/2015-Vegetable-Weed-Control.pdf (accessed 9 Sept. 2021).

- J. Neal, *Weed Management Resources* (North Carolina, 2021), https:// weeds.ces.ncsu.edu/weeds-management-resources/ (accessed 9 Sept. 2021).

- Several authors, *Weed Management Handbook* (2021), https:// pnwhandbooks.org/weed/about (accessed 9 Sept. 2021).

- *Weed Control Guide for Vegetable Crops* (Michigan, 2021), https://shop. msu.edu/products/bulletin-e0433 (accessed 9 Sept. 2021).

- *Cornell Integrated Crop and Past Management Guidelines for Commercial Vegetable Production*, http://veg-guidelines.cce.cornell. edu/ (accessed 9 Sept. 2021).

- Florida Vegetable Weed Management can be found at https://edis.ifas. ufl.edu/entity/topic/vegetable_weeds (accessed 9 Sept. 2021).

Vegetable growing techniques never remain the same for long. As growers we can think back how new technologies have replaced the old ways we did things on the farm. Some that come to my mind include hybrid varieties, safer sprayers and pesticides, plastic mulch, drip irrigation, mechanical harvesters, smartphones, and many more. New technologies are becoming available perhaps at an even increasing rate today, compared with yesterday. The demands on vegetable growers to grow vegetables of high quality that are inexpensive for the consumer are growing. Farmers are increasingly asked to produce under increasing demands for protecting the environment and keen attention to food safety. Labor is a big challenge for vegetable growers: the cost, quality of the worker, and the availability. Many of these challenges are being addressed by new technologies that address the need for increased efficiencies. This part in the handbook is meant to introduce vegetable farmers to this new era of technology, and to demonstrate how these technologies can apply to small farms.

Increased efficiency refers to controlling more of the variability in farming. Some refer to this as *Precision Vegetable Farming*. This chapter's goal is to summarize some the new technologies that have come into use on vegetable farms over the last decade and will become commonplace on vegetable farms into the future. This list is not meant to be all-inclusive, but to paint a picture of what could be of value to you as a vegetable grower today and into your farm's future. Many growers are already adopting new technologies described herein and more growers are interested in learning more. More technologies will undoubtedly come onto the scene in the near future.

Computers have advanced the availability and usefulness of new technologies like no other force in the past. A computer that took up an entire room a few decades ago now fits in the palm of our hand. Computers

Knott's Handbook for Vegetable Growers, Sixth Edition. George J. Hochmuth and Rebecca G. Sideman.
© 2023 John Wiley & Sons Ltd. Published 2023 by John Wiley & Sons Ltd.
Companion website: https://www.wiley.com/go/hochmuth/vegetablegrowers6e

hooked up with satellites have opened an entire world of opportunity for seeing and analyzing our farms, fields, and machinery.

Artificial intelligence (AI). The general area of using computers to help us (or our machines) make decisions can be referred to as *artificial intelligence*. AI depends on robust data sets collected relative to the questions we need answers to and decisions we want to make. One example might be sensors. We might be interested in the health of our crops where a change in color may be related to a disease, or relative maturity—important for determining timely harvesting. Sensors feed information to the grower or to a computer for analyses. Another area of AI would be using computers to control a harvesting machine, a fertilizer applicator, or robot. Satellites are often involved in gathering and sharing data on farm activities.

Precision Agriculture (PA), sometimes called "satellite farming," is the use of temporal, spatial, and individual data sets to help make management decisions where some degree of uncertainty is present. PA often involves the use of satellite information and computers as tools to assist the farmer in making decisions on the farm. The decisions could be anything from basic production technologies such as planter guidance in the field to managing the interconnected aspects of multilayers of decision-making, such as soil chemical conditions, soil type, and fertilizer application. PA helps the farmer make decisions that are in the best interests of efficiency, economy, and the environment.

Global Positioning Satellite (GPS). As its name implies, GPS helps position activities on the farm. GPS tells the farmer where something was measured and can track activities as they move across a field. GPS is important for grid-sampling for soil testing activities. From the GPS information, a map can be created showing how soil test variables like pH or extractable phosphorus change across the field. This information can help a variable-rate fertilizer or lime spreader apply the right amount of material in the right places.

GPS helps inform the grower about yield in a field. Yields of small grains or corn are measured as the combine harvests the crop. The yields across a field can then be related to prior fertilizer applications to see if the variable rate fertilizer resulted in expected or increased yields and thus paid for the grid sampling and reduced fertilizer applications (which helps the environment). Yield monitoring is easily adapted to vegetable crops that are mechanically harvested, like sweet corn, green beans, and potato, but soon other vegetables will be adapted to yield monitoring technologies.

GPS can help autonomous vehicles move across a field in desired patterns. GPS will be important for "driving" autonomous vehicles in a field.

468

Fields can be marked where certain activities were carried out last year. For example, GPS can be used to avoid wheel tracks (and the soil compaction) from previous trips across a field. Sensors can be placed on satellites to measure many variables important to vegetable growers. These variables may include placement of rows and roads, mapping irrigation systems, identifying poor performing spots in a field, providing for accurate swathing in pesticide placement, and more. Satellites with civilian signals are becoming more commonplace. **Graphical Information Systems (GIS)** is another important component of precision agriculture. GIS are sophisticated computer programs that provide for the easy managing of large data sets to provide the answers farmers need from any new technology. For example, GIS helps analyze data from yield and soil conditions such as pH and fertility, so the farmer can see how soil conditions affect yield across a field more precisely. GIS and GPS work hand-in-hand to provide more information to the farmer for making decisions affecting profitability as well as affecting aspects like environmental health.

EXAMPLES OF PRECISION AGRICULTURE SYSTEMS AND TOOLS

Grid sampling. Soil sampling is a time and energy requiring process and a sample only represents an average of the sampled area (usually 20 acres) sampled. The area sampled will normally have variations in soil fertility. Since the sample is an average, this means there are spots in the sampled area that are lower or higher in extractable nutrients. It follows that the fertilizer recommendation is an average for the sampled area. There may be spots that do not receive adequate fertilization, some that receive more than needed, and some that received the correct amounts. Grid sampling breaks a field into smaller sampling units, perhaps 2.5 acres, so that a clearer picture can be obtained about the variation in soil fertility across the entire field. The sampling grids are provided GPS coordinates, so the farmer can know the variation in the field. The results from the lab can be matched with the field coordinates and a computer can draw a map of the field and its fertility status. With this information and the appropriate equipment, fertilizer or lime applications can be tailor-applied across the field. Grid sampling and variable-rate technology can save the farmer money (as fertilizer is expensive), improve yields, and help reduce the environmental footprint for the spots that had been receiving excessive fertilizer.

Yield monitoring. At the end of the example above, a farmer can use PA to monitor yield, on the run, during harvesting and sending the data immediately to a computer via satellite. Yield results can be analyzed to see if the variable-rate fertilizer application increased the field average yield. Yield monitoring and grid sampling and relating the two require the GPS and GIS capabilities with modern computers.

Soil moisture sensing. Another important aspect of farming is proper irrigation—too much water wastes resources and could leach nutrients causing environmental problems, and too little can reduce yield and quality of vegetables. We discussed soil moisture sensors in Part 7 for irrigation management in this handbook. Certain sensor technologies can be used to gather real-time soil moisture data and use it through PA to apply the right amount of water at the right time. Coupling irrigation management with real-time weather information via various agricultural weather apps, can help the farmer manage the irrigation system. Though an app on the smartphone, a farmer can turn off an irrigation system from some distance if there is rain imminent at the location.

Autonomous vehicles. Farming requires considerable labor inputs for managing equipment on the farm. Along with human management comes issues with quality of workmanship and availability. Autonomous vehicles have come a long way in recent years for their development and practicality on farms. Using the computer technology discussed above farmers can have machinery that does field work without a human driver. Some autonomous farm vehicles are being commercialized, for example, sprayers where the first use will probably be in tree crops. Already several agricultural equipment manufacturers are prototyping autonomous (and electric) tractors and equipment.

Robots. Another area of rapid change is in robotics. Already there are many examples where robots are playing roles in agriculture, for example, in packing houses. Another area is in fruit harvesting. Some of the crops, for which it is most difficult to maintain timely availability of quality human harvesters, such as strawberry, are seeing the most rapid development of robotic harvesters. Computer-controlled robots allow for reductions in human harvesting and also allow for harvesting to proceed 24 hours a day. Robotic harvesters rely on computer artificial vision to determine the ripeness and position of the fruit on the plant. Individual fruits can be clipped from the plant without the machine touching the fruit. Soon we will see the widespread commercialization of all kinds of robotic harvesters. Vegetable growers stand to gain the biggest benefit because of the challenges associated with obtaining high quality harvesting labor in a timely manner. Rules and regulations with human labor can be burdensome, particularly for small farmers. Robotic technology will continue to get better and less expensive, and the robots can work around the clock.

Unmanned Aerial Vehicles (UAVs). Probably the best-known example of computerized technology would be the unmanned aerial vehicles, otherwise known as drones. This technology has found its way into everyday life for many people, including farmers. Once a curiosity for flying with a camera over fields to look at crops, the technology now has advanced to a considerable level of utility. Drones use GPS technology to feed data to the computer for analyses on all kinds of important aspects of crop culture. At the simplest level, drones can use a camera to make videos of a crop as it flies overhead and transmit those data to the farmer, who can see how the crop is doing, especially in hard-to-reach areas on the farm. Optical and thermal sensors can be used to determine crop condition, diseases, or nutrient deficiencies. Drones are not invasive to the crop. Heavy-duty drones can even carry a payload of pesticides or nutrients for spraying on predetermined spots in the field. Drones have great utility in assisting crop consultants and scouts to be able to cover much more area in the same amount of time as they once did by riding portions of a field. Drones can help preview a field so the scout can ride to the spots in the field needing the most hands-on attention. There are many benefits to drones for vegetable growers and the drone technology is probably the best example of a modern computer-based technology readily available to small-scale vegetable growers.

Smart sprayers and weeders. Another important innovation in computer use on the farm is with smart sprayers or weeders. Using computer vision technology, sprayers can be controlled to apply herbicides only to certain areas around the plant and not a general over-the-top application. In plastic mulch systems sprayers can direct the spray to the plant hole and not generally over the entire mulched bed. The same idea can be achieved with mechanical weeders that see the weed and remove the weed and not the crop plant. These technologies accomplish the goal of weeding with using much less chemical or none at all, and also with less labor involved, a real plus for a vegetable grower.

Research is being conducted to evaluate ultraviolet light in controlling disease organisms in the field. Soon we will see robotic and smart machines treating fields for diseases or insects without human intervention and without chemicals.

Crop health meters. Handheld meters are becoming more prevalent on the farm. For example, handheld chlorophyll meters can estimate crop health from a measure of chlorophyll content of the leaves.

Smartphones. Our smartphones play an important role in managing these new technologies. Downloading an app to our phone gives us continual access to the new technology. For example, we can manage irrigation systems from a distance by watching the real-time soil moisture data graphics. We can watch the weather in our locale and make decisions about irrigation if rain is on the way.

Where do small farmers fit in? Many small farmers think the new technologies are primarily for large farmers who may have more resources and would benefit from the increased efficiency, but technologies can play an equally valuable role in the small-farm management. You may not follow on-the-go grain yield monitoring, but small farms can take advantage of GPS and grid soil testing to increase efficiency in managing fertilizer. Small farmers can take advantage of computers and soil moisture sensors to save on pumping and water use, including the reduction in potential leaching of nutrients. Handheld or drone-based chlorophyll meters may help a farmer follow the crop health and predict needed fertilizer applications. Smart sprayers will have a role to play on small vegetable farms. Drones are not expensive anymore and they can help a small farmer scout a crop and diagnose crop problems from the air. As the technology becomes more developed, it will become less expensive and more easily adapted to small farms. Vegetable farmers should keep an eye on the technology developments—there are many companies who see this kind of technology being the norm in the future. Agriculture is changing rapidly and successful farmers in the future will be the ones that keep abreast of the changes.

Further Reading

P. Kaushik, *Precision Vegetable Farming Technologies: An Update ONLINEFIRST* (2021). DOI: 10.5772/intechopen.97805, https://www.intechopen.com/online-first/76674 (accessed 25 Sep. 2021).

GPS.gov., "Agriculture," *Applications of GPS in Agriculture*, https://www.gps.gov/applications/agriculture/ (accessed 10 Oct. 2021).

T. Hammonds, *Use of GIS in Agriculture* (Cornell Small Farms Program, 2017), https://smallfarms.cornell.edu/about/ (accessed 10 Oct. 2021).

Geographical Information Systems (GIS). University of Florida Extension, https://edis.ifas.ufl.edu/entity/topic/gis (accessed 10 Oct. 2021).

J. Fletcher and A. Singh, *Applications of Unmanned Aerial Systems in Agricultural Operation Management: Part 1: Overview* (University of Florida Extension, 2020), https://edis.ifas.ufl.edu/publication/AE541 (accessed 10 Oct. 2021).

A. Singh and J. Fletcher, *Applications of Unmanned Aerial Systems in Agricultural Operation Management: Part 11: Platforms and Payloads* (University of Florida Extension, 2021), https://edis.ifas.ufl.edu/publication/ae552 (accessed 10 Oct. 2021).

Y. Ampatzidas, *Applications of Artificial Intelligence for Precision Agriculture* (University of Florida Extension, 2018), https://edis.ifas.ufl.edu/publication/AE529 (accessed 10 Oct. 2021).

Knott's Handbook for Vegetable Growers, Sixth Edition. George J. Hochmuth and Rebecca G. Sideman.
© 2023 John Wiley & Sons Ltd. Published 2023 by John Wiley & Sons Ltd.
Companion website: https://www.wiley.com/go/hochmuth/vegetablegrowers6e

477

01

FOOD SAFETY

FOOD SAFETY IN VEGETABLE PRODUCTION

Microbial contamination of produce can lead to foodborne illness outbreaks, and the key to reducing risks is preventing contamination before it happens. Reviewing and strengthening current good agricultural practices (GAPs) and good manufacturing practices (GMPs) in packing facilities can reduce microbial risks. There are many possible ways for produce to become contaminated by pathogenic microorganisms during production, harvest and handling, including through:

- Soil
- Irrigation water
- Animal manure
- Inadequately composted manure
- Wild and domestic animals
- Inadequate field worker hygiene
- Harvesting equipment
- Transport containers (field to packing facility)
- Wash and rinse water
- Unsanitary handling during sorting and packaging, in wholesale or retail operations, and at home
- Equipment used to soak, pack or cut produce
- Ice
- Cooling units (hydrocoolers)
- Transport vehicles
- Improper storage conditions (temperatures)
- Improper packaging
- Cross contamination in storage, display and preparation

Adapted from website of: Food Safety, *Cornell Vegetable Program* (Ithaca, NY: Cornell University), https://cvp.cce.cornell.edu/food_safety.php (accessed 12 Mar. 2021).

MINIMIZING RISKS OF MICROBIAL CONTAMINATION

Human Pathogens That May Be Associated With Fresh Vegetables

- Soil-associated pathogenic bacteria (*Clostridium botulinum, Listeria moncytogenes*)
- Fecal-associated pathogenic bacteria (*Salmonella* spp., *Shigella* spp., *E. coli* O157:H7 and others)
- Pathogenic parasites (*Cryptosporidium, Cyclospora*)
- Pathogenic viruses (Hepatitis, Norwalk virus and others)

Basic Principles of Good Agricultural Practices (GAPs)

- Prevention of microbial contamination of fresh produce is favored over reliance on corrective actions once contamination has occurred.
- To minimize microbial food safety hazards in fresh produce, growers or packers should use GAPs in those areas over which they have a degree of control while not increasing other risks to the food supply or the environment.
- Anything that comes in contact with fresh produce has the potential of contaminating it. For most foodborne pathogens associated with produce, the major source of contamination is associated with human or animal feces.
- Whenever water comes in contact with fresh produce, its source and quality dictate the potential for contamination.
- Practices using manure or municipal biosolid wastes should be closely managed to minimize the potential for microbial contamination of fresh produce.
- Worker hygiene and sanitation practices during production, harvesting, sorting, packing and transport play a critical role in minimizing the potential for microbial contamination of fresh produce.
- Follow all applicable local, state and Federal laws and regulations, or corresponding or similar laws, regulations, or standards for operators outside the U.S. for agricultural practices.
- Accountability at all levels of the agricultural environment (farms, packing facility, distribution center and transport operation) is important to a successful food safety program. There must be qualified personnel and effective monitoring to ensure that all elements of the program function correctly and to help track produce back through the distribution channels to the producer.

Adapted from J. R. Gorny and D. Zagory, "Food Safety," in *The Commercial Storage of Fruits, Vegetables, and Florist and Nursery Stocks* (USDA Agriculture Handbook 66, 2016), https://www.ars.usda.gov/arsuserfiles/oc/np/commercialstorage/commercialstorage.pdf.

GOOD AGRICULTURAL PRACTICES (GAPs) AND THE FOOD SAFETY MODERNIZATION ACT (FSMA)

The principles of Good Agricultural Practices (GAPs) were first introduced by the US Food and Drug Administration (FDA) in 1998. In response to this guidance, the United States Department of Agriculture (USDA) formally implemented the Good Agricultural Practices & Good Handling Practices (GAPs and GHPs) audit verification program.

In response to growing food safety issues, the Food Safety Modernization Act (FSMA) was passed by Congress and signed into law in January 2011. The new law requires companies to implement a food safety program that significantly minimizes potential hazards and risks of foodborne illness.

While GAPs programs have largely been voluntary, imposed by the industry or buyers, all non-exempt operations must follow the new FSMA federal guidelines, except for exempt commodities (outlined in the regulation) and for those producers exporting to foreign countries. In those circumstances, voluntary GAPs programs may still be required by buyers or trade organizations.

The Produce Safety Rule (PSR), one of the foundational rules of the FSMA, establishes standards to ensure safe growing, harvesting, packing, and holding of covered produce on farms. Various parts of the PSR set requirements to prevent pathogens from agricultural water, biological soil amendments of animal origin, human waste, and biological hazards from domesticated or wild animals from contaminating produce. Whether covered or exempt, taking immediate steps to implement field sanitation GAPs will benefit a company's financial viability and overall produce safety.

Currently, the PSR is required for all non-exempt operations. A flow chart with details on coverage and exemptions is available on the FDA website for operations to help determine whether they need to comply with the PSR (https://www.fda.gov/media/94332/download) (accessed 12 Mar. 2021). Considering that the FSMA regulations are science-based and recommend minimum standards for safe food practices, growers must be aware of the new rules so they can comply with the more stringent regulations.

For growers who must comply with the PSR, the Produce Safety Alliance (a collaboration between Cornell University, the FDA and the USDA) has developed a training curriculum and many resources to help growers develop and implement a written farm food safety plan. For more information, visit: https://producesafetyalliance.cornell.edu/ (accessed 12 Mar. 2021).

Adapted from J. A. Lepper, J. De, C. R. Pabst, A. Sreedharan, R. M. Goodrich-Schneider, and K. R. Schneider, *Food Safety on the Farm: Good Agricultural Practices and Good Handling Practices—Field Sanitation* (2019), https://edis.ifas.ufl.edu/publication/FS160 (accessed 12 Mar. 2021).

Additional information on agricultural food safety can be found at:

- https://onfarmfoodsafety.org/ (accessed 12 Mar. 2021)
- https://producesafetyalliance.cornell.edu/ (accessed 12 Mar. 2021)
- http://www.gaps.cornell.edu/ (accessed 12 Mar. 2021)
- https://newfarmers.usda.gov/food-safety (accessed 12 Mar. 2021)
- http://www.foodsafety.gov/ (accessed 12 Mar. 2021)
- http://www.extension.iastate.edu/foodsafety/ (accessed 12 Mar. 2021)
- https://www.fda.gov/food/guidance-regulation-food-and-dietary-supplements/food-safety-modernization-act-fsma (accessed 12 Mar. 2021)

CLEANING AND SANITIZING

The Food Safety Modernization Act Produce Safety Rule requires cleaning and sanitizing harvest containers, and all food contact surfaces. Remember that cleaning and sanitizing are two different things, and cleaning always happens first. Cleaning is the physical removal of dirt and organic matter, whereas sanitizing is the treatment of a surface to reduce or remove microorganisms.

Cleaning and sanitizing is a four-step process:

1. Remove any obvious dirt or debris from the surface
2. Apply detergent and scrub. Detergents should be appropriate for use on food contact surfaces.
3. Rinse the surface with clean water to remove soil and detergent.
4. Apply sanitizer approved for use on food contact surfaces. Rinse if required by label.

Sanitizers may also be used in water that comes in contact with fruits and vegetables during or after harvest. In this case, sanitizers are used as water treatments to prevent the spread of contamination in harvest and postharvest systems such as dump tanks and flumes, reducing the risk of cross-contamination by human pathogens. The attributes of several sanitizers used for this purpose are describe in Table 11.1 Sanitizers fall under the EPA definition of an antimicrobial pesticide, and growers should use EPA-labelled sanitizers whose labelled uses are consistent with intended uses, for example: sanitizing food contact surfaces or treating fruit or vegetable washwater.

481

TABLE 11.1. SANITIZING CHEMICALS FOR PACKINGHOUSES

Compound	Advantages	Disadvantages
Chlorine (most widely used sanitizer in packinghouse water systems)	Relatively inexpensive. Broad spectrum—effective on many different microbes. Practically no residue left on the commodity.	Corrosive to equipment. Sensitive to pH. Below 6.5 or above 7.5 reduces activity or increases noxious odors. Can irritate skin and damage mucous membranes.
Chlorine dioxide	Activity is much less pH dependent than chlorine.	Must be generated on-site. Greater human exposure risk than chlorine. Off-gassing of noxious gases is common. Concentrated gases can be explosive.
Peroxyacetic acid	No known toxic residues or byproducts. Produces very little off-gassing. Less affected by organic matter than chlorine. Low corrosiveness to equipment.	Activity is reduced in the presence of metal ions. Concentrated product is very toxic to humans. Sensitive to pH. Greatly reduced activity at pH above 7–8.

**TABLE 11.1.   SANITIZING CHEMICALS FOR PACKINGHOUSES
(Continued)**

Compound	Advantages	Disadvantages
Ozone	Very strong oxidizer/ sanitizer. Can reduce pesticide residues in the water. Less sensitive to pH than chlorine (but breaks down much faster above ~pH 8.5). No known toxic residues or byproducts.	Must be generated on-site. Ozone gas is toxic to humans. Off-gassing can be a problem. Treated water should be filtered to remove particulates and organic matter. Very corrosive to equipment (including rubber and some plastics). Highly unstable in water— half-life ~15 min; may be less than 1 min in water with organic matter or soil.

Note: Although quaternary ammonia is an effective sanitizer with useful properties and can be used to sanitize equipment, it is not registered for contact with food.

An up-to-date list of labeled sanitizers for produce is maintained by the Produce Safety Alliance, and is available here: https://producesafetyalliance.cornell.edu/resources/general-resource-listing/ (accessed 16 Sept. 2021).

See also: D. P. Clements, G. Wall, D. Stoeckel, C. Fisk, K. Woods, and E. Bihn, *Introduction to Selecting an EPA-Labeled Sanitizer* (2018), https://producesafetyalliance.cornell.edu/sites/ producesafetyalliance.cornell.edu/files/shared/documents/Sanitizer-Factsheet.pdf (accessed 21 ept. 2021); Food Safety on the Farm Series. University of Florida Extension, https://edis.ifas.ufl. edu/entity/topic/series_food_safety_on_the_farm (accessed 4 Dec. 2021).

02

GENERAL POSTHARVEST HANDLING PROCEDURES

The basic steps in postharvest handling for four groups of vegetables are presented in Tables 11.2 (leafy, floral and succulent vegetables), 11.3 (underground storage organs), 11.4 (immature fruits), and 11.5 (mature fruits). Additional information on postharvest handling can be found on these websites:

Food and Agriculture Organization (FAO) Information Network on Post-harvest Operations (InPhO), www.fao.org/inpho (accessed 4 Dec. 2021)

North Carolina State University Cooling and Handling Publications, https://content.ces.ncsu.edu/proper-postharvest-cooling-and-handling-methods (accessed 4 Dec. 2021)

UC Davis postharvest Technology Research & Information Center, http://postharvest.ucdavis.edu (accessed 4 Dec. 2021)

USDA/Agricultural Marketing Service/Fruit & Vegetable Division, https://www.ams.usda.gov/market-news/fruits-vegetables (accessed 4 Dec. 2021)

USDA/Economic Research Service, http://www.ers.usda.gov/ (accessed 4 Dec. 2021)

USDA Food Safety and Inspection Service, www.fsis.usda.gov (accessed 4 Dec. 2021)

University of Florida Postharvest Programs, http://postharvest.ifas.ufl.edu (accessed 4 Dec. 2021)

The Commercial Storage of Fruits, Vegetables, and Florist and Nursery Stocks. USDA Agriculture Handbook 66 (2016). https://www.ars.usda.gov/arsuserfiles/oc/np/commercialstorage/commercialstorage.pdf (accessed 4 Dec. 2021)

TABLE 11.2. LEAFY, FLORAL, AND SUCCULENT VEGETABLES

Step	Function

1. Harvesting mostly by hand; some harvesting aids are in use
2. Transport to packinghouse and unloading if not field packed
3. Cutting and trimming (by harvester or by different worker on mobile packing line or in packinghouse)
4. Sorting and manual sizing (as above)
5. Washing or rinsing
6. Wrapping (e.g., cauliflower, head lettuce) or bagging (e.g., celery)
7. Packing in shipping containers (waxed fiberboard or plastic to withstand water or ice exposure for cooling)
8. Palletization of shipping containers
9. Cooling methods
   - Vacuum cooling: lettuce
   - Hydrovacuum cooling: cauliflower, celery
   - Hydrocooling: artichoke, celery, green onion, leaf lettuce, leek, spinach
   - Package ice: broccoli, parsley, spinach
   - Room cooling: artichoke, cabbage
10. Transport, destination handling, retail handling

TABLE 11.3.   UNDERGROUND STORAGE ORGAN VEGETABLES

Step	Function

1. Mechanical harvest (digging, lifting) except hand harvesting of sweet potato
2. Curing (in field) of potato, onion, garlic, and tropical crops
3. Field storage of potatoes and tropical storage organ vegetables in pits, trenches or clamps
4. Collection into containers or into bulk trailers
5. Transport to packinghouse and unloading
6. Cleaning by dry brushing or with water
7. Sorting to eliminate defects
8. Sizing
9. Packing in bags or cartons; consumer packs placed within master containers
10. Palletization of shipping containers
11. Cooling methods
    - Hydrocooling of temperate storage roots and tubers
    - Room cooling of potato, onion, garlic, and tropical storage organs
12. Curing
    - Forced-air drying (onion and garlic)
    - High temperature and relative humidity (potato and tropical storage organs)
13. Storage
    - Ventilated storage of potato, onion, garlic, and sweet potato in cellars and warehouses
    - Temporary storage of temperate storage organ vegetables
    - Long-term storage of potato, onion, garlic, and tropical crops following curing
14. Fungicide treatment (sweet potato); sprout inhibitor (potato)
15. Transport, destination handling, retail handling

TABLE 11.4.   IMMATURE FRUIT VEGETABLES

Step	Function

1. Harvesting mostly by hand into buckets or trays; some harvesting aids are in use.
   - Sweet corn, snap bean, and pea are also harvested mechanically.
   - Field-packed vegetables are usually not washed, but may be wiped with a moist cloth or spray-washed on a mobile packing line.
2. For packinghouse operations, stacking buckets or trays on trailers or transferring to shallow pallet bins.
3. Transporting harvested vegetables to packinghouse.
4. Unloading by dry or wet dump.
5. Washing or rinsing.
6. Sorting to eliminate defects.
7. Waxing cucumber and pepper.
8. Sizing.
9. Packing in shipping containers by weight or count.
10. Palletizing shipping containers.
11. Cooling methods:
    - Hydrocooling: bean, pea, sweet corn.
    - Forced-air cooling: chayote, cucumber, eggplant, okra, summer squash.
    - Slush-ice cooling and vacuum cooling: sweet corn.
12. Temporary storage.
13. Transporting, destination handling, retail handling.

TABLE 11.5. MATURE FRUIT VEGETABLES

Step	Function

1. Harvesting
2. Hauling to the packinghouse or processing plant
3. Cleaning
4. Sorting to eliminate defects
5. Waxing (tomato, pepper)
6. Sizing and sorting into grades
7. Packing—shipping containers
8. Palletization and unitization
   - Curing of winter squash and pumpkin
   - Ripening of melon and tomato with ethylene
9. Cooling (hydrocooling, room cooling, forced-air cooling)
10. Temporary storage
11. Loading into transport vehicles
12. Destination handling (distribution centers, wholesale markets, etc.)
13. Delivery to retail
14. Retail handling

TABLE 11.6. APPROXIMATE TIME FROM VEGETABLE PLANTING TO MARKET MATURITY UNDER OPTIMUM GROWING CONDITIONS

Time to Market Maturity[1] (Days)

Vegetable	Early Variety	Late Variety	Common Variety
Bean, broad	—	—	120
Bean, bush	48	60	—
Bean, edible soy	84	115	—
Bean, pole	58	70	—
Bean, lima, bush	65	80	—
Bean, lima, pole	80	88	—
Beet	50	65	—
Broccoli[2]	60	85	—
Broccoli raab	—	—	70
Brussels sprouts[2]	90	185	—
Cabbage	70	120	—
Cardoon	—	—	120
Carrot	50	95	—
Cauliflower[2]	55	90	—
Celeriac	—	—	110
Celery[2]	90	125	—
Chard, Swiss	50	60	—
Chervil	—	—	60
Chicory	65	150	—
Chinese cabbage	50	80	—
Chive	—	—	90
Collards	70	85	—
Corn, sweet	60	95	—
Corn salad	—	—	60
Cress	—	—	45
Cucumber, pickling	48	58	—
Cucumber, slicing	55	70	—
Dandelion	—	—	85
Eggplant[2]	60	80	—

	Time to Market Maturity[1] (Days)		
Vegetable	Early Variety	Late Variety	Common Variety
Endive	80	100	—
Florence fennel	—	—	100
Kale	—	—	55
Kohlrabi	50	60	—
Leek	—	—	120
Lettuce, butterhead	55	70	—
Lettuce, cos	65	75	—
Lettuce, head	70	85	—
Lettuce, leaf	40	50	—
Melon, cantaloupe	75	105	—
Melon, casaba	—	—	110
Melon, honeydew	90	110	—
Melon, Persian	—	—	110
Okra	50	60	—
Onion, dry	90	150	—
Onion, green	45	70	—
Parsley	65	75	—
Parsley root	—	—	90
Parsnip	—	—	120
Pea	56	75	—
Pea, edible-podded	60	75	—
Pepper, hot[2]	60	95	—
Pepper, sweet[2]	65	80	—
Potato	90	120	—
Pumpkin	85	120	—
Radicchio	90	95	—
Radish	22	30	—
Radish, winter	50	65	—
Roselle	—	—	175
Rutabaga	—	—	90
Salsify	—	—	150

| | Time to Market Maturity[1] (Days) | | |
Vegetable	Early Variety	Late Variety	Common Variety
Scolymus	—	—	150
Scorzonera	—	—	150
Sorrel	—	—	60
Southernpea	65	85	—
Spinach	37	45	—
Squash, summer	40	50	—
Squash, winter	80	120	—
Sweet potato	120	150	—
Swiss chard	—	—	60
Tomatillo	—	—	80
Tomato[2]	60	85	—
Tomato, processing	118	130	—
Turnip	35	50	—
Watercress	—	—	180
Watermelon	75	95	—

[1] Maturity may vary depending on season, latitude, production practices, variety, and other factors.
[2] Time from transplanting.

TABLE 11.7. APPROXIMATE TIME FROM POLLINATION OF FRUITING VEGETABLES TO MARKET MATURITY UNDER WARM GROWING CONDITIONS

Vegetable	Time to Market Maturity[1] (Days)
Bean	7–10
Cantaloupe	42–46
Corn, fresh market[2]	18–23
Corn, processing[2]	21–27
Cucumber, pickling (¾–1⅛ in. in diameter)	4–5
Cucumber, slicing	15–18
Eggplant (⅔ maximum size)	25–40
Okra	4–6
Pepper, green stage (about maximum size)	45–55
Pepper, red stage	60–70
Pumpkin, Connecticut field	80–90
Pumpkin, Dickinson	90–110
Pumpkin, Small Sugar	65–75
Squash, summer, crookneck[3]	6–7
Squash, summer, straightneck[3]	5–6
Squash, summer, scallop[3]	4–5
Squash, summer, zucchini[3]	3–4
Squash, winter, banana	70–80
Squash, winter, Boston Marrow	60–70
Squash, winter, buttercup	60–70
Squash, winter, butternut	60–70
Squash, winter, Golden Delicious	60–70
Squash, winter, hubbard	80–90
Squash, winter, Table Queen or Acorn	55–60
Strawberry	25–42
Tomato, mature green stage	35–45
Tomato, red ripe stage	45–60
Watermelon	42–45

[1] Maturity may vary depending on season, latitude, production practices, variety, and other factors.
[2] Days from 50% silking.
[3] For a weight of ¼–½ lb.

ESTIMATING YIELDS OF CROPS

Predicting crop yields before harvest aids in scheduling harvests of various fields for total yields, and to obtain highest yields of a particular grade or stage of maturity. To estimate yields, follow these steps:

1. Select and measure a typical 10-ft section of a row. If the field is variable or large, you may want to select several 10-ft sections.
2. Harvest the crop from the measured section or sections.
3. Weigh the entire sample for total yields or grade the sample and weigh the graded sample for yield of a particular grade.
4. If you have harvested more than one 10-ft section, divide the yield by the number of sections harvested.
5. Multiply the sample weight by the conversion factor in the table for your row spacing. The value obtained will equal hundredweight (cwt) per acre.

TABLE 11.8. CONVERSION FACTORS FOR ESTIMATING YIELDS

Row Spacing (in.)	Multiply Sample Weight (lb) by Conversion Factor to Obtain cwt/acre
12	43.6
15	34.8
18	29.0
20	26.1
21	24.9
24	21.8
30	17.4
36	14.5
40	13.1
42	12.4
48	10.9

Example 1: A 10-ft sample of carrots planted in 12-in. rows yields 9 lb of No. 1 carrots.

$$9 \times 43.6 = 392.4 \, \text{cwt} \, / \, \text{acre}$$

Example 2: The average yield of three 10-ft samples of No. 1 potatoes planted in 36-in. rows is 26 lb.

$$26 \times 14.5 = 377 \, \text{cwt} \, / \, \text{acre}$$

493

TABLE 11.9. YIELDS OF VEGETABLE CROPS

Vegetable	Approximate Average Yield in the United States (cwt/acre)	Good Yield (cwt/acre)
Artichoke	140	160
Asparagus	35	45
Bean, fresh market	62	100
Bean, processing	75	135
Bean, lima, processing	28	40
Beet, fresh market	140	200
Beet, processing	320	400
Broccoli	155	200
Brussels sprouts	180	200
Cabbage, fresh market	320	450
Cabbage, processing	500	800
Carrot, fresh market	350	800
Carrot, processing	520	700
Cauliflower	200	200
Celeriac	—	200
Celery	550	750
Chard, Swiss	—	150
Chinese cabbage, napa	—	400
Chinese cabbage, bok-choy	—	300
Collards	120	200
Corn, fresh market	110	200
Corn, processing	150	200
Cucumber, fresh market	185	300
Cucumber, processing	110	300
Eggplant	250	350
Endive, escarole	180	200
Garlic	165	200
Horseradish	—	80
Lettuce, head	360	400
Lettuce, leaf	200	325
Lettuce, Romaine	300	350
Melon, cantaloupe	250	250
Melon, honeydew	285	250
Melon, Persian	130	160

TABLE 11.9. YIELDS OF VEGETABLE CROPS (*Continued*)

Vegetable	Approximate Average Yield in the United States (cwt/acre)	Good Yield (cwt/acre)
Okra	60	150
Onion	545	650
Pea, fresh market	40	60
Pea, processing (shelled)	30	45
Pepper, bell	325	375
Pepper, chili (fresh and dried)	185	230
Pepper, pimiento	—	100
Potato	315	400
Pumpkin	220	400
Radish	90	200
Rhubarb	—	200
Rutabaga	—	400
Snowpea	—	80
Southernpea	—	35
Spinach, fresh market	160	225
Spinach, processing	180	220
Squash, summer	160	300
Squash, winter	—	400
Sweet potato	145	300
Strawberry	400	600
Tomato, fresh market	290	400
Tomato, processing	895	900
Tomato, cherry	—	600
Turnip	—	400
Watermelon	365	500

TABLE 11.10. COMPARISON OF TYPICAL PRODUCT EFFECTS AND COST FOR COMMON COOLING METHODS

Product Effect	Forced-Air	Hydro	Vacuum	Water Spray	Ice	Room
Cooling time (hr)	1–10	0.1–1.0	0.3–2.0	0.3–2.0	0.1–0.3[1]	20–100
Product moisture loss (%)	0.1–2.0	0–0.5	2.0–4.0	No data	No data	0.1–2.0
Water contact with product	No	Yes	No	Yes	Yes, unless bagged	No
Potential for decay contamination	Low	High[2]	None	High[2]	Low	Low
Capital cost	Low	Low	Medium	Medium	High	High
Energy efficiency	Low	High	High	Medium	Low	Low
Water-resistant packaging needed	No	Yes	No	Yes	Yes	No
Portable	Sometimes	Rarely done	Common	Common	Common	No
Feasibility of in-line cooling	Rarely done	Yes	No	No	Rarely done	No

Adapted from J. F. Thompson, "Pre-cooling and Storage Facilities," in *The Commercial Storage of Fruits, Vegetables, and Florist and Nursery Stocks* (USDA Agriculture Handbook 66, 2016), https://www.ars.usda.gov/arsuserfiles/oc/np/commercialstorage/commercialstorage.pdf (accessed 4 Dec. 2021)

[1]Top icing can take much longer.

[2]Recirculated water must be constantly sanitized to minimize accumulation of decay pathogens.

TABLE 11.11.   GENERAL COOLING METHODS FOR VEGETABLES

Method[1]	Vegetable	Comments
Room cooling	All vegetables	Too slow for most perishable commodities. Cooling rates vary extensively within loads, pallets, and containers
Forced-air cooling (pressure cooling)	Strawberry, fruit-type vegetables, tubers, cauliflower	Much faster than room cooling; cooling rates very uniform if properly used. Container venting and stacking requirements are critical to effective cooling
Hydrocooling	Stems, leafy vegetables, some fruit-type vegetables	Very fast cooling; uniform cooling in bulk if properly used, but may vary extensively in packed shipping containers; daily cleaning and sanitation measures essential; product must tolerate wetting; water-tolerant shipping containers may be needed
Package-icing	Roots, stems, some flower-type vegetables, green onion, Brussels sprouts	Fast cooling; limited to commodities that can tolerate water–ice contact; water-tolerant shipping containers are essential
Vacuum cooling	Leafy vegetables; some stem and flower-type vegetables	Commodities must have a favorable surface-to-mass ratio for effective cooling. Causes about 1% weight loss for each 10°F cooled. Adding water during cooling prevents this weight loss but equipment is more expensive, and water-tolerant shipping containers are needed

TABLE 11.11. GENERAL COOLING METHODS FOR VEGETABLES (*Continued*)

Method[1]	Vegetable	Comments
Transit cooling: mechanical refrigeration	All vegetables	Cooling in most available equipment is too slow and variable; generally not effective
Top-icing and channel-icing	Some roots, stems, leafy vegetables, cantaloupe	Slow and irregular, top-ice weight reduces net pay load; water-tolerant shipping containers needed

Adapted from A. A. Kader (ed.), *Postharvest Technology of Horticultural Crop*, 2nd ed. (University of California, Division of Agriculture and Natural Resources Publication 3311, 1992). Copyright © 1992 Regents of the University of California. See also: J. A. Watson, D. Treadwell, S. A. Sargent, J. K. Brecht, and W. Pelletier, *Postharvest Storage, Packaging and Handling of Specialty Crops: A Guide for Florida Small Farm Producers* (University of Florida Extension, 2019).

[1] For these methods to be effective, the product must be cooled continuously until reaching the consumer.

TABLE 11.12. SPECIFIC COOLING METHODS FOR VEGETABLES

Vegetable	Size of Operation		Remarks
	Large	Small	
Leafy Vegetables			
Cabbage	VC, FA	FA	
Iceberg lettuce	VC	FA	
Kale, collards	VC, R, WVC	FA	
Leaf lettuces, spinach, endive, escarole, Chinese cabbage, bok choy, romaine	VC, FA, WVC, HC	FA	
Root Vegetables			
With tops	HC, PI, FA	HC, FA	Carrots can be VC
Topped	HC, PI	HC, PI, FA	
Irish potato	R w/evap coolers		With evap coolers, facilities should be adapted to curing
Sweet potato	HC	R	
Stem and Flower Vegetables			
Artichoke	HC, PI	FA, PI	
Asparagus	HC	HC	
Broccoli, Brussels sprouts	HC, FA, PI	FA, PI	
Cauliflower	FA, VC	FA	

TABLE 11.12. SPECIFIC COOLING METHODS FOR VEGETABLES (*Continued*)

Vegetable	Size of Operation		Remarks
	Large	Small	
Celery, rhubarb	HC, WVC, VC	HC, FA	
Green onion, leek	PI, HC, WVC	PI	
Mushroom	FA, VC	FA	
Pod Vegetables			
Bean	HC, FA	FA	
Pea	FA, PI, VC	FA, PI	
Bulb Vegetables			
Dry onion	R	R, FA	Should be adapted to curing
Garlic	R		
Fruit-type Vegetables			
Cucumber, eggplant	R, FA, FA-EC	FA, FA-EC	Fruit type vegetables are chilling sensitive but at varying temperature.
Melons			
Cantaloupe	HC, FA, PI	FA, FA-EC	
Crenshaw, honeydew, casaba	FA, R	FA, FA-EC	
Watermelon	FA, HC	FA, FA-EC	
Pepper	R, FA, FA-EC, VC	FA, FA-EC	

500

Summer squash, okra	R, FA, FA-EC	FA, FA-EC	
Sweet corn	HC, VC, PI	HC, FA, PI	
Tomatillo	R, FA, FA-EC	FA, FA-EC	
Tomato	R, FA, FA-EC	R	
Winter squash	R		
Fresh Herbs			
Not packaged	HC, FA	FA, R	Can be easily damaged by water beating in HC
Packaged	FA	FA, R	
Cactus			
Leaves (nopalitos)	R	FA	
Fruit (tunas or prickly pears)	R	FA	

Adapted from A. A. Kader (ed.), *Postharvest Technology of Horticultural Crops*, 3rd ed. (University of California, Division of Agriculture and Natural Resources Publication 3311, 2002). Copyright © 2002 Regents of the University of California.
R = Room Cooling; FA = Forced-Air Cooling; HC = Hydrocooling; VC = Vacuum Cooling; WVC = Water Spray Vacuum Cooling; FA-EC = Forced-Air Evaporative Cooling; PI = Package Icing.

TABLE 11.13. RELATIVE PERISHABILITY AND POTENTIAL STORAGE LIFE OF FRESH VEGETABLES IN AIR AT NEAR OPTIMUM STORAGE TEMPERATURE AND RELATIVE HUMIDITY

Potential Storage Life (weeks)

<2	2–4	4–8	8–16
Asparagus	Artichoke	Beet	Garlic
Bean sprouts	Green bean	Carrot	Onion
Broccoli	Brussels sprouts	Potato (immature)	Potato (mature)
Cantaloupe	Cabbage	Radish	Pumpkin
Cauliflower	Celery		Winter squash
Green onion	Eggplant		Sweet potato
Leaf lettuce	Head lettuce		Taro
Mushroom	Mixed melons		Yam
Pea	Okra		
Spinach	Pepper		
Sweet corn	Summer squash		
Tomato (ripe)	Tomato (partially ripe)		
Fresh-cut vegetables			

Adapted from A. A. Kader (ed.), *Postharvest Technology of Horticultural Crops*, 3rd ed. (University of California, Division of Agriculture and Natural Resources Publication 3311, 2002). Copyright © 2002 Regents of the University of California.

TABLE 11.14. RECOMMENDED TEMPERATURE AND RELATIVE
HUMIDITY CONDITIONS AND APPROXIMATE
STORAGE LIFE OF FRESH VEGETABLES

For more detailed information about specific commodities, see the
following resources:
The Commercial Storage of Fruits, Vegetables, and Florist and Nursery
Stocks. USDA Agriculture Handbook 66 (2016). https://www.ars.usda.
gov/arsuserfiles/oc/np/commercialstorage/commercialstorage.pdf

| | Storage Conditions | | |
Vegetable	Temperature (°F)	Relative Humidity (%)	Approximate Storage Life
Amaranth	32–36	95–100	10–14 days
Anise	32–36	90–95	2–3 weeks
Artichoke, globe	32	95–100	2–3 weeks
Artichoke, Jerusalem	31–32	90–95	4 months
Asparagus	36	95–100	2–3 weeks
Bean, fava	32	90–95	1–2 weeks
Bean, lima	37–41	95	5–7 days
Bean, snap	40–45	95	7–10 days
Bean, yardlong	40–45	95	7–10 days
Beet, bunched	32	98–100	10–14 days
Beet, topped	32	98–100	4 months
Bitter melon	50–54	85–90	2–3 weeks
Bok choy	32	95–100	3 weeks
Boniato	55–60	85–90	4–5 months
Broccoli	32	95–100	10–14 days
Brussels sprouts	32	95–100	3–5 weeks
Cabbage, early	32	98–100	3–6 weeks
Cabbage, late	32	95–100	5–6 months
Cabbage, Chinese	32	95–100	2–3 months
Cactus, leaves	41–50	90–95	2–3 weeks
Cactus, pear	41	90–95	3 weeks
Calabaza	50–55	50–70	2–3 months
Carrot, bunched	32	98–100	10–14 days
Carrot, topped	32	98–100	6–8 months
Cassava	32–41	85–90	1–2 months

503

TABLE 11.14. RECOMMENDED TEMPERATURE AND RELATIVE
HUMIDITY CONDITIONS AND APPROXIMATE
STORAGE LIFE OF FRESH VEGETABLES
(*Continued*)

Vegetable	Storage Conditions		
	Temperature (°F)	Relative Humidity (%)	Approximate Storage Life
Cauliflower	32	95–98	3–4 weeks
Celeriac	32	98–100	6–8 months
Celery	32	98–100	1–2 months
Chard	32	95–100	10–14 days
Chayote	45	85–90	4–6 weeks
Chicory, witloof	36–38	95–98	2–4 weeks
Chinese broccoli	32	95–100	10–14 days
Collards	32	95–100	10–14 days
Cucumber, slicing	50–54	85–90	10–14 days
Cucumber, pickling	40	95–100	7 days
Daikon	32–34	95–100	4 months
Eggplant	50	90–95	1–2 weeks
Endive, escarole	32	95–100	2–4 weeks
Garlic	32	65–70	6–7 months
Ginger	55	65	6 months
Greens, cool-season	32	95–100	10–14 days
Greens, warm-season	45–50	95–100	5–7 days
Horseradish	30–32	98–100	10–12 months
Jicama	55–65	85–90	1–2 months
Kale	32	95–100	2–3 weeks
Kohlrabi	32	98–100	2–3 months
Leek	32	95–100	2 months
Lettuce	32	98–100	2–3 weeks
Malanga	45	70–80	3 months
Melon			
Cantaloupe	36–41	95	2–3 weeks
Casaba	45–50	85–90	3–4 weeks
Crenshaw	45–50	85–90	2–3 weeks

| Vegetable | Storage Conditions | | Approximate Storage Life |
	Temperature (°F)	Relative Humidity (%)	
Honeydew	41–50	85–90	3–4 weeks
Persian	45–50	85–90	2–3 weeks
Watermelon	50–59	90	2–3 weeks
Mushroom	32	90	7–14 days
Mustard greens	32	90–95	7–14 days
Okra	45–50	90–95	7–10 days
Onion, dry	32	65–70	1–8 months
Onion, green	32	95–100	3 weeks
Parsley	32	95–100	8–10 weeks
Parsnip	32	98–100	4–6 months
Pea, English, snow, snap	32–34	90–98	1–2 weeks
Pepino	41–50	95	4 weeks
Pepper, chili	41–50	85–95	2–3 weeks
Pepper, sweet	45–50	95–98	2–3 weeks
Potato, early[1]	50–59	90–95	10–14 days
Potato, late[2]	40–54	95–98	5–10 months
Pumpkin	54–59	50–70	2–3 months
Radicchio	32–34	95–100	3–4 weeks
Radish, spring	32	95–100	1–2 months
Radish, winter	32	95–100	2–4 months
Rhubarb	32	95–100	2–4 weeks
Rutabaga	32	98–100	4–6 months
Salsify	32	95–98	2–4 months
Scorzonera	32–34	95–98	6 months
Shallot	32–36	65–70	—
Southernpea	40–41	95	6–8 days
Spinach	32	95–100	10–14 days
Sprouts			
Alfalfa	32	95–100	7 days
Bean	32	95–100	7–9 days
Radish	32	95–100	5–7 days

TABLE 11.14. RECOMMENDED TEMPERATURE AND RELATIVE
HUMIDITY CONDITIONS AND APPROXIMATE
STORAGE LIFE OF FRESH VEGETABLES
(*Continued*)

| Vegetable | Storage Conditions | | Approximate Storage Life |
	Temperature (°F)	Relative Humidity (%)	
Squash, summer	45–50	95	1–2 weeks
Squash, winter[3]	54–59	50–70	2–3 months
Strawberry	32	90–95	5–7 days
Sweet corn, shrunken	32	90–95	10–14 days
Sweet corn, sugary	32	95–98	5–8 days
Sweet potato[4]	55–59	85–95	4–10 months
Tamarillo	37–40	85–95	10 weeks
Taro	45–50	85–90	4 months
Tomatillo	45–55	85–90	3 weeks
Tomato, mature green	50–55	90–95	2–5 weeks
Tomato, firm ripe	46–50	85–90	1–3 weeks
Turnip greens	32	95-100	10-14 days
Turnip root	32	95	4-5 months
Water chestnut	32–36	85–90	2–4 months
Watercress	32	95–100	2–3 weeks
Yam	59	70-80	2-7 months

Adapted from A. A. Kader (ed.), *Postharvest Technology of Horticultural Crops*, 3rd ed. (University of California, Division of Agriculture and Natural Resources Publication 3311, 2002). Copyright © 2002 Regents of the University of California.

[1] Winter, spring, or summer-harvested potatoes are usually not stored. However, they can be held 4–5 months at 40°F if cured 4 or more days at 60–70°F before storage. Potatoes for chips should be held at 70°F or conditioned for best chip quality.

[2] Fall-harvested potatoes should be cured at 50–60°F and high relative humidity for 10–14 days. Storage temperatures for table stock or seed should be lowered gradually to 38–40°F. Potatoes intended for processing should be stored at 50–55°F; those stored at lower temperatures or with a high reducing sugar content should be conditioned at 70°F for 1–4 weeks, or until cooking tests are satisfactory.

[3] Winter squash varieties differ in storage life.

[4] Sweet potatoes should be cured immediately after harvest by holding at 85°F and 90–95% relative humidity for 4–7 days.

TABLE 11.15. POSTHARVEST HANDLING OF FRESH CULINARY HERBS

Herb	Storage Conditions			
	Temperature (°F)	Relative Humidity (%)	Ethylene Sensitivity	Approximate Storage Life
Basil	50	90	High	7 days
Chives	32	95–100	Medium	—
Cilantro	32–34	95–100	High	2 weeks
Dill	32	95–100	High	1–2 weeks
Epazote	32–41	90–95	Medium	1–2 weeks
Mint	32	95–100	High	2–3 weeks
Oregano	32–41	90–95	Medium	1–2 weeks
Parsley	32	95–100	High	1–2 months
Perilla	50	95	Medium	7 days
Sage	32–50	90–95	Medium	2–3 weeks
Thyme	32	90–95	—	2–3 weeks

Adapted from A. A. Kader (ed.), *Postharvest Technology of Horticultural Crops*, 3rd ed. (University of California, Division of Agriculture and Natural Resources Publication 3311, 2002). Copyright © 2002 Regents of the University of California.

TABLE 11.16. RESPIRATION RATES OF FRESH CULINARY HERBS

	Respiration Rates (mg/kg/hr of CO_2)		
Herb	32°F	50°F	68°F
Basil	36	71	167
Chervil	12	80	170
Chinese chive	54	99	432
Chives	22	110	540
Coriander	22	nd	nd
Dill	22	103	nd
Fennel	19[1]	nd	32
Ginger	nd	nd	6[2]
Ginseng	6	15	nd
Marjoram	28	68	nd
Mint	20	76	252
Oregano	22	101	176
Sage	36	103	157
Tarragon	40	94	234
Thyme	12	25	52
Tarragon	38	82	203

Adapted from *The Commercial Storage of Fruits, Vegetables, and Florist and Nursery Stocks* (USDA Agriculture Handbook 66, 2016), https://www.ars.usda.gov/arsuserfiles/oc/np/commercialstorage/commercialstorage.pdf (accessed 4 Dec. 2021).

[1] At 36°F.
[2] At 55°F.

TABLE 11.17. AVERAGE RESPIRATION RATES OF VEGETABLES AT VARIOUS TEMPERATURES

Respiration Rate (mg/kg/hr of CO_2)[1]

Vegetable	32°F	41°F	50°F	59°F	68°F	77°F
Artichoke, globe	30	43	71	110	193	nd[2]
Artichoke, Jerusalem	10	12	19	50	nd	nd
Asparagus[33]	60	105	215	235	270	nd
Bean, snap	20	34	58	92	130	nd
Bean, long	40	46	92	202	220	nd
Beet	5	11	18	31	60	nd
Bok choy	6	11	20	39	56	nd
Broccoli	21	34	81	170	300	nd
Brussels sprouts	40	70	147	200	276	nd
Cabbage	5	11	18	28	42	62
Cabbage, Chinese	10	12	18	26	39	nd
Carrot, topped	15	20	31	40	25	nd
Cauliflower	17	21	34	46	79	92
Celeriac	7	13	23	35	45	nd
Celery	15	20	31	40	71	nd
Chicory	3	6	13	21	37	nd
Cucumber	nd	nd	26	29	31	37
Eggplant, American	nd	nd	nd	69[4]	nd	nd
Eggplant, Japanese	nd	nd	nd	131[4]	nd	nd
Eggplant, white egg	nd	nd	nd	113[4]	nd	nd

TABLE 11.17. AVERAGE RESPIRATION RATES OF VEGETABLES AT VARIOUS TEMPERATURES (*Continued*)

Vegetable	Respiration Rate (mg/kg/hr of CO_2)[1]						
	32°F	41°F	50°F	59°F	68°F	77°F	
Endive	45	52	73	100	133	200	
Garlic	8	16	24	22	20	nd	
Jicama	6	11	14	nd	6	nd	
Kohlrabi	10	16	31	46	nd	nd	
Leek	15	25	60	96	110	115	
Lettuce, head	12	17	31	39	56	82	
Lettuce, leaf	23	30	39	63	101	147	
Luffa	14	27	36	63	79	nd	
Melon							
Cantaloupe	6	10	15	37	55	67	
Honeydew	nd	8	14	24	30	33	
Watermelon	nd	4	8	nd	21	nd	
Mushroom	35	70	97	nd	264	nd	
Napalito	nd	18	40	56	74	nd	
Okra	21[3]	40	91	146	261	345	
Onion, dry	3	5	7	7	8	nd	
Parsley	30	60	114	150	199	274	
Parsnip	12	13	22	37	nd	nd	
Pea, English	38	64	86	175	271	313	

510

Pea, edible pod	39	64	89	176	273	nd
Pepper	nd	7	12	27	34	nd
Potato, cured	nd	12	16	17	22	nd
Prickly pear	nd	nd	nd	nd	32	nd
Radicchio	8	13[5]	23[6]	nd	nd	45
Radish, topped	16	20	34	74	130	172
Radish, with tops	6	10	16	32	51	75
Rhubarb	11	15	25	40	49	nd
Rutabaga	5	10	14	26	37	nd
Salsify	25	43	49	nd	193	nd
Spinach	21	45	110	179	230	nd
Squash, summer	25	32	67	153	164	nd
Sweet corn	41	63	105	159	261	359
Swiss chard	19[7]	nd	nd	nd	29	nd
Southernpea, pods	24[7]	25	nd	nd	148	nd
Southernpea, peas	29[7]	nd	nd	nd	126	nd
Tomatillo	nd	13	16	nd	32	nd
Tomato	nd	nd	15	22	35	43
Turnip, topped	8	10	16	23	25	nd
Watercress	22	50	110	175	322	377

Adapted from *The Commercial Storage of Fruits, Vegetables, and Florist and Nursery Stocks* (USDA Agriculture Handbook 66, 2016), https://www.ars.usda.gov/arsuserfiles/oc/np/commercialstorage/commercialstorage.pdf (accessed 4 Dec. 2021).

[1] To calculate heat production, multiply mg/kg/hr by 220 to get BTU/ton/day or by 61 to get kcal/ton/day.
[2] nd = not determined.
[3] 1 day after harvest.
[4] At 55°F.
[5] At 43°F.
[6] At 45°F.
[7] At 36°F.

TABLE 11.18. RECOMMENDED CONTROLLED ATMOSPHERE OR MODIFIED ATMOSPHERE CONDITIONS DURING TRANSPORT AND/OR STORAGE OF SELECTED VEGETABLES

	Temperature[1] (°F)		Controlled Atmosphere[2] (%)		
Vegetable	Optimum	Range	O_2	CO_2	Potential for Application
Artichoke	32	32–41	2–3	2–3	Moderate
Asparagus	36	34–41	Air	10–14	High
Bean, green	46	41–50	2–3	4–7	Slight
Bean, processing	46	41–50	8–10	20–30	Moderate
Broccoli	32	32–41	1–2	5–10	High
Brussels sprouts	32	32–41	1–2	5–7	Slight
Cabbage	32	32–41	2–3	3–6	High
Cantaloupe	37	36–41	3–5	10–20	Moderate
Cauliflower	32	32–41	2–3	3–4	Slight
Celeriac	32	32–41	2–4	2–3	Slight
Celery	32	32–41	1–4	3–5	Slight
Chinese cabbage	32	32–41	1–2	0–5	Slight
Cucumber, slicing	54	46–54	1–4	0	Slight
Cucumber, processing	39	34–39	3–5	3–5	Slight
Herbs[3]	34	32–41	5–10	4–6	Moderate
Leek	32	32–41	1–2	2–5	Slight
Lettuce, crisphead	32	32–41	1–3	0	Moderate
Lettuce, cut or shredded	32	32–41	1–5	5–20	High
Lettuce, leaf	32	32–41	1–3	0	Moderate
Mushroom	32	32–41	3–21	5–15	Moderate
Okra	50	45–54	air	4–10	Slight
Onion, dry	32	32–41	1–2	0–10	Slight
Onion, green	32	32–41	2–3	0–5	Slight
Parsley	32	32–41	8–10	8–10	Slight
Pea, sugar	32	32–50	2–3	2–3	Slight
Pepper, bell	46	41–54	2–5	2–5	Slight

TABLE 11.18. RECOMMENDED CONTROLLED ATMOSPHERE OR
MODIFIED ATMOSPHERE CONDITIONS DURING
TRANSPORT AND/OR STORAGE OF SELECTED
VEGETABLES (*Continued*)

| Vegetable | Temperature[1] (°F) | | Controlled Atmosphere[2] (%) | | Potential for Application |
	Optimum	Range	O[2]	CO[2]	
Pepper, chili	46	41–54	3–5	0–5	Slight
Pepper, processing	41	41–54	3–5	10-20	Moderate
Radish, topped	32	32–41	1–2	2–3	Slight
Spinach	32	32–41	7–10	5–10	Slight
Sweet corn	32	32–41	2–4	5–10	Slight
Tomato mature-green	54	54–59	3–5	3–5	Moderate
Tomato ripe	50	50–59	3–5	3–5	Moderate
Witloof chicory	32	32–41	3–4	4–5	Slight

Adapted from M. Saltveit, "A summary of CA requirements and recommendations for vegetables,"
Acta Horticulturae 600:723–727 (2003).

[1] Usual and/or recommended range. A relative humidity of 90–98% is recommended.

[2] Specific CA recommendations depend on varieties, temperature, and duration of storage.

[3] Herbs: chervil, chives, coriander, dill, sorrel, and watercress.

TABLE 11.19. OPTIMUM CONDITIONS FOR CURING ROOT, TUBER, AND BULB VEGETABLES PRIOR TO STORAGE

Vegetable	Temperature (°F)	RH (%)	Duration (days)
Cassava	86–95	85–90	4–7
Malanga	86–95	90–95	7
Potato			
Early crop	59–68	90–95	4–5
Late crop	50–59	90–95	10–15
Sweet potato	84–90	80–90	4–7
Taro	93–97	95	5
Water chestnut	86–90	95–100	3
Yam	86–95	85-95	4-7
Garlic and onion	Ambient (75–90 best) 93–113	60–75	5–10 (field drying) 0.5–3 (forced heated air)

TABLE 11.20.   MOISTURE LOSS FROM VEGETABLES

High	Medium	Low
Broccoli	Artichoke	Eggplant
Cantaloupe	Asparagus	Garlic
Chard	Bean, snap	Ginger
Green onion	Beet[1]	Melons
Kohlrabi	Brussels sprouts	Onion
Leafy greens	Cabbage	Potato
Mushroom	Carrot[1]	Pumpkin
Oriental vegetables	Cassava[2]	Winter squash
Parsley	Cauliflower	
Strawberry	Celeriac[1]	
	Celery	
	Sweet corn[3]	
	Cucumber[2]	
	Endive	
	Escarole	
	Leek	
	Lettuce	
	Okra	
	Parsnip[1]	
	Pea	
	Pepper	
	Radish[1]	
	Rutabaga[1,2]	
	Sweet potato	
	Summer squash	
	Tomato[2]	
	Yam	

Adapted from B. M. McGregor, *Tropical Products Transport Handbook* (USDA Agricultural Handbook 668, 1987).

[1] Root crops with tops have a high rate of moisture loss.
[2] Waxing reduces the rate of moisture loss.
[3] Husk removal reduces water loss.

TABLE 11.21. STORAGE SPROUT INHIBITORS

Sprout inhibitors are most effective when used in conjunction with good storage; their use cannot substitute for poor storage or poor storage management. However, storage temperatures may be somewhat higher when sprout inhibitors are used than when they are not. Finding alternatives to CIPC is the subject of considerable research effort. Several compounds that suppress sprouting have been identified, such as ethylene, S-carvone, 1,4-dimethyl naphthalene, essential oils (e.g., caraway oil, clove oil, spearmint oil) and chemical components of essential oils. Not all are commercially available. A selection of materials with commercial labels are listed below. In all cases, follow label instructions.

Vegetable	Material	Application
Potato (do not use on seed potatoes)	Maleic hydrazide	When most tubers are 1½–2 in. in diameter. Vines must remain green for several weeks after application
	Chlorpropham (CIPC)	In storage, 2–3 weeks after harvest as an aerosol treatment. Do not store seed potatoes in a treated storage.
		During washing, as an emulsifiable concentrate added to wash water to prevent sprouting during marketing
	3-Decen-2-one	In storage, application by thermal fogging. Not for application on seed potatoes.
Onion	Maleic hydrazide	Apply when 50% of the tops are down, the bulbs are mature, the necks soft, and 5–8 leaves are still green

For more information, see: M. J. Frazer, N. Olsen and G. Kleinkopf, *Organic and Alternative Methods for Potato Sprout Control in Storage* (University of Idaho Publication CIS 1120, 2004); N. Gumbo, L. S. Magwaza, and N. Z. Ngobese, "Evaluating ecologically acceptable sprout suppressants for enhancing dormancy and potato storability: A review," *Plants* 10:2307 (2021).

TABLE 11.22. SUSCEPTIBILITY OF VEGETABLES TO CHILLING INJURY

Vegetable	Approximate Lowest Safe Temperature (°F)	Appearance When Stored Between 32°F and Safe Temperature[1]
Asparagus	32–36	Dull, gray-green, and limp tips
Bean, lima	34–40	Rusty-brown specks, spots, or areas
Bean, snap	45	Pitting and russeting
Chayote	41–50	Dull brown discoloration, pitting, flesh darkening
Cucumber	45	Pitting, water-soaked spots, decay
Eggplant	45	Surface scald, alternaria rot, blackening of seeds
Ginger	45	Softening, tissue breakdown, decay
Jicama	55–65	Surface decay, discoloration
Melon		
Cantaloupe	36–41	Pitting, surface decay
Casaba, Crenshaw, Persian	45–50	Pitting, surface decay, failure to ripen
Honeydew	45–50	Reddish-tan discoloration, pitting, surface decay, failure to ripen
Watermelon	40	Pitting, objectionable flavor
Okra	45	Discoloration, water-soaked areas, pitting, decay
Pepper, sweet	45	Sheet pitting, alternaria rot on fruit and calyx, darkening of seed
Potato	38	Mahogany browning, sweetening

517

TABLE 11.22.   SUSCEPTIBILITY OF VEGETABLES TO CHILLING INJURY (*Continued*)

Vegetable	Approximate Lowest Safe Temperature (°F)	Appearance When Stored Between 32°F and Safe Temperature[1]
Pumpkin and hard-shell squash	50	Decay, especially alternaria rot
Sweet potato	55	Decay, pitting, had core when cooked
Tamarillo	37–40	Surface pitting, discoloration
Taro	50	Internal browning, decay
Tomato, ripe	45–50	Water-soaking and softening, decay
Tomato, mature green	55	Poor color when ripe, alternaria rot
Watermelon	40	Pitting, objectionable flavor

Adapted from C. Y. Wang, "Chilling and Freezing Injury," in *The Commercial Storage of Fruits, Vegetables, and Florist and Nursery Stocks* (USDA Agriculture Handbook 66, 2016), https://www.ars.usda.gov/arsuserfiles/oc/np/commercialstorage/commercialstorage.pdf (accessed 4 Dec. 2021).
[1] Severity of injury is related to temperature and time.

TABLE 11.23. RELATIVE SUSCEPTIBILITY OF VEGETABLES TO FREEZING INJURY

Most Susceptible	Moderately Susceptible	Least Susceptible
Asparagus	Broccoli	Beet
Bean, snap	Carrot	Brussels sprouts
Cucumber	Cauliflower	Cabbage, mature, and savoy
Eggplant	Celery	Kale
Lettuce	Onion, dry	Kohlrabi
Okra	Parsley	Parsnip
Pepper, sweet	Pea	Rutabaga
Potato	Pumpkin	Salsify
Squash, summer	Radish	Turnip
Sweet potato	Spinach	
Tomato	Squash, winter	

Adapted from C. Y. Wang, "Chilling and Freezing Injury," in *The Commercial Storage of Fruits, Vegetables, and Florist and Nursery Stocks* (USDA Agriculture Handbook 66, 2016), https://www.ars.usda.gov/arsuserfiles/oc/np/commercialstorage/commercialstorage.pdf (accessed 4 Dec. 2021).

TABLE 11.24. **CHILLING THRESHOLD TEMPERATURES AND VISUAL SYMPTOMS OF CHILLING INJURY FOR SOME SUBTROPICAL AND TROPICAL STORAGE ORGAN VEGETABLES**

Vegetable	Chilling Threshold (°F)	Symptoms
Cassava	41–46	Internal breakdown, increased water loss, failure to sprout, increased decay, and loss of eating quality
Ginger	54	Accelerated softening and shriveling, oozes moisture from the surface, decay
Jicama	55–59	External decay, rubbery and translucent flesh with grown discoloration, increased water loss
Malanga	45	Tissue breakdown and internal discoloration, increased water loss, increased decay and undesirable flavor changes
Potato	39	Mahogany browning: reddish-brown areas in the flesh; adverse effects on cooking quality
Sweet potato	54	Internal brown-black discoloration, adverse effects on cooled quality, "hard core," and accelerated decay
Taro	45–50	Tissue breakdown and internal discoloration, increased water loss, increased decay and undesirable flavor changes
Yam	55	Tissue softening, internal discoloration (grayish flecked with reddish brown), shriveling, and decay

Copyright 2003 from *Postharvest Physiology and Pathology of Vegetables*, 2nd ed., by J. R. Bartz and J. K. Brecht (eds.). Reproduced by permission of Routledge/Taylor & Francis Group, LLC.

TABLE 11.25. SYMPTOMS OF FREEZING INJURY ON SOME VEGETABLES

Vegetable	Symptoms
Artichoke	Epidermis becomes detached and forms whitish to light-tan blisters. When blisters are broken, underlying tissue turns brown
Asparagus	Tip becomes limp and dark; the rest of the spear is water-soaked. Thawed spears become mushy
Beet	External and internal water-soaking and sometimes blackening of conducting tissue
Broccoli	The youngest florets in the center of the curd are most sensitive to freezing injury. They turn brown and give strong off-odors upon thawing
Cabbage	Leaves become water-soaked, translucent, and limp upon thawing; separated epidermis
Carrot	A blistered appearance; jagged lengthwise cracks. Interior becomes water-soaked and darkened upon thawing
Cauliflower	Curds turn brown and have a strong off-odor when cooked
Celery	Leaves and petioles appear wilted and water-soaked upon thawing. Petioles freeze more readily than leaves
Garlic	Thawed cloves appear water-soaked, grayish-yellow
Lettuce	Blistering, dead cells of the separated epidermis on outer leaves become tan; increased susceptibility to physical damage and decay
Onion	Thawed bulbs are soft, grayish-yellow, and water-soaked in cross section; often limited to individual scales
Pepper, bell	Dead, water-soaked tissue in part of or all pericarp surface; pitting, shriveling, and decay follow thawing
Potato	Freezing injury may not be externally evident, but shows as gray or bluish-gray patches beneath the skin. Thawed tubers become soft and watery
Radish	Thawed tissues appear translucent; roots soften and shrivel
Sweet potato	A yellowish-brown discoloration of the vascular ring, and a yellowish-green water-soaked appearance of other tissues. Roots soften and become very susceptible to decay
Tomato	Water-soaked and soft upon thawing. In partially frozen fruits, the margin between healthy and dead tissue is distinct, especially in green fruits
Turnip	Small water-soaked spots or pitting on the surface. Injured tissues appear tan or gray and give off an objectionable odor

Adapted from A. A. Kader, J. M. Lyons, and L. L. Morris, "Postharvest Responses of Vegetables to Preharvest Field Temperature," *HortScience* 9:523–527 (1974).

TABLE 11.26. **SOME POSTHARVEST PHYSIOLOGICAL DISORDERS OF VEGETABLES, ATTRIBUTABLE DIRECTLY OR INDIRECTLY TO PREHARVEST FIELD TEMPERATURES**

Vegetable	Disorder	Symptoms and Development
Asparagus	Feathering	Bracts of the spears are partly spread as a result of high temperature
Brussels sprouts	Black leaf speck	Becomes visible after storage for 1–2 weeks at low temperature. Has been attributed in part to cauliflower mosaic virus infection in the field, which is influenced by temperature and other environmental factors
	Tip burn	Leaf margins turn light tan to dark brown
Cantaloupe	Vein tract browning	Discoloration of unnetted longitudinal stripes; related partly to high temperature and virus diseases
Garlic	Waxy breakdown	Enhanced by high temperature during growth; slightly sunken, light-yellow areas in fleshy cloves, then the entire clove becomes amber, slightly translucent, and waxy but still firm
Lettuce	Tip burn	Light-tan to dark-brown margins of leaves. Has been attributed to several causes, including field temperature; it can lead to soft rot development during postharvest handling
	Rib discoloration	More common in lettuce grown when day temperatures exceed 81°F or when night temperatures are between 55 and 64°F than in lettuce grown during cooler periods
	Russet spotting	Small tan, brown, or olive spots randomly distributed over the affected leaf; a postharvest disorder of lettuce induced by ethylene. Lettuce is more susceptible to russet spotting when harvested after high field temperatures (above 86°F) for 2 days or more during the 10 days before harvest

TABLE 11.26. **SOME POSTHARVEST PHYSIOLOGICAL DISORDERS OF VEGETABLES, ATTRIBUTABLE DIRECTLY OR INDIRECTLY TO PREHARVEST FIELD TEMPERATURES** (*Continued*)

Vegetable	Disorder	Symptoms and Development
	Rusty-brown discoloration	Rusty-brown discoloration has been related to internal rib necrosis associated with lettuce mosaic virus infection, which is influenced by field temperature and other environmental factors
Onion	Translucent scale	Grayish water-soaked appearance of the outer two or three fleshy scales of the bulb; translucency makes venation very distinct. In severe cases, the entire bulb softens and off-odors may develop
Potato	Blackheart	May occur in the field during excessively hot weather in waterlogged soils. Internal symptom is dark-gray to purplish or black discoloration usually in the center of the tuber
Radish	Pithiness	Textured white spots or streaks in cross section, large air spaces near the center, tough and dry roots. Results from high temperature

Adapted from A. A. Kader, J. M. Lyons, and L. L. Morris, "Postharvest Responses of Vegetables to Preharvest Field Temperature," *HortScience* 9:523–527 (1974).

TABLE 11.27. **SYMPTOMS OF SOLAR INJURY ON SOME VEGETABLES**

Vegetable	Symptoms
Bean, snap	Very small brown or reddish spots on one side of the pod coalesce and become water-soaked and slightly shrunken
Cabbage	Blistering of some outer leaves which leads to a bleached papery appearance. Desiccated leaves are susceptible to decay
Cauliflower	Discoloration of curds from yellow to brown to black (solar browning)
Cantaloupe	Sunburn: dry, sunken, and white to light tan areas. In milder sunburn, ground color is green or spotty brown
Lettuce	Papery areas on leaves, especially the cap leaf, develop during clear weather when air temperatures are higher than 77°F; affected areas become focus for decay
Honeydew melon	White to gray area at or near the top, may be slightly wrinkled, undesirable flavor. Brown blotch: tan to brown discolored areas caused by death of epidermal cells due to excessive ultraviolet radiation
Muskmelon	Dry, sunken, and white to light-tan areas. In milder sunburn, grown color is green or spotty brown.
Onion and garlic	Sunburn: dry scales are wrinkled and this may extend to one or two fleshy scales, injured area may be bleached depending on the color of the bulb
Pepper, bell	Dry and papery areas, yellowing, and sometimes wilting
Potato	Sunscald: water and blistered areas on the tuber surface. Injured areas become sunken and leathery and subsurface tissue rapidly turns dark-brown to black when exposed to air
Tomato	Sunburn (solar yellowing): affected areas on the fruit become whitish, translucent, thin walled, a netted appearance may develop. Mild solar injury might not be noticeable at harvest, but becomes more apparent after harvest as uneven ripening

Adapted from A. A. Kader, J. M. Lyons, and L. L. Morris, "Postharvest Responses of Vegetables to Preharvest Field Temperature," *HortScience* 9:523–527 (1974).

TABLE 11.28. CLASSIFICATION OF HORTICULTURAL COMMODITIES ACCORDING TO ETHYLENE PRODUCTION RATES

Very Low	Low	Moderate	High	Very High
Artichoke	Blackberry	Banana	Apple	Cherimoya
Asparagus	Blueberry	Fig	Apricot	Mammee apple
Cauliflower	Casaba melon	Guava	Avocado	Passion fruit
Cherry	Cranberry	Honeydew melon	Cantaloupe	Sapote
Citrus	Cucumber	Mango	Feijoa	
Grape	Eggplant	Plantain	Kiwi fruit (ripe)	
Jujube	Okra	Tomato	Nectarine	
Leafy vegetables	Olive		Papaya	
Most cut flowers	Pepper		Peach	
Pomegranate	Persimmon		Pear	
Potato	Pineapple		Plum	
Root vegetables	Pumpkin			
Strawberry	Raspberry			
	Tamarillo			
	Watermelon			

Adapted from A. A. Kader (ed.), *Postharvest Technology of Horticultural Crops*, 3rd ed. (University of California, Division of Agriculture and Natural Resources Publication 3311, 2002). Copyright © 2002 Regents of the University of California. Used by permission.

TABLE 11.29. COMPATIBLE PRODUCE FOR LONG-DISTANCE TRANSPORT

Produce in the same temperature column can be safely mixed. Ethylene-sensitive vegetables should not be mixed with ethylene-producing fruits and vegetables. Dry vegetables should not be mixed with other fruits and vegetables.

Produce	Recommended Storage and Transport Temperatures			
	32–36°F	40–45°F	45–50°F	55–65°F
Dry vegetables	Dry onion[1,2,3] Garlic			Ginger[4] Pumpkin Squash, winter Potato, early crop Tomato, mature green
Ethylene-sensitive vegetables†	Arugula* Asparagus Belgian endive Bok choy Broccoflower Broccoli* Brussels sprouts Cabbage[1] Carrot[1,2] Cauliflower Celery[1,2,3] Chard* Chicory Chinese cabbage Collards* Cut vegetables Endive Escarole Green onion[6] Herbs (not basil) Kailon* Kale* Leek[7] Lettuce Mushroom*[6] Mustard greens* Parsnip Snow pea* Spinach* Sweet pea* Turnip greens Watercress	Beans, snap, etc.*[5] Cactus leaves Cucumber* Pepper, chili Potato, late crop[1] Southern peas* Tomatillo	Basil* Chayote Eggplant*[5] Kiwano Long bean Okra Squash, summer Watermelon	

Vegetables (not ethylene-sensitive)	Amaranth*	Radicchio	Calabaza	Cassava	
	Anise	Radish	Haricot vert	Jicama	
	Artichoke	Rhubarb[6]	Pepper, bell[5]	Sweet potato (boniato)	
	Bean sprouts*	Rutabaga	Winged bean	Taro (malanga)	
	Beet	Salsify	Luffa*‡	Yam	
	Celeriac	Scorzonera		Tomato, ripe*‡	
	Daikon	Shallot			
	Horseradish	Sweet corn[6]			
	Jerusalem artichoke	Swiss chard			
	Kohlrabi	Turnip			
	Lo bok	Water chestnut			
Fruits and melons (very low ethylene-producing)	Barbados cherry	Date	Cactus pear (tuna)	Babaco	Lemon[8]	Breadfruit
	Bitter melon	Gooseberry	Jujube	Calamondin*	Lime[8]	Canistel
	Blackberry*	Grape[6,7,9]	Kumquat	Carambola	Pineapple[5,10]	Grapefruit (CA, AZ)[8]
	Blueberry	Longan	Mandarin[8]	Casaba melon	Pummelo[8]	Jaboticaba*
	Caimito	Loquat	Olive	Cranberry	Tamarillo[8]	
	Cashew apple	Lychee	Orange (CA, AZ)[8]	Grapefruit[8]	Tangelo[8]	
	Cherry	Orange (FL)[8]	Pepino	Juan Canary melon	Ugli fruit	
	Coconut	Raspberry*	Persimmon			
	Currant	Strawberry*	Pomegranate			
			Tamarind			
			Tangerine[8]			

TABLE 11.29. COMPATIBLE PRODUCE FOR LONG-DISTANCE TRANSPORT (Continued)

Produce in the same temperature column can be safely mixed. Ethylene-sensitive vegetables should not be mixed with ethylene-producing fruits and vegetables. Dry vegetables should not be mixed with other fruits and vegetables.

Produce	Recommended Storage and Transport Temperatures				
Ethylene-producing fruits and melons	Apple[1,2,3]	Nectarine	Durian	Avocado, unripe	Atemoya
	Apricot	Peach	Feijoa	Crenshaw melon	Banana
	Avocado, ripe	Pear, Asian	Guava	Custard apple	Cherimoya
	Cantaloupe	Pear, European[1,3]	Honeydew melon	Passion fruit	Jackfruit
	Cut fruits	Plum	Persian melon		Mamey sapote
	Fig[1,6,7]	Prune			Mango
	Kiwifruit	Quince			Mangosteen
					Papaya
					Plantain
					Sapote
					Soursop*

Adapted from A. A. Kader (ed.), Postharvest Technology of Horticultural Crops, 3rd ed. (University of California, Division of Agriculture and Natural Resources Publication 3311, 2002). Copyright © 2002 Regents of the University of California. Used by permission.

*Less than 14-day shelf life at recommended temperature and normal atmosphere conditions.
[1]Ethylene should be kept below 1 ppm in storage areas for sensitive vegetables.
[2]Produces moderate amounts of ethylene and should be treated as an ethylene-producing fruit.
[1]Odors from apples and pears are absorbed by cabbage, carrots, celery, figs, onions, and potatoes.
[2]Celery absorbs odor from onions, apples, and carrots.
[3]Onion odor is absorbed by apples, celery, pears, and citrus.
[4]Ginger odor is absorbed by eggplant.
[5]Pepper odor is absorbed by beans, pineapples, and avocados.
[6]Green onion odor is absorbed by figs, grapes, mushrooms, rhubarb, and corn.
[7]Leek odor is absorbed by figs and grapes.
[8]Citrus absorbs odor from strongly scented fruits and vegetables.
[9]Sulfur dioxide released from pads used with table grapes will damage other produce.
[10]Avocado odor is absorbed by pineapple.

INTEGRATED CONTROL OF POSTHARVEST DISEASES

Effective and consistent control of storage diseases depends on integrating the following practices:

- Select disease resistant cultivars where possible.
- Maintain correct crop nutrition by use of leaf and soil analysis.
- Irrigate based on crop requirements and avoid overhead irrigation.
- Apply pre-harvest treatments to control insects and diseases.
- Harvest the crop at the correct maturity for storage.
- Apply postharvest treatments to disinfest and control diseases and disorders on produce.
- Maintain good sanitation in packing areas and keep wash water free of contamination.
- Store produce under conditions least conducive to growth of pathogens.

TABLE 11.30. IMPORTANT POSTHARVEST DISEASES OF VEGETABLES

Vegetable	Disease	Causal Agent	Fungal Class/Type
Artichoke	Gray mol	*Botrytis cinerea*	Hyphomycete
	Watery soft rot	*Sclerotinia sclerotiorum*	Discomycete
Asparagus	Bacterial soft rot	*Erwinia* or *Pseudomonas* spp.	Bacteria
	Fusarium rot	*Fusarium* spp.	Hyphomycete
	Phytophthora rot	*Phytophthora* spp.	Oomycete
	Purple spot	*Stemphylium* spp.	Hyphomycete
Bulbs	Bacterial soft rot	*Erwinia carotovora*	Bacterium
	Black rot	*Aspergillus niger*	Hyphomycete
	Blue mold rot	*Penicillium* spp.	Hyphomycete
	Fusarium basal rot	*Fusarium oxysporum*	Hyphomycete
	Neck rot	*Botrytis* spp.	Hyphomycete
	Purple blotch	*Alternaria porri*	Hyphomycete
	Sclerotium rot	*Sclerotium rolfsii*	Agonomycete
	Smudge	*Colletotrichum circinans*	Coelomycete
Carrot	Bacterial soft rot	*Erwinia* spp.	Bacteria
		Pseudomonas spp.	
	Black rot	*A. radicina*	Hyphomycete
	Cavity spot	Disease complex	Soil fungi
	Chalaropsis rot	*Chalara* spp.	Hyphomycetes
	Crater rot	*R. carotae*	Agonomycete
	Gray mold rot	*B. cinerea*	Hyphomycete
	Sclerotium rot	*S. rolfsii*	Agonomycete
	Watery soft rot	*Sclerotinia* spp.	Discomycete

530

Celery	Bacterial soft rot	*Erwinia* or *Pseudomonas* spp.	Bacteria
	Brown spot	*Cephalosporium apii*	Hyphomycete
	Cercospora spot	*Cercospora apii*	Hyphomycete
	Gray mold	*Botrytis cinerea*	Hyphomycete
	Licorice rot	*Mycocentrospora acerina*	Hyphomycete
	Phoma rot	*Phoma apiicola*	Coelomycete
	Pink rot	*Sclerotinia* spp.	Discomycete
	Septoria spot	*Septoria apiicola*	Coelomycete
Crucifers	Alternaria leaf spot	*Alternaria* spp.	Hyphomycete
	Bacterial soft rot	*E. carotovora*	Bacterium
	Black rot	*Xanthomonas campestris*	Bacterium
	Downy mildew	*Peronospora parasitica*	Oomycete
	Rhizoctonia rot	*Rhizoctonia solani*	Agonomycete
	Ring spot	*Mycosphaerella brassicicola*	Loculoascomyete
	Virus diseases	Cauliflower mosaic virus	Virus
		Turnip mosaic virus	
Cucurbits	Watery soft rot	*Sclerotinia* spp.	Discomycete
	White blister	*Albugo candida*	Oomycete
	Anthracnose	*Colletotrichum* spp.	Coelomycete
	Bacterial soft rot	*Erwinia* spp.	Bacterium
	Black rot	*Didymella bryoniae*	Loculoascomycete
	Botryodiplodia rot	*Botryodiplodia theobromae*	Coelomycete
	Charcoal rot	*Macrophomina phaseolina*	Coelomycete
	Fusarium rot	*Fusarium* spp.	Hyphomycete
	Leak	*Pythium* spp.	Oomycete
	Rhizopus rot	*Rhizopus* spp.	Zygomycete

TABLE 11.30. IMPORTANT POSTHARVEST DISEASES OF VEGETABLES (Continued)

Vegetable	Disease	Causal Agent	Fungal Class/Type
	Sclerotium rot	Sclerotium rolfsii	Agonomycete
	Soil rot	R. solani	Agonomycete
Legumes	Alternaria blight	A. alternata	Hyphomycete
	Anthracnose	Colletotrichum spp.	Coelomycete
	Ascochyta pod spot	Ascochyta spp.	Coelomycetes
	Bacterial blight	Pseudomonas spp.	Bacteria
		Xanthomonas spp.	
	Chocolate spot	B. cinerea	Hyphomycete
	Cottony leak	Pythium spp.	Oomycete
		Mycosphaerella blight	Loculoascomycete
		M. pinodes	
	Rust	Uromyces spp.	Hemibasidiomycete
	Sclerotium rot	S. rolfsii	Agonomycete
	Soil rot	R. solani	Agonomycete
	White mold	Sclerotinia spp.	Discomycete
Lettuce	Bacterial rot	Erwinia, Pseudomonas,	Bacteria
		Xanthomonas spp.	
	Gray mold rot	B. cinerea	Hyphomycete
	Rhizoctonia rot	R. solani	Agonomycete
	Ringspot	Microdochium panattonianum	Hyphomycete
	Septoria spot	S. lactucae	Coelomycete
	Stemphylium spot	Stemphylium herbarum	Hyphomycete

	Disease	Pathogen	Class
Potato	Watery soft rot	*Sclerotinia* spp.	Discomycete
	Bacterial soft rot	*Erwinia* spp.	Bacteria
	Blight	*Phytophthora infestans*	Oomycete
	Charcoal rot	*S. bataticola*	Agonomycete
	Common scab	*Streptomyces scabies*	Actinomycete
	Fusarium rot	*Fusarium* spp.	Hyphomycete
	Gangrene	*Phoma exigua*	Coelomycete
	Ring rot	*Clavibacter michiganensis*	Bacterium
	Sclerotium rot	*S. rolfsii*	Agonomycete
	Silver scurf	*Helminthosporium solani*	Hyphomycete
	Watery wound rot	*Pythium* spp.	Oomycete
Solanaceous fruits	Alternaria rot	*A. alternata*	Hyphomycete
	Anthracnose	*Colletotrichum* spp.	Coelomycete
	Bacterial canker	*C. michiganensis*	Bacterium
	Bacterial speck	*Pseudomonas syringae*	Bacterium
	Bacterial spot	*X. campestris*	Bacterium
	Fusarium rot	*Fusarium* spp.	Hyphomycetes
	Gray mold rot	*B. cinerea*	Hyphomycete
	Light blight	*P. infestans*	Oomycete
	Phoma rot	*Phoma lycopersici*	Hyphomycete
	Phomopsis rot	*Phomopsis* spp.	Coelomycetes
	Phytophthora rot	*Phytophthora* spp.	Oomycete
	Pleospora rot	*Stemphylium herbarum*	Hyphomycete
	Rhizopus rot	*Rhizopus* spp.	Zygomycetes
	Sclerotium rot	*S. rolfsii*	Agonomycete
	Soil rot	*R. solani*	Agonomycete
	Sour rot	*Geotrichum candidum*	Hyphomycete
	Watery soft rot	*Sclerotinia* spp.	Discomycetes

TABLE 11.30. IMPORTANT POSTHARVEST DISEASES OF VEGETABLES (Continued)

Vegetable	Disease	Causal Agent	Fungal Class/Type
Sweet potato	Black rot	*Ceratocystis fimbriata*	Pyrenomycete
	Fusarium rot	*Fusarium* spp.	Hyphomycete
	Rhizopus rot	*Rhizopus* spp.	Zygomycetes
	Soil rot	*Streptomyces ipomoeae*	Actinomycete
	Scurf	*Monilochaetes infuscans*	Hyphomycete

Adapted from P. L. Sholberg and W. S. Conway, "Postharvest Pathology," in *The Commercial Storage of Fruits, Vegetables, and Florist and Nursery Stocks* (USDA Agriculture Handbook 66, 2016), https://www.ars.usda.gov/arsuserfiles/oc/np/commercialstorage/commercialstorage.pdf.

08

VEGETABLE QUALITY

The word *quality* is used in reference to fresh fruits and vegetables in several ways including *market* quality, *edible* quality, *dessert* quality, *shipping* quality, *table* quality, *nutritional* quality, *internal* quality, and *appearance* quality. Quality of fresh vegetables is a combination of characteristics, attributes, and properties that give the vegetables value to humans for food and enjoyment. Producers are concerned that their commodities have good appearance and few visual defects, but useful varieties must also score high on yield, disease resistance, ease of harvest, and shipping quality. To receivers and distributors, appearance quality is most important; they are also keenly interested in firmness and long storage life. Consumers consider good-quality vegetables to be those that look good, are firm, and offer good flavor and nutritive value. Although consumers buy on the basis of appearance, their satisfaction and repeat purchases depend on good edible quality. From harvest through distribution, careful attention and record-keeping can help ensure crop quality for the final consumer. Key quality assurance records for vegetable crops are provided in Table 11.31, and several measurable parameters of quality are indicated in Table 11.32.

TABLE 11.31.   QUALITY ASSURANCE RECORDS FOR VEGETABLES

Field packing
Maturity/ripeness stage and uniformity
Harvest method (hand or mechanical)
Temperature of product (harvest during cool times of the day and keep
  product shaded)
Uniformity of packs (size, trimming, maturity)
Well-constructed boxed; palletization and unitization
Condition of field boxes or bins (no rough or dirty surfaces)
Cleaning and sanitization of bins and harvest equipment

Packinghouse
Time from harvest to arrival
Shaded receiving area
Uniformity of harvest (size, trimming, maturity)
Washing/hydrocooling operation (sanitization)
Change water daily and maintain constant sanitizer levels in dump
  tanks

TABLE 11.31.   QUALITY ASSURANCE RECORDS FOR VEGETABLES (*Continued*)

Decay control chemical usage
Sorting for size, color, quality, etc.
Discard product that falls on the floor
Check culls for causes of rejection and for sorting accuracy
Sanitize facilities and equipment regularly
Well-constructed boxes; palletization and unitization.

Cooler
Time from harvest to cooler
Time from arrival to start of cooling
Package design (ventilation)
Speed of cooling and final temperature
Temperature of product after cooling
Temperature of holding room
Time from cooling to loading

Loading trailer
First-in, first-out truck loading
Temperature of product
Boxes palletized and unitized
Truck condition (clean, undamaged, precooled)
Loading pattern; palletization and unitization
Duration of transport
Temperature during transport (thermostat setting and use of recorders)

Arrival at distribution center
Transit time
Temperature of product
Product condition and uniformity
    Uniformity of packs (size, trimming, maturity)
    Ripeness stage, firmness
    Decay incidence/type
Maintain refrigeration during cooling

TABLE 11.32. QUALITY COMPONENTS OF FRESH VEGETABLES

Main Factors	Components
Appearance (visual)	*Size*: dimensions, weight, volume
	Shape and form: diameter/depth ratio, compactness, uniformity
	Color: uniformity, intensity
	Gloss: nature of surface wax
	Defects, external and internal: morphological, physical and mechanical, physiological, pathological, entomological
Texture (feel)	Firmness, hardness, softness
	Crispness
	Succulence, juiciness
	Mealiness, grittiness
	Toughness, fibrousness
Flavor (taste and smell)	Sweetness
	Sourness (acidity)
	Astringency
	Bitterness
	Aroma (volatile compounds)
	Off-flavors and off-odors
Nutritional value	Carbohydrates (including dietary fiber)
	Proteins
	Lipids
	Vitamins
	Essential elements
Safety	Naturally occurring toxicants
	Contaminants (chemical residues, heavy metals)
	Mycotoxins
	Microbial contamination

Adapted from A. A. Kader (ed.), *Postharvest Technology of Horticultural Crops*, 3rd ed. (University of California, Division of Agriculture and Natural Resources Publication 3311, 2002). Copyright © 2002 Regents of the University of California. Used by permission.

09

U.S. STANDARDS FOR GRADE OF VEGETABLES

U.S. STANDARDS

Grade standards issued by the U.S. Department of Agriculture (USDA) are currently in effect for most vegetables for fresh market and for processing. Some standards have been unchanged since they became effective, whereas others have been revised quite recently. Tables 11.33 and 11.34 provide a brief summary of standards for some vegetable and fruit crops.

For many specialty vegetable products, U.S. grade standards do not exist, but the USDA Agricultural Marketing Service publishes inspection instructions. Such instructions exist for many specialty products including Asparagus bean, batatas, bok choy, cardoon, chayote squash, Chinese cabbage, Chinese chives, coriander, daikon, dandelion, ginger root, horseradish root, Japanese eggplant, Jerusalem artichoke, jicama, rocket (arugula), salsify, scorzonera, tomatillo, and yam, among many others.

For copies of the U.S. Standards for Grades of Fresh and Processing Vegetables, refer to:

https://www.ams.usda.gov/grades-standards/vegetables (accessed 4 Dec. 2021)

TABLE 11.33. QUALITY FACTORS FOR FRESH VEGETABLES IN THE U.S. STANDARDS FOR GRADES

Vegetable	Effective Date	Quality Factors[1]
Anise, sweet	2016	Firmness, size, tenderness, trimming, blanching, and freedom from decay and damage caused by growth cracks, pithy branches, wilting, freezing, and seedstems
Artichoke	2006	Stem length, trimming, shape, overmaturity, uniformity of size, compactness, and freedom from decay and defects
Asparagus	2006	Freshness (turgidity), trimming, straightness, freedom from damage and decay, diameter of stalks, percent green color
Bean, lima	2016	Uniformity, maturity, size, freshness, shape, and freedom from soft decay, sprouted beans, worm holes, and from damage caused by dirt, russeting, scars, or other
Bean, snap	1990	Uniformity, size, shape, maturity, firmness, tenderness, cleanness, and freedom from soft rot and damage caused by leaves, leafstems, and hail
Beet, bunched or topped	2016	Root shape, trimming of rootlets, firmness (turgidity), smoothness, cleanness, minimum size (diameter), and freedom from damage from cuts, freezing, growth cracks, and rodents
Beet, greens	2008	Freshness, cleanness, tenderness, and freedom from decay, other kinds of leaves, discoloration, insects, mechanical injury, and freezing injury
Broccoli (Italian sprouting)	2006	Color, maturity, stalk diameter and length, compactness, base cut, and freedom from defects and decay
Brussels sprouts	2016	Color, maturity (firmness), size (diameter and length), and freedom from soft decay and seedstems, free from damage caused by discoloration, dirt, freezing

539

TABLE 11.33. QUALITY FACTORS FOR FRESH VEGETABLES IN THE U.S. STANDARDS FOR GRADES *(Continued)*

Vegetable	Effective Date	Quality Factors[1]
Cabbage	2016	Uniformity, solidity (maturity or firmness), trimming, color, freedom from soft rot, seedstems, and damage caused by discoloration, freezing
Cantaloupe	2008	Internal quality (soluble solids content = >11% very good, >9% good), uniformity of size, shape, ground color and netting; maturity and turgidity; and freedom from "wet slip," sunscald; freedom from damage caused by liquid in the seed cavity, sunburn, hail, dirt, surface mold, and bruises
Carrot, bunched	2020	Root shape, color, cleanness, smoothness, firmness, root diameter, free from soft rot, free from damage caused by freezing, growth cracks, sunburn, pithiness, woodiness, internal discoloration, oil spray, dry rot, or other means. Tops: length and fullness, and freedom from decay and damage caused by freezing, seedstems, discoloration
Carrot, topped	2020	Uniformity, firmness, trimming, color, shape, size, cleanness, smoothness, and freedom from defect and damage caused by growth cracks, freezing, pithiness, woodiness, internal discoloration, oil spray, dry rot
Carrots with short trimmed tops	2020	Roots: size, firmness, cleanliness, color, smoothness, form, and freedom from soft rot and damage caused by freezing, growth cracks, sunburn, pithiness, woodiness, internal discoloration, oil spray, dry rot, and decay; leaves: (cut to <4 in.) freedom from yellowing or other discoloration, disease, insects, and seedstems

540

Commodity	Year	Quality factors
Cauliflower	2017	Curds: cleanliness, compactness, white color, size (diameter), freshness and trimming of jacket leaves, and freedom from defect and decay, and damage due to bruising, cuts, discoloration, enlarged bracts, fuzziness, hollow stem, insects, mold, riciness, wilting, and other means
Celery	1959	Stalk form, uniformity, compactness, cleanliness, color, trimming, length of stalk and midribs, freedom from blackheart, brown stem, soft rot, doubles, and free from damage caused by freezing, growth cracks, horizontal cracks, pithy branches, seedstems, suckers, wilting, blight
Collard greens or Broccoli greens	2016	Freshness, tenderness, cleanliness, trimming, color, and freedom from decay and damage caused by coarse stalks and seedstems, discoloration, freezing
Corn, sweet	1992	Uniformity of color and size, freshness, milky kernels, cob length, freedom from insect injury, discoloration, and other defects, coverage with fresh husks
Cucumber	2016	Color, shape, freshness, firmness, maturity, size (diameter and length), and freedom from decay, sunscald, and injury caused by scars, yellowing, sunburn, dirt, freezing, mosaic or other disease, cuts, and bruises
Cucumber, greenhouse	1985	Freshness, shape, firmness, color, size (length of 11 in. or longer), and freedom from decay, cuts, bruises, scars, insect injury, and other defects
Dandelion greens	1955	Freshness, cleanliness, tenderness, and freedom from damage caused by seed stems, discoloration, freezing
Eggplant	2013	Color, turgidity, cleanness, shape, freedom from decay and wormholes and from injury caused by scars, freezing
Endive, Escarole, or Chicory	2016	Freshness, trimming, color (blanching), freedom from decay and damage caused by seedstems, broken, bruised, spotted or discolored leaves, wilting, dirt
Garlic	2016	Maturity, curing, compactness, well-filled cloves, covering by outer sheath, bulb size, and freedom from mold, decay, shattered cloves, and damage caused by dirt or staining, sunburn, sunscald, cuts, sprouts, tops, roots

TABLE 11.33.   QUALITY FACTORS FOR FRESH VEGETABLES IN THE U.S. STANDARDS FOR GRADES (*Continued*)

Vegetable	Effective Date	Quality Factors[1]
Honeydew and Honey Ball melons	1967	Maturity, firmness, shape, and freedom from decay and defect (sunburn, bruising, hail spots, and mechanical injuries)
Horseradish roots	1936	Uniformity of shape and size, firmness, smoothness, and freedom from hollow heart, other defects, and decay
Kale	2005	Uniformity of growth and color, trimming, freshness, and freedom from stunting, defect, decay, and damage
Lettuce, crisp-head	1975	Turgidity, color, maturity (firmness), trimming (number of wrapper leaves), and freedom from tip burn, other physiological disorders, mechanical damage, seedstems, other defects, and decay
Lettuce, field-grown leaf	2006	Development, trimming, and freedom from decay, freezing, insects, worms or excreta, and freedom from injury by broken midribs, bruising, dirt, discoloration, downy mildew, tipburn, seedstems, watersoaking, yellowing
Lettuce, greenhouse leaf	2016	Development, trimming, and freedom from coarse stems, bleached or discolored leaves, wilting, spray burn, dirt, freezing
Lettuce, romaine	2016	Freshness, development, trimming, and freedom from decay and damage caused by seedstems, broken, bruised, or discolored leaves, tipburn, wilting, freezing, dirt
Mushroom	2016	Maturity, shape, trimming, size, firmness, and freedom from open veils, disease, spots, insect injury, and decay

542

Commodity	Year	Quality factors
Mustard greens and turnip greens	2016	Freshness, tenderness, cleanness, and freedom from damage caused by seedstems, discoloration, freezing, disease, insects, or mechanical means; roots (if attached): firmness, size, and freedom from damage
Okra	2013	Freshness, tenderness, uniformity of shape and color, not misshapen, and freedom from defect, decay, and damage
Onion, dry Creole	2014	Maturity, firmness, shape, size (diameter), and freedom from soft rot, doubles, bottlenecks, and from damage caused by splits, seedstems, tops, roots, moisture, sunburn, sunscald, cuts, staining, dirt
Bermuda-Granex-Grano	2014	Maturity, firmness, shape, size (diameter) and freedom from decay, wet sunscald doubles, bottlenecks, seedstems, splits, dry sunken areas, sunburn, sprouting, staining, dirt, mechanical damage, tops, roots, translucent scales, watery scales, moisture, and other defects
Other varieties	2014	Maturity, firmness, shape, size (diameter), and freedom from decay, wet sunscald, doubles, bottlenecks, scallions, and freedom from damage caused by seedstems, splits, tops, roots, dry sunken areas, sunburn, sprouts, freezing, peeling, cracked fleshy scales, watery scales, dirt or staining, or translucent scales
Onion, common green	2016	Turgidity, color, tenderness, length, diameter, form, cleanness, bulb trimming, freedom from damage caused by seedstems, roots, or foreign material
Onion sets	2016	Size, maturity, firmness, and freedom from decay and damage caused by sprouts, tops, freezing, mold, moisture, dirt, or chaff
Parsley	2007	Freshness, uniformity, green color, and freedom from decay and from damage caused by seedstems, yellow or discolored leaves, wilting, freezing, or dirt
Parsnip	2016	Trimming, size, cleanness, smoothness, firmness, shape, freedom from woodiness, soft rot or wet breakdown, and from damage caused by discoloration, bruises, cuts, rodents, growth cracks, pithiness

TABLE 11.33.   QUALITY FACTORS FOR FRESH VEGETABLES IN THE U.S. STANDARDS FOR GRADES (*Continued*)

Vegetable	Effective Date	Quality Factors[1]
Pea pods (snow peas)	2007	Cleanness, shape, freshness, immaturity, length, tenderness, color, and free from broken, decayed, flabby and molded pods, and from injury by blistering, bruising, cracks, cuts, dirt, discoloration, freezing, scars, or shriveling
Pea, fresh	2016	Maturity, size, shape, freshness, firmness, degree filled, freedom from decay and watersoaking, and from damage caused by black calyxes, freezing, splitting, hail, dirt, leaves, and mildew
Pepper, sweet	2005	Maturity, color, shape, size, firmness, and freedom from defects (sunburn, sunscald, freezing injury, hail, scars) and decay affecting calyxes, walls, or stems
Pepper, other	2007	Maturity, color, firmness, shape, size, freedom from defects (blossom end rot, crushed/broken, freezing injury, hail, scars, sunscald) and decay affecting calyxes, walls, or stems
Potato	2011	Uniformity, maturity, firmness, cleanness, shape, size, and freedom from freezing, blackheart, late blight, southern bacterial wilt and ring rot, soft rot and wet breakdown, freedom from damage by any other cause
Radish	1968	Tenderness, cleanness, smoothness, shape, size, and freedom from pithiness and other defects. Bunched radishes have tops that are fresh and free from damage

Commodity	Year	Factors
Rhubarb	2016	Color, freshness, tenderness, straightness, trimming, cleanness, stalk diameter and length, and freedom from decay and damage caused by scars, freezing, or other means
Shallot, bunched	2016	Firmness, overall length, root diameter, form, tenderness, trimming, cleanness, and freedom from decay and damage caused by seedstems; tops: freshness, green color, and no mechanical damage
Southernpea (cowpea, field pea)	2016	Maturity, pod fill, pod shape, and freedom from decay and worm holes, and from damage caused by stems, leaves and trash, stings or other insect injury, scars, discoloration, or wilting
Spinach bunches	1987	Similar varietal characteristics, form, freshness, cleanness, trimming, and freedom from damage caused by coarse stalks or seedstems, flower buds, discoloration, wilting, freezing, and foreign material
Spinach leaves	1946	Color, turgidity, cleanness, trimming, and freedom from seedstems, coarse stalks, and other defects
Squash, summer	2016	Immaturity, tenderness, shape, presence of stems, firmness, and freedom from decay, discoloration, freezing, cuts, bruises, scars, and other defects
Squash, winter and Pumpkin	1983	Maturity, firmness, freedom from discoloration, cracking, dry rot; uniformity of size
Strawberry	2006	Maturity (>1/2 or >3/4 of surface showing red or pink color, depending on grade), firmness, attached calyx, size, and freedom from mold and from damage caused by dirt and moisture.
Sweet potato	2005	Firmness, smoothness, cleanness, shape, size, and freedom from freezing injury, internal breakdown, black rot, growth cracks, secondary rootlets, sprouts, cuts, bruises, scars, scurf, pox, wireworms, and weevils.
Tomato	1991	Maturity and ripeness (color chart), firmness, shape, size, and freedom from puffiness, freezing injury, sunscald, scars, catfaces, and growth cracks

Vegetable	Effective Date	Quality Factors[1]
Tomato, greenhouse	2007	Maturity, firmness, shape, size, cleanness, and freedom from sunscald, freezing injury, bruises, cuts, shriveling, puffiness, catfaces, growth cracks, scars, disease, moldy stems, and skin checks
Tomato, on the vine	2008	Maturity, firmness, shape, attachment to stems/vines, freedom from sunscald, freezing injury, and damage due to bruises, cuts, shriveling, puffiness, catfaces, growth cracks, and scars. Vines: non-brittle, free from mold
Turnip and rutabaga	2016	Roots: uniformity of color, size, and shape, smoothness, cleanness, trimming, freshness, and freedom from defects and damage (discoloration, cuts, growth cracks, pithiness, woodiness, water core, dry rot). Tops: Freshness, freedom from discoloration and freezing
Watermelon	2021	Maturity and ripeness (optional internal quality criteria: soluble solids content = >10% very good, >8% good), shape, not overripe, freedom from anthracnose, sunscald, and whiteheart

Adapted from the *U.S. Standards for Grades of Fresh and Processing Vegetables* (USDA Agricultural Marketing Service), https://www.ams.usda.gov/grades-standards/vegetables (accessed Oct. 2021).

[1] In addition to the factors specified, quality factors include freedom from decay and damage due to disease, insects, and mechanical or other means.

TABLE 11.34. QUALITY FACTORS FOR PROCESSING VEGETABLES IN THE U.S. STANDARDS FOR GRADES

Vegetable	Date Issued	Quality Factors
Asparagus, green	1972	Freshness, shape, green color, size (spear length), and freedom from defect (freezing damage, dirt, disease, insect injury, and mechanical injuries) and decay
Bean, shelled lima	1953	Tenderness green color, and freedom from decay and from injury caused by discoloration, shriveling, sunscald, freezing, heating, disease, insects, or other means
Bean, snap	1959	Freshness, tenderness, shape, size, and freedom from decay and from damage caused by scars, rust, disease, insects, bruises, punctures, broken ends, or other means
Beet	1945	Firmness, tenderness, shape, size, and freedom from soft rot, cull material, growth cracks, internal discoloration, white zoning, rodent damage, disease, insects, and mechanical injury
Broccoli	1959	Freshness, tenderness, green color, compactness, trimming, and freedom from decay and damage caused by discoloration, freezing, pithiness, scars, dirt, or mechanical means
Cabbage	1944	Firmness, trimming, and freedom from soft rot, seedstems, and from damage caused by bursting, discoloration, freezing, disease, birds, insects, or mechanical or other means
Carrot	1984	Firmness, color, shape, size (root length), smoothness, not woody, and freedom from soft rot, cull material, and from damage caused by growth cracks, sunburn, green core, pithy core, water core, internal discoloration, disease, or mechanical means

547

TABLE 11.34. QUALITY FACTORS FOR PROCESSING VEGETABLES IN THE U.S. STANDARDS FOR GRADES (Continued)

Vegetable	Date Issued	Quality Factors
Cauliflower	1959	Freshness, compactness, color, and freedom from jacket leaves, stalks, and other cull material, decay, and damage caused by discoloration, bruising, fuzziness, enlarged bracts, dirt, freezing, hail, or mechanical means
Corn, sweet	1962	Maturity, freshness, and freedom from damage by freezing, insects, birds, disease, cross-pollination, or fermentation
Cucumber, pickling	1936	Color, shape, freshness, firmness, maturity, and freedom from decay and from damage caused by dirt, freezing, sunburn, disease, insects, or mechanical or other means
Mushroom	1964	Freshness, firmness, shape, and freedom from decay, disease spots, and insects, and from damage caused by insects, bruising, discoloration, or feathering
Okra	1965	Freshness, tenderness, color, shape, and freedom from decay and insects, and from damage caused by scars, bruises, cuts, punctures, discoloration, dirt or other means
Onion	1944	Maturity, firmness, and freedom from decay, sprouts, bottlenecks, scallions, seedstems, sunscald, roots, insects, and mechanical injury
Pea, fresh shelled for canning/freezing	1946	Tenderness, succulence, color, and freedom from decay, scald, rust, shriveling, heating, disease, and insects
Pepper, sweet	1948	Firmness, color, shape, and freedom from decay, insects, and damage by any means that results in 5–20% trimming (by weight) depending on grade

548

Potato	1983	Shape, smoothness, freedom from decay and defect (freezing injury, blackheart, sprouts), size, specific gravity, glucose content, and fry color
Potato for chipping	1978	Firmness, cleanness, shape, freedom from defect (freezing, blackheart, decay, insect injury, and mechanical injury), size; optional tests for specific gravity and fry color are included
Southernpea	1965	Pods: maturity, freshness, and freedom from decay; seeds: freedom from scars, insects, decay, discoloration, splits, cracked skin, and other defects
Spinach	2016	Freshness, freedom from decay, mildew, grass weeds, wood, roots, worms, and other foreign material, and freedom from damage caused by seedstems, discoloration, seedbuds, coarse stalks and stems, spray residue, insects, dirt, or mechanical means
Sweet potato for canning/freezing	1959	Firmness, shape, color, size, and freedom from decay and defect (freezing injury, scald, cork, internal discoloration, bruises, cuts, growths cracks, pithiness, stringiness, and insect injury)
Sweet potato for dicing/pulping	1951	Firmness, shape size, and freedom from decay and defect (scald, freezing injury, cork, internal discoloration, pithiness, growth cracks, insect damage, and stringiness)
Tomato	1983	Firmness, ripeness (color as determined by a photoelectric instrument), and freedom from insect or mechanical damage, freezing, decay, growth cracks, sunscald, gray wall, and blossom-end rot
Tomato, green	1950	Firmness, color (green), and freedom from decay and defect (growth cracks, scars, catfaces, catfaces, sunscald, disease, insects, or mechanical damage)
Tomato, Italian type for canning	1957	Firmness, color uniformity, and freedom from decay and defect (growth cracks, sunscald, freezing, disease, insects, or mechanical injury)

Adapted from the *U.S. Standards for Grades of Fresh and Processing Vegetables* (USDA Agricultural Marketing Service), https://www.ams.usda.gov/grades-standards/vegetables (accessed Oct. 2021).

INTERNATIONAL STANDARDS

Standards for vegetables being traded internationally are published by the Organization for Economic Cooperation and Development (OECD). These standards are available from the OECD Bookshop at http://www.oecdbookshop.org in the series International Standardization of Fruit and Vegetables.

https://www.oecd-ilibrary.org/agriculture-and-food/international-standards-for-fruit-and-vegetables_19935668 (accessed 4 Dec. 2021)

10
MINIMALLY PROCESSED VEGETABLES

Examples of minimally processed or fresh-cut products are shredded lettuce, sliced tomatoes, salad mixes (raw vegetable salads), peeled baby carrots, broccoli florets, cauliflower florets, cut celery stalks, shredded cabbage, and cut melon. Basic requirements for preparation of minimally processed products are provided in Table 11.35, and storage characteristics of some of these products are provided in Table 11.36.

The U.S. Food and Drug Administration issued guidance for industry (Guide to Minimize Microbial Food Safety Hazards of Fresh-Cut Fruits and Vegetables) in 2008, these guidelines were updated as recently as 2018. These guidelines are available at https://www.fda.gov/regulatory-information/search-fda-guidance-documents/guidance-industry-guide-minimize-microbial-food-safety-hazards-fresh-cut-fruits-and-vegetables (accessed Oct. 2021).

TABLE 11.35. BASIC REQUIREMENTS FOR PREPARATION OF MINIMALLY PROCESSED VEGETABLES

- High-quality raw material (correct variety, production practices, harvest and storage conditions)
- Strict hygiene and good manufacturing practices, use of Hazard Analysis and Critical Control Points (HACCP) principles
- Low temperatures during processing
- Careful cleaning and/or washing before and after peeling
- Good-quality water (sensory, microbiological, pH) for washing
- Use of mild processing aids (e.g., chlorine, ozone, antioxidants, or calcium salts) in wash water for disinfection or to prevent browning and texture loss
- Gentle draining, spin-or air-drying following washing
- Gentle peeling, cutting, slicing, and/or shredding
- Correct packing materials and packaging methods
- Correct temperature during distribution and handling

Adapted from R. Ahvenainen, "New approaches in improving the shelf life of minimally processed fruit and vegetables," *Trends in Food Science & Technology* 7:79–86 (1996).

TABLE 11.36. PHYSIOLOGY AND STORAGE CHARACTERISTICS OF FRESH-CUT VEGETABLES (ALL PRODUCTS SHOULD BE STORED AT 32–41°F)

Vegetable	Fresh-Cut Product	Respiration Rate in Air at 41°F (mL CO_2/kg/hr)*	Common Quality Defects (Other Than Microbial Growth)	Beneficial Atmosphere	
				%O_2	%CO_2
Asparagus tips	Trimmed spears	40	Browning, softening	10–20	10–15
Beans, snap	Cut	15–18	Browning	2–5	3–12
Beet	Cubed	5	Leakage; color loss	5	5
Broccoli	Florets, Slaw from stems	20–35	Yellowing, off-odors	3–10	5–10
Cabbage	Shredded	13–20	Browning	3–7	5–15
Carrot	Sticks, shredded, coins, "baby carrots"	7–10; 12–15	Surface drying ("white blush"), leakage	0.5–5	10
Cauliflower	Florets	—	Discoloration; off odors	5–10	<5
Celery	Sticks	2–3	Browning, surface drying	—	—
Cucumber	Sliced	5	Leakage	—	—
Garlic	Peeled clove	20	Sprout growth, discoloration	3	5–10
Jicama	Sticks	5–10	Browning; texture loss	3	10
Leek	Sliced	25	Discoloration	5	5
Lettuce, iceberg	Chopped, shredded	6; 10	Browning of cut edges	<0.5–3	10–15
Lettuce, other	Chopped	10–13	Browning of cut edges	1–3	5–10
Melons Cantaloupe	Cubed	5–8	Leakage; softening; glassiness (translucency)	3–5	5–15

552

Commodity	Cut form		Physiological deterioration		
Honeydew	Cubed	2–4	Leakage; softening; glassiness (translucency)	2–3	5–15
Watermelon	Cubed	2–4	Leakage; softening	3–5	5–15
Onion, bulb	Sliced, diced	8–12	Texture, juice loss, discoloration	2–5	10–15
Onion, green	Chopped	25–30	Discoloration, growth; leakage	—	—
Pepper	Sliced, diced	3; 6	Texture loss, browning	3	5–10
Potato	Sticks, peeled	4–8	Browning, drying	1–3	6–9
Rutabaga	Cubed	10	Discoloration, drying	5	5
Spinach	Cleaned, cut	6–12	Off-odors; rapid deterioration of small pieces	1–3	8–10
Squash, summer	Cubed, sliced	12–24	Browning; leakage	1	—
Strawberry	Sliced; topped	12; 6	Loss of texture, juice, color	1–2	5–10
Tomato	Sliced	3	Leakage	3	3

Adapted from A. A. Kader (ed.), *Postharvest Technology of Horticultural Crops*, 3rd ed. (University of California, Division of Agriculture and Natural Resources Publication 3311, 2002). Copyright © 2002 Regents of the University of California. Used by permission. See also: *The Commercial Storage of Fruits, Vegetables, and Florist and Nursery Stocks* (USDA Agriculture Handbook 66, 2016), https://www.ars.usda.gov/arsuserfiles/oc/np/commercialstorage/commercialstorage.pdf.

11
PACKAGING OF FRESH VEGETABLES

While some industry standards exist for packing several fruit and vegetable crops, the standards can vary from region to region, and by market. The list provided in Table 11.37 is not exhaustive, and because different buyers and sellers use countless container types and sizes, communications with buyers is essential.

TABLE 11.37. SHIPPING CONTAINERS FOR FRESH VEGETABLES

Vegetable	Container[1]	Approximate Net Weight (lb)
Asparagus	24 × 1 lb banded bunches (upright), 1 1/9 bushel box	24
Beans, green/ snap	1 1/9 bushel box; loose or in 5 × 5 lb bags	25–30
Beets	Topped: poly bag	25
	Topped: 1 1/9 bushel waxed carton	40
	Bunched with tops: 12 bunches in 1 1/9 bushel waxed carton	
	Bunched with tops: 24 bunches per case in a leafy greens waxed box	
Pepper, green bell	1 1/9 bushel box (36 count minimum)	25
Pepper, colored bell	5/9 bushel box	
Pepper, hot	½ bushel or 1 1/9 bushel box	
Baby salad greens	½ bushel box or crate	8–10
Basil	12 or 24 bunches per box	
Broccoli	Bunched (2–3 heads): 14–18 bunches in 21-lb waxed box	21
	Crown cuts: loose in 20-lb box	20
	Side shoots: 11-lb plastic bag	11

TABLE 11.37. **SHIPPING CONTAINERS FOR FRESH VEGETABLES**
(Continued)

Vegetable	Container[1]	Approximate Net Weight (lb)
Brussels sprouts	Bulk-pack cartons or boxes	25
Cabbage	$1\frac{2}{3}$, $1\frac{3}{4}$, or $1\frac{7}{8}$ bushel waxed cabbage containers, 16–18 count	45–50
	Cabbage bag	50
Cantaloupe	Cartons or bins, by count (6, 9 or 15)	25–40
Carrots	Jumbo: loose in 25 or 50 lb poly bags	25, 50
	Baby with tops: bunched: 12 bunches (5 carrots per bunch) in a $1\frac{1}{9}$ bushel wax box	
Cauliflower	$\frac{1}{2}$ bushel wax box, 6 heads per box (9 and 12 ct. also accepted)	
Celeriac	Cartons	20
Cilantro	12 or 24 bunches per $\frac{1}{2}$ bushel box	
Corn, sweet	1 bushel wood crate, 48 ears per crate	
Cucumber	$1\frac{1}{9}$ bushel wax box, 65–70 count	40
	$\frac{5}{9}$ waxed carton	20
	24-count or 36-count boxes	
Eggplant, specialty	$\frac{1}{2}$ bushel box	15
Eggplant, classic	$\frac{1}{2}$ bushel	15
	$1\frac{1}{9}$ bushel box	30–35
Garlic	Carton or in bulk	10
Garlic scapes	Bushel box or crate, loose or 12–24 bunches per box	

TABLE 11.37. **SHIPPING CONTAINERS FOR FRESH VEGETABLES** *(Continued)*

Vegetable	Container[1]	Approximate Net Weight (lb)
Greens, cooking (kale, collards, chard, mustard greens, etc.)	24 × 1-lb bunches in 1⅑ bushel waxed leafy greens box	
Ground cherries	Carton	25
Kohlrabi	Film or mesh bags	25 or 50
	Carton	24
Lettuce	Leaf lettuce: 24-count cartons	24–28
	Iceberg lettuce: 24-count cartons	24–28
	Butterhead/Boston: 24-count cartons	24–28
	Bibb/greenhouse grown: 12-count in clamshells	
Leek	½ bushel box or basket, in bunches of 2–3	
Onions, green	½ bushel box (24 bunches), 6–9 per bunch	20
	1⅑ bushel box (48 bunches), 6–9 per bunch	
Onions, storage	Bag	25, 50
Parsnip	Carton	20
Pea tendrils	Bushel crate or cartons; loose or in bunches	20–25
Pea, snap	½ bushel box or carton	10
Potato	1⅑ bushel, loose	50
	50-lb bags	50
	5 or 10-lb plastic or paper bags in box or bin	50
Pumpkin, pie	½ or ⅝ or 1⅑ bushel crate	45–50

TABLE 11.37. SHIPPING CONTAINERS FOR FRESH VEGETABLES (*Continued*)

Vegetable	Container[1]	Approximate Net Weight (lb)
Pumpkin, large	By count or weight in bulk bins	
Radish	½ bushel box with 20 bunches of 8–12	
Squash, summer	$\frac{5}{9}$ bushel wax box	20
Squash, winter	$1\frac{1}{9}$ bushel	35–45
	Bulk bin/container	800–900
Sweet potato	½ bushel or $1\frac{1}{9}$ bushel box	40–50
Tomatoes	Field-grown: loose, in 25-lb box	25
	Greenhouse-grown/heirlooms: single layer in 10-lb flat box	10
Turnips	Topped: $\frac{5}{9}$ bushel waxed carton with polyethylene liners	25
	Topped: $1\frac{1}{9}$ bushel waxed carton	40
	Bunched with tops: 12 bunches in $1\frac{1}{9}$ bushel carton	
	Bunched with tops: 24 bunches per case in a leafy greens waxed box	
Watermelons	By count or weight in bulk bins	

[1] Other containers are being developed and used in the marketplace. The requirements of each market should be determined.

This table was adapted from the following resources: Wholesale and retail product specifications: Guidance and best practices for fresh produce for small farms and food hubs. P. Tripp and L. G. Wilson (2014), https://www.ncagr.gov/smallfarms/documents/wholesale-and-retail-product-specs-2.pdf (accessed 4 Dec. 2021); *Growing for Wholesale: Grading and Packing Guidelines by Crop* (Cornell Cooperative Extension Vegetable Program, 2019), https://cvp.cce.cornell.edu/submission.php?id=503 (accessed 4 Dec. 2021); Adapted from M. Boyette, D. C. Sanders, and G. A. Rutledge, *Packaging Requirements for Fresh Fruits and Vegetables* (1996), https://content.ces.ncsu.edu/packaging-requirements-for-fresh-fruits-and-vegetables (accessed 4 Dec. 2021).

TABLE 11.38. STANDARDIZED SHIPPING CONTAINER DIMENSIONS DESIGNED FOR A STANDARD 40 × 48 IN. PALLET

	Outside Base Dimensions	
Inches	Containers Per Layer	Layers Can Be Cross-Stacked
15¾ × 11¾	10	Yes
19¾ × 11¾	8	No
19¾ × 15¾	6	No
23¾ × 15¾	5	Yes

Adapted from S. A. Sargent, M. A. Ritenour and J. K. Brecht, "Handling, Cooling, and Sanitation Techniques for Maintaining Postharvest Quality," in *Vegetable Production Handbook* (University of Florida, 2005–2006).

TABLE 11.39. TRANSPORT EQUIPMENT INSPECTION

Most carriers check their transport equipment before presenting it to the shipper for loading. The condition of the equipment is critical to maintaining the quality of the products. Therefore, the shipper also should check the equipment to ensure it is in good working order and meets the needs of the product. Carriers provide guidance on checking and operating the refrigeration systems.

All transportation equipment should be checked for:

- cleanliness—the load compartment should be regularly steam-cleaned
- damage—walls, floors, doors, ceilings should be in good condition
- temperature control—refrigerated units should be recently calibrated and supply continuous air circulation for uniform product temperatures

Shippers should insist on clean equipment. A load of products can be ruined by:

- odors from previous shipments
- toxic chemical residues
- insects nesting in the equipment
- decaying remains of agricultural products
- debris blocking drain openings or air circulation along the floor

TABLE 11.39. TRANSPORT EQUIPMENT INSPECTION
(Continued)

Shippers should insist on well-maintained equipment and check for the following:

- damage to walls, ceilings, or floors which can let in the outside heat, cold moisture, dirt, and insects
- operation and condition of doors, ventilation openings, and seals
- provisions for load locking and bracing

For refrigerated trailers and van containers, the following additional checks are important:

- with the doors closed, have someone inside the cargo area check for light—the door gaskets must seal. A smoke generator also can be used to detect leaks
- the refrigeration unit should cycle from high to low speed when the desired temperature is reached and then back to high speed
- determine the location of the sensing element which controls the discharge air temperature. If it measures return air temperature, the thermostat may have to be set higher to avoid chilling injury or freezing injury of the products
- a solid return air bulkhead should be installed at the front of the trailer
- a heating device should be available for transportation in areas with extreme cold weather
- equipment with a top air delivery system must have a fabric air chute or metal ceiling duct in good condition

Adapted from B. M. McGregor, *Tropical Products Handbook* (USDA Agricultural Handbook 668, 1987).

TABLE 11.40. CHARACTERISTICS OF DIRECT MARKETING ALTERNATIVES FOR FRESH VEGETABLES

Marketing Strategy:	Opportunities	Challenges
Pick-your-own	• No harvest labor or transportation expenses • No post-harvest storage needed • Provides entertainment: agritourism potential • Grower determines hours of operation	• Field attendants are needed • Promotion/advertisement needed • Liability with visitors to the farm • Need to provide parking, instruction, restroom/washing facilities, containers • Can be difficult to balance availability with demand • Weather-dependent
Roadside market or farm stand	• Minimal transportation costs • Can provide entertainment: agritourism potential • Grower determines hours of operation • Farmer determines packaging/ marketing strategy	• Checkout attendants are needed • Liability with visitors to the farm • Need to provide parking, containers • Need to invest in building/stand infrastructure, and adequate post-harvest storage facilities • Variety of offerings is helpful in attracting customers

**TABLE 11.40.   CHARACTERISTICS OF DIRECT MARKETING
ALTERNATIVES FOR FRESH VEGETABLES (*Continued*)**

Marketing Strategy:	Opportunities	Challenges
Farmer's market	• No investment in infrastructure needed • Minimal advertising costs assumed • Farmer determines packaging/ marketing strategy	• Checkout attendants are needed • There may be competition from other vendors • Need adequate post-harvest storage facilities and transportation to market • Variety of offerings is helpful in attracting customers • Not all farmers' markets are well-attended
Community Supported Agriculture (CSA)	• Minimal investment in infrastructure needed • Income received early in the season • Minimal investment in packaging and transportation • Farmer determines packaging/ marketing strategy	• Pick-up or distribution strategy is needed • Advertisement needed to build clientele; there may be competition in some regions • Need adequate post-harvest storage facilities and transportation to market • Variety of offerings is essential

Also refer to *Direct Marketing Options, Section 26 in the 2021 Guide to Farming* (The Cornell Small Farms Program), https://smallfarms.cornell.edu/guide/guide-to-farming/direct-marketing-options/ (accessed 4 Dec. 2021).

TABLE 11.41. CHARACTERISTICS OF SOME WHOLESALE MARKETING STRATEGIES FOR FRESH VEGETABLES

Marketing Strategy:	Opportunities	Challenges
Terminal market	• Can move very large quantities at one time • Terminal markets provide good market information	• Buyer sets the price • Large quantities are usually needed • Transportation and specialized containers are needed • Must meet buyer's standards or U.S. grades
Cooperative and private packing facilities	• Harvesting and packing equipment may be provided • Transportation may be provided • Relatively low market investment required • Firms may provide technical assistance with growing and selling	• Must meet buyer's standards or U.S. grades • Prices received depend on market prices, costs, and revenues
Grocers and restaurants	• Grower sets price, or grower and buyer compromise on price • Required volumes vary	• Need adequate post-harvest storage facilities and transportation to market • Difficult to enter market and develop customers • Individual buyers may require third-party certifications and liability insurance • Must meet buyer's standards

TABLE 11.41. CHARACTERISTICS OF SOME WHOLESALE MARKETING STRATEGIES FOR FRESH VEGETABLES (Continued)

Marketing Strategy:	Opportunities	Challenges
Wholesale/ broker	• Relatively low market investment required • Good wholesaler/ brokers can sell produce quickly at good prices	• Need adequate post-harvest storage facilities • Transportation and packing costs vary • Buyer sets the price • Must meet buyer's standards or U.S. grades • Large quantities are usually needed

Adapted from *Cucurbit Production and Pest Management* (Oklahoma Cooperative Extension Circular E-853, 1986). Also refer to: *Direct Marketing Options, Section 26 in the 2021 Guide to Farming* (The Cornell Small Farms Program), https://smallfarms.cornell.edu/guide/guide-to-farming/direct-marketing-options/ (accessed 4 Dec. 2021); M. Leroux, *Guide to Marketing Channel Selection: How to sell through wholesale and direct marketing channels* (Cornell Cooperative Extension, 2010).

PART **12**

APPENDIX

Knott's Handbook for Vegetable Growers, Sixth Edition. George J. Hochmuth
and Rebecca G. Sideman.
© 2023 John Wiley & Sons Ltd. Published 2023 by John Wiley & Sons Ltd.
Companion website: https://www.wiley.com/go/hochmuth/vegetablegrowers6e

SOURCES OF INFORMATION ON VEGETABLE PRODUCTION

The Land Grant University Agricultural Experiment Station and Cooperative Extension Service in each state has been the principal source of information on vegetables in printed form. Most of this information is now on the Internet because of the high cost of producing, storing, and distributing printed information. An easy Google Search can take the reader to numerous university Extension websites containing information on vegetable production to supplement information in this Knott's Handbook.

SOME PERIODICALS FOR VEGETABLE GROWERS

American Vegetable Grower
Greenhouse Grower
Productores de Hortalizas
Meister Media Worldwide
3377 Euclid Avenue
Willoughby, OH 44094-5992
Ph (800) 572-7740
http://www.meistermedia.com

Florida Grower
AgNet Media, Inc.
27206 SW 22nd Place
Newberry, FL 32669
Ph 352-671-1909

Great American Publishing
P.O. Box 128
Sparta, MI 49345
Ph (616) 887-9008
Fax (616) 887-2666

Growing for Market
P.O. Box 3747
Lawrence, KS 66046
Ph (785) 748-0605
Fax (785) 748-0609
http://www.growingformarket.com

Potato Country
https://potatocountry.com/

Potato Grower Magazine
520 Park Avenue
Idaho Falls, ID 83402
Ph (208) 524-7000
http://www.potatogrower.com

Specialty Crop Industry
AgNet Media, Inc.
27206 SW 22nd Place
Newberry, FL 32669
Ph 352-671-1909

Spudman
75 Applewood Drive, Suite A
P.O. Box 128
Sparta, MI 49345
Ph 616-520-2137
https://www.spudman.com/

The Grower
105-355 Elmira Road N.
Guelph, ON N1K 1S5
Ph 866-898-8488 ext. 221

The Packer
https://www.thepacker.com/

The Produce News
P.O. Box 971401
Boca Raton, FL 33497
Ph (888) 986-7990
http://www.theproducenews.com

Vegetable Growers News
75 Applewood Drive, Suite A
P.O. Box 128
Sparta, MI 49345
Ph 616-520-2137
http://www.vegetablegrowersnews.com

Vegetables West Grower and PCA
https://vegetableswest.com/

Western Grower and Shipper Magazine
https://www.wga.com/magazine

U.S. UNITS OF MEASUREMENT

Length
1 foot = 12 inches
1 yard = 3 feet
1 yard = 36 inches
1 rod = 16.5 feet
1 mile = 5,280 feet

Area
1 acre = 43,560 square feet
1 section = 640 acres
1 section = 1 square mile

Volume
1 liquid pint = 16 liquid ounces
1 liquid quart = 2 liquid pints
1 liquid quart = 32 liquid ounces
1 gallon = 8 liquid pints
1 gallon = 4 liquid quarts
1 gallon = 128 liquid ounces
1 peck = 16 pints (dry)
1 peck = 8 quarts (dry)
1 bushel = 4 pecks
1 bushel = 64 pints (dry)
1 bushel = 32 quarts (dry)

Mass or Weight
1 pound = 16 ounces
1 hundredweight = 100 pounds
1 ton = 20 hundredweight
1 ton = 2000 pounds
1 unit (fertilizer) = 1% ton = 20 pounds

CONVERSION FACTORS FOR U.S. UNITS

Multiply	By	To Obtain
Length		
feet	12	inches
feet	0.33333	yards
inches	0.08333	feet
inches	0.02778	yards
miles	5,280	feet
miles	63,360	inches
miles	1,760	yards
rods	16.5	feet
yards	3	feet
yards	36	inches
yards	0.000568	miles
Area		
acres	43,560	square feet
acres	160	square rods
acres	4,840	square yards
square feet	144	square inches
square feet	0.11111	square yards
square inches	0.00694	square feet
square miles	640	acres
square miles	27,878,400	square feet
square miles	3,097,600	square yards
square yards	0.0002066	acres
square yards	9	square feet
square yards	1,296	square inches
Volume		
bushels	2,150.42	cubic inches
bushels	4	pecks
bushels	64	pints
bushels	32	quarts
cubic feet	1,728	cubic inches

Multiply	By	To Obtain
cubic feet	0.03704	cubic yards
cubic feet	7.4805	gallons
cubic feet	59.84	pints (liquid)
cubic feet	29.92	quarts (liquid)
cubic yards	27	cubic feet
cubic yards	46,656	cubic inches
cubic yards	202	gallons
cubic yards	1,616	pints (liquid)
cubic yards	807.9	quarts (liquid)
gallons	0.1337	cubic feet
gallons	231	cubic inches
gallons	128	ounces (liquid)
gallons	8	pints (liquid)
gallons	4	quarts (liquid)
gallons water	8.3453	pounds water
pecks	0.25	bushels
pecks	537.605	cubic inches
pecks	16	pints (dry)
pecks	8	quarts (dry)
pints (dry)	0.015625	bushels
pints (dry)	33.6003	cubic inches
pints (dry)	0.0625	pecks
pints (dry)	0.5	quarts (dry)
pints (liquid)	28.875	cubic inches
pints (liquid)	0.125	gallons
pints (liquid)	16	ounces (liquid)
pints (liquid)	0.5	quarts (liquid)
quarts (dry)	0.03125	bushels
quarts (dry)	67.20	cubic inches
quarts (dry)	2	pints (dry)
quarts (liquid)	57.75	cubic inches
quarts (liquid)	0.25	gallons
quarts (liquid)	32	ounces (liquid)
quarts (liquid)	2	pints (liquid)

Multiply	By	To Obtain
Mass or Weight		
ounces (dry)	0.0625	pounds
ounces (liquid)	1.805	cubic inches
ounces (liquid)	0.0078125	gallons
ounces (liquid)	0.0625	pints (liquid)
ounces (liquid)	0.03125	quarts (liquid)
pounds	16	ounces
pounds	0.0005	tons
pounds of water	0.01602	cubic feet
pounds of water	27.68	cubic inches
pounds of water	0.1198	gallons
tons	32,000	ounces
tons	20	hundredweight
tons	2,000	pounds
Rate		
feet per minute	0.01667	feet per second
feet per minute	0.01136	miles per hour
miles per hour	88	feet per minute
miles per hour	1.467	feet per minute

METRIC UNITS OF MEASUREMENT

Length
1 millimeter = 1,000 microns
1 centimeter = 10 millimeters
1 meter = 100 centimeters
1 meter = 1,000 millimeters
1 kilometer = 1,000 meters

Area
1 hectare = 10,000 square meters

Volume
1 liter = 1,000 milliliters

Mass or Weight
1 gram = 1,000 milligrams
1 kilogram = 1,000 grams
1 quintal = 100 kilograms
1 metric ton = 1,000 kilograms
1 metric ton = 10 quintals

CONVERSION FACTORS FOR SI AND NON-SI UNITS

To Convert Column 1 into Column 2, Multiply by	Column 1 SI Unit	Column 2 Non-SI Unit	To Convert Column 2 into Column 1, Multiply by
Length			
0.621	kilometer, km (10^3 m)	mile, mi	1.609
1.094	meter, m	yard, yd	0.914
3.28	meter, m	foot, ft	0.304
1.0	micrometer, μ (10^{-6} m)	micron, μ	1.0
3.94×10^{-2}	millimeter, mm (10^{-3} m)	inch, in	25.4
10	nanometer, nm (10^{-9} m)	Angstrom, Å	0.1
Area			
2.47	hectare, ha	Acre	0.405
247	square kilometer, km^2 (10^3 m)^2	Acre	4.05×10^{-3}
0.386	square kilometer, km^2 (10^3 m)^2	square mile, mi^2	2.590
2.47×10^{-4}	square meter, m^2	Acre	4.05×10^{-3}
10.76	square meter, m^2	square foot, ft^2	9.29×10^{-2}
1.55×10^{-3}	square millimeter, mm^2 (10^{-6} m)^2	square inch, in^2	645

574

Volume

6.10×10^4	cubic meter, m³	cubic inch, in³	1.64×10^{-5}
2.84×10^{-2}	liter, L (10^{-3} m³)	bushel, bu	35.24
1.057	liter, L (10^{-3} m³)	quart (liquid), qt	0.946
3.53×10^{-2}	liter, L (10^{-3} m³)	cubic foot, ft³	28.3
0.265	liter, L (10^{-3} m³)	Gallon	3.78
33.78	liter, L (10^{-3} m³)	ounce (fluid), oz	2.96×10^{-2}
2.11	liter, L (10^{-3} m³)	pint (fluid), pt	0.473
9.73×10^{-3}	cubic meter, m³	acre-inch	102.8
35.3	cubic meter, m³	cubic foot, ft³	2.83×10^{-2}

Mass

2.20×10^{-3}	gram, g (10^{-3} kg)	pound, lb	454
3.52×10^{-2}	gram, g	ounce (avdp), oz	28.4
2.205	kilogram, kg	pound, lb	0.454
10^2	kilogram, kg	quintal (metric), q	10^2
1.10×10^{-3}	kilogram, kg	ton (2,000 lb), ton	907
1.102	megagram, Mg (tonne)	ton (U.S.), ton	0.907

Yield and Rate

0.893	kilogram per hectare, kg/ha	pound per acre, lb/acre	1.12
7.77×10^{-2}	kilogram per cubic meter, kg/m³	pound per bushel, lb/bu	12.87
1.49×10^{-2}	kilogram per hectare, kg/ha	bushel per acre, 60 lb	67.19
1.59×10^{-2}	kilogram per hectare, kg/ha	bushel per acre, 56 lb	62.71

To Convert Column 1 into Column 2, Multiply by	Column 1 SI Unit	Column 2 Non-SI Unit	To Convert Column 2 into Column 1, Multiply by
1.86×10^{-2}	kilogram per hectare, kg/ha	bushel per acre, 48 lb	53.75
0.107	liter per hectare, l/ha	gallon per acre, gal/acre	9.35
893	megagram per hectare, Mg/ha	pound per acre, lb/acre	1.12×10^{-3}
0.446	megagram per hectare, Mg/ha	ton (2,000 lb) per acre, ton/acre	2.24
2.24	meter per second, m/s	mile per hour	0.477

Specific Surface

10	square meter per kilogram, m²/kg	square centimeter per gram, cm²/g	0.1
10^3	square meter per kilogram, m²/kg	square millimeter per gram, mm²/g	10^{-3}

Pressure

9.90	megapascal, MPa (10^6 Pa)	atmosphere	0.101
10	megapascal, MPa (10^6 Pa)	bar	0.1
1.00	megagram per cubic meter, Mg/m³	gram per cubic centimeter, g/cm³	1.00
2.09×10^{-2}	pascal, Pa	pound per square foot, lb/ft²	47.9
1.45×10^{-4}	pascal, Pa	pound per square inch, lb/in²	6.90×10^3

Temperature

$1.00\,(K-273)$	Kelvin, K	Celsius, °C	$1.00\,(°C+273)$
$(9/5\,°C)+32$	Celsius, °C	Fahrenheit, °F	$\tfrac{5}{9}(°F-32)$

Energy, Work, Quantity of Heat

9.48×10^{-4}	joule, J	British thermal unit, Btu	1.05×10^{3}
0.239	joule, J	calorie, cal	4.19
10^{7}	joule, J	erg	10^{-7}
0.735	joule, J	foot-pound	1.36
2.387×10^{-5}	joule per square meter, J/m²	calorie per square centimeter (langley)	4.19×10^{4}
10^{5}	newton, N	dyne	10^{-5}
1.43×10^{-3}	watt per square meter, W/m²	calorie per square centimeter minute (irradiance), cal/cm²/min	698

Transpiration and Photosynthesis

3.60×10^{-2}	milligram per square meter per second, mg/m²/s	gram per square decimeter per hour, g/dm²/hr	27.8
5.56×10^{-3}	milligram (H_2O) per square meter per second, mg/m²/s	micromole (H_2O) per square centimeter per second, μmol/cm²/s	180
10^{-4}	milligram per square meter per second, mg/m²/s	milligram per square centimeter per second, mg/cm²/s	10^{4}
35.97	milligram per square meter per second, mg/m²/s	milligram per square decimeter per hour, mg/dm²/hr	2.78×10^{-2}

To Convert Column 1 into Column 2, Multiply by	Column 1 SI Unit	Column 2 Non-SI Unit	To Convert Column 2 into Column 1, Multiply by
Angle			
57.3	radian, rad	degrees (angle), °	1.75×10^{-2}
Water Measurement			
9.73×10^{-3}	cubic meter, m^3	acre-inches, acre-in	102.8
9.81×10^{-3}	cubic meter per hour, m^3/hr	cubic feet per second, ft^3/s	101.9
4.40	cubic meter per hour, m^3/hr	U.S. gallons per minute, gal/min	0.227
8.11	hectare-meters, ha-m	acre-feet, acre-ft	0.123
97.28	hectare-meters, ha-m	acre-inches, acre-in	1.03×10^{-2}
8.1×10^{-2}	hectare-centimeters, ha-cm	acre-feet, acre-ft	12.33
Concentrations			
1	centimole per kilogram, cmol/kg (ion exchange capacity)	milliequivalents per 100 grams, meq/100 g	1
0.1	gram per kilogram, g/kg	percent, %	10
1	megagram per cubic meter, Mg/m^3	gram per cubic centimeter, g/cm^3	1
1	milligram per kilogram, mg/kg	parts per million, ppm	1

CONVERSIONS FOR RATES OF APPLICATION

1 ton per acre = 20.8 grams per square foot

1 ton per acre = 1 pound per 21.78 square feet

1 ton per acre furrow slice (6-inch depth) = 1 gram per 1,000 grams soil

1 gram per square foot = 96 pounds per acre

1 pound per acre = 0.0104 grams per square foot

1 pound per acre = 1.12 kilograms per hectare

100 pounds per acre = 0.2296 pounds per 100 square feet

grams per square foot × 96 = pounds per acre

kilograms per 48 square feet = tons per acre

pounds per square feet × 21.78 = tons per acre

WATER AND SOIL SOLUTION CONVERSION FACTORS

Concentration

1 decisiemens per meter (dS/m) = 1 millimho per centimeter (mmho/cm)
1 decisiemens per meter (dS/m) = approximately 640 milligrams per liter salt
1 part per million (ppm) = 1/1,000,000
1 percent = 0.01 or 1/100
1 ppm × 10,000 = 1 percent
ppm × 0.00136 = tons per acre-foot of water
ppm = milligrams per liter
ppm = 17.12 × grains per gallon
grains per gallon = 0.0584 × ppm
ppm = 0.64 × micromhos per centimeter (in range of 100–5,000 micromhos per centimeter)
ppm = 640 × millimhos per centimeter (in range of 0.1–5.0 millimhos per centimeter)
ppm = grams per cubic meter
mho = reciprocal ohm
millimho = 1000 micromhos
millimho = approximately 10 milliequivalents per liter (meq/liter)
milliequivalents per liter = equivalents per million
millimhos per centimeter = $EC \times 10^3$ ($EC \times 1,000$) at 25 °C (EC = electrical conductivity)
micromhos per centimeter = $EC \times 10^6$ ($EC \times 1,000,000$) at 25 °C
millimhos per centimeter = 0.1 siemens per meter
millimhos per centimeter = ($EC \times 10^3$) = decisiemens per meter (dS/m)
1,000 micromhos per centimeter = approximately 700 ppm
1,000 micromhos per centimeter = approximately 10 milliequivalents per liter
1,000 micromhos per centimeter = 1 ton of salt per acre-foot of water
milliequivalents per liter = 0.01 × ($EC \times 10^6$) (in range of 100–5,000 micromhos per centimeter)
milliequivalents per liter = 10 × ($EC \times 10^3$) (in range of 0.1–5.0 millimhos per centimeter)

Pressure and Head

1 atmosphere at sea level = 14.7 pounds per square inch
1 atmosphere at sea level = 29.9 inches of mercury
1 atmosphere at sea level = 33.9 feet of water
1 atmosphere = 0.101 megapascal (MPa)

1 bar = 0.10 megapascal (MPa)
1 foot of water = 0.8826 inch mercury
1 foot of water = 0.4335 pound per square inch
1 inch of mercury = 1.133 feet water
1 inch of mercury = 0.4912 pound per square inch
1 inch of water = 0.07355 inch mercury
1 inch of water = 0.03613 pound per square inch
1 pound per square inch = 2.307 feet water
1 pound per square inch = 2.036 inches mercury
1 pound per square foot = 47.9 pascals

Weight and Volume (U.S. Measurements)
1 acre-foot of soil = about 4,000,000 pounds
1 acre-foot of water = 43,560 cubic feet
1 acre-foot of water = 12 acre-inches
1 acre-foot of water = about 2,722,500 pounds
1 acre-foot of water = 325,851 gallons
1 cubic-foot of water = 7.4805 gallons
1 cubic foot of water at 59 °F = 62.37 pounds
1 acre-inch of water = 27,154 gallons
1 gallon of water at 59 °F = 8.337 pounds
1 gallon of water = 0.1337 cubic foot or 231 cubic inches

Flow (U.S. Measurements)
1 cubic foot per second = 448.8 gallons per minute
1 cubic foot per second = about 1 acre-inch per hour
1 cubic foot per second = 23.80 acre-inches per hour
1 cubic foot per second = 3600 cubic feet per hour
1 cubic foot per second = about $7\frac{1}{2}$ gallons per second
1 gallon per minute = 0.00223 cubic feet per second
1 gallon per minute = 0.053 acre-inch per 24 hours
1 gallon per minute = 1 acre-inch in $4\frac{1}{2}$ hours
1000 gallons per minute = 1 acre-inch in 27 minutes
1 acre-inch per 24 hours = 18.86 gallons per minute
1 acre-foot per 24 hours = 226.3 gallons per minute
1 acre-foot per 24 hours = 0.3259 million gallons per 24 hours

U.S.-Metric Equivalents

1 cubic meter = 35.314 cubic feet
1 cubic meter = 1.308 cubic yards
1 cubic meter = 1000 liters
1 liter = 0.0353 cubic feet
1 liter = 0.2642 U.S. gallon
1 liter = 0.2201 British or Imperial gallon
1 cubic centimeter = 0.061 cubic inch
1 cubic foot = 0.0283 cubic meter
1 cubic foot = 28.32 liters
1 cubic foot = 7.48 U.S. gallons
1 cubic foot = 6.23 British gallons
1 cubic inch = 16.39 cubic centimeters
1 cubic yard = 0.7645 cubic meter
1 U.S. gallon = 3.7854 liters
1 U.S. gallon = 0.833 British gallon
1 British gallon = 1.201 U.S. gallons
1 British gallon = 4.5436 liters
1 acre-foot = 43,560 cubic feet
1 acre-foot = 1,233.5 cubic meters
1 acre-inch = 3,630 cubic feet
1 acre-inch = 102.8 cubic meters
1 cubic meter per second = 35.314 cubic feet per second
1 cubic meter per hour = 0.278 liter per second
1 cubic meter per hour = 4.403 U.S. gallons per minute
1 cubic meter per hour = 3.668 British gallons per minute
1 liter per second = 0.0353 cubic feet per second
1 liter per second = 15.852 U.S. gallons per minute
1 liter per second = 13.206 British gallons per minute
1 liter per second = 3.6 cubic meters per hour
1 cubic foot per second = 0.0283 cubic meter per second
1 cubic foot per second = 28.32 liters per second
1 cubic foot per second = 448.8 U.S. gallons per minute
1 cubic foot per second = 373.8 British gallons per minute
1 cubic foot per second = 1 acre-inch per hour (approximately)
1 cubic foot per second = 2 acre-feet per day (approximately)
1 U.S. gallon per minute = 0.06309 liter per second
1 British gallon per minute = 0.07573 liter per second

Power and Energy
1 horsepower = 550 foot-pounds per second
1 horsepower = 33,000 foot-pounds per minute
1 horsepower = 0.7457 kilowatts
1 horsepower = 745.7 watts
1 horsepower-hour = 0.7457 kilowatt-hour
1 kilowatt = 1.341 horsepower
1 kilowatt-hour = 1.341 horsepower-hours
1 acre-foot of water lifted 1 foot = 1.372 horsepower-hours of work
1 acre-foot of water lifted 1 foot = 1.025 kilowatt-hours of work

HEAT AND ENERGY EQUIVALENTS AND DEFINITIONS

1 joule = 0.239 calorie

1 joule = Nm (m^2 kg s^{-2})

temperature of maximum density of water = 3.98 °C (about 39 °F)

1 British thermal unit (Btu) = heat needed to change 1 pound water at maximum density of 1 °F

1 Btu = 1.05506 kilojoules (kJ)

1 Btu/lb = 2.326 kJ/kg

1 Btu per minute = 0.02356 horsepower

1 Btu per minute = 0.01757 kilowatts

1 Btu per minute = 17.57 watts

1 horsepower = 42.44 Btu per minute

1 horsepower-hour = 2547 Btu

1 kilowatt-hour = 3415 Btu

1 kilowatt = 56.92 Btu per minute

1 pound water at 32 °F changed to solid ice requires removal of 144 Btu

1 pound ice in melting takes up to 144 Btu

1 ton ice in melting takes up to 288,000 Btu

USEFUL WEBSITES FOR UNITS AND CONVERSIONS

Weights, Measures, and Conversion Factors for Agricultural Commodities and Their Products (USDA Agricultural Handbook AH-697, 1992), https://www.ers.usda.gov/publications/pub-details/?pubid=41881.

Metric Conversions (2013), https://www.extension.iastate.edu/agdm/wholefarm/html/c6-80.html.

A Dictionary of Units – Part 1 (2004), http://www.cleavebooks.co.uk/dictunit/dictunit1.htm.

A Dictionary of Units – Part 2 (2004), http://www.cleavebooks.co.uk/dictunit/dictunit2.htmU.S.

U.S. Metric Association (2022), https://usma.org/.

INDEX

Knott's Handbook for Vegetable Growers, Sixth Edition. George J. Hochmuth and Rebecca G. Sideman.
© 2023 John Wiley & Sons Ltd. Published 2023 by John Wiley & Sons Ltd.
Companion website: https://www.wiley.com/go/hochmuth/vegetablegrowers6e